W. R. Gocht H. Zantop R. G. Eggert

International Mineral Economics

Mineral Exploration,
Mine Valuation,
Mineral Markets,
International Mineral Policies

Springer-Verlag

W. R. Gocht   H. Zantop   R. G. Eggert

International Mineral Economics

Mineral Exploration, Mine Valuation, Mineral Markets, International Mineral Policies

With 46 Figures

Springer-Verlag
Berlin Heidelberg New York
London Paris Tokyo

WERNER R. GOCHT, Dr. rer. nat., Dr. rer. pol.,
Professor and Director of the Research Institute
for International Technical and Economic Cooperation
Aachen University of Technology
D-5100 Aachen, FRG

HALF ZANTOP, Ph. D.,
Associate Professor of Economic Geology
Dartmouth College
Hanover, NH 03755, USA

RODERICK G. EGGERT, Ph. D.,
Assistant Professor of Mineral Economics
Colorado School of Mines
Golden, CO 80401, USA

ISBN 3-540-18749-9 Springer-Verlag Berlin Heidelberg New York
ISBN 0-387-18749-9 Springer-Verlag New York Berlin Heidelberg

Library of Congress Cataloging-in-Publication Data. Gocht, Werner. International mineral economics. Bibliography: p. Includes index. 1. Mineral industries. I. Eggert, Roderick G. II. Zantop, H. (Half), 1938–. III. Title. HD9506.A2G63 1988   338.2   88-04892

This work is subject to copyright. All rights are reserved, whether the whole or part of the material is concerned, specifically the rights of translation, reprinting, re-use of illustrations, recitation, broadcasting, reproduction of microfilms or in other ways, and storage in data banks. Duplication of this publication or parts thereof is only permitted under the provisions of the German Copyright Law of September 9, 1965, in its version of June 24, 1985, and a copyright fee must always be paid. Violations fall under the prosecution act of the German Copyright Law.

© Springer-Verlag Berlin Heidelberg 1988
Printed in Germany

The use of registered names, trademarks etc. in this publication does not imply, even in the absence of a specific statement, that such names are exempt from the relevant protective laws and regulations and therefore free for general use.

Typesetting: Fotosatz & Design, Berchtesgaden
Offsetprinting and Bookbinding: Druckhaus Beltz, Hemsbach/Bergstraße

Preface

Exploration and mining geologists, while concerned mostly with the geologic aspects of mineral resources, must be able to judge the technical feasibility and to estimate the economic potential of a mineral deposit when recommending detailed exploration. Likewise, mineral economists must recognize the significance of the geologic, technical, and political influences on mining and mineral processing when evaluating the economic potential of a prospective deposit. Mining engineers must understand the ways in which geologic, political and economic factors affect the feasibility of a project.

In *International Mineral Economics,* we would like to introduce professionals and students to the essential concepts of mineral exploration, mine valuation and mineral market analysis, and international mineral policies. Although each of these topics is covered in detail in the specialized literature, we have attempted to integrate their most important aspects here, to provide those involved in mineral resources with a broad, interdisciplinary overview.

This project grew out of an earlier book by Werner Gocht, in German, *Wirtschaftsgeologie und Rohstoffpolitik* (1983). We have expanded the coverage of the geologic aspects of mineral occurrence and exploration methods; thoroughly revised the treatment of the economic analysis of mineral projects and markets; and updated and expanded the analysis of international mineral policies.

Werner Gocht, of the Aachen University of Technology, coordinated the project. Half Zantop, of Dartmouth College, wrote the chapters on economic geology, mineral exploration, and mineral development. Roderick G. Eggert, of the Colorado School of Mines, was responsible for the chapters on the economic, institutional and legal framework of mineral development; the economic evaluation of mineral deposits; and mineral market models and forecasts. Werner Gocht wrote the chapters on mineral market analysis and on international mineral policies. Each of us, however, integrated his work with that of the others. All of us are thoroughly familiar with the entire manuscript and feel responsible for it as a team.

The book would not have been possible without the assistance and encouragement of many individuals and organizations we wish to acknowledge. Our home institutions provided financial assistance. Bruce Gemmell, Diana K. Ten Eyck, and John Tilton read parts of the manuscript. Comments from all of these readers contributed greatly to its quality and presentation. Judith Jütte-Rauhut provided able research assistance in Aachen. Richard Williams assisted in translating parts of the text on international mineral policies from German to English.

We alone, of course, are responsible for any remaining errors or faulty analysis. We encourage comments and criticism from our readers, with the hope that the next edition of the book will benefit from these suggestions.

February 1988

WERNER R. GOCHT
HALF ZANTOP
RODERICK G. EGGERT

Contents

Introduction 1

Part I   **Economic Geology, Mineral Exploration, and Mineral Development**

1   **Mineral Deposits and Metallogenic Concepts** 5
1.1   Composition of the Earth and Types of Mineral Resources 8
1.2   Geologic Cycles and the Formation of Mineral Deposits . 10
1.2.1   Mineral Deposits Associated with Magmatic Processes . 11
1.2.2   Mineral Deposits Associated with Sedimentary Processes 14
1.2.3   Mineral Deposits Associated with Metamorphic Processes 16
1.3   Geologic Models of Mineral Deposits 17

2   **Exploration Methods** 22
2.1   Development Phases of Exploration and Mining Projects 23
2.2   Exploration Strategies, Procedures, and Stages 26
2.2.1   Program Design 26
2.2.2   Reconnaissance Exploration 27
2.2.3   Detailed Follow-up Exploration 27
2.3   Exploration Methods 29
2.3.1   Remote Sensing 30
2.3.2   Aerial Photography 32
2.3.3   Geologic Exploration 34
2.3.4   Geophysical Exploration 34
2.3.4.1   Magnetic Surveys 37
2.3.4.2   Electric Surveys 38
2.3.4.3   Electromagnetic Surveys 40
2.3.4.4   Radiometric Surveys 41
2.3.4.5   Gravimetric Surveys 41
2.3.4.6   Seismic Surveys 43
2.3.4.7   Airborne and Ground Geophysical Surveys 44
2.3.5   Geochemical Exploration 45

2.3.5.1	Stream Sediment Geochemical Surveys	47
2.3.5.2	Hydrogeochemical Surveys	48
2.3.5.3	Soil Geochemical Surveys	48
2.3.5.4	Biogeochemical Surveys	49
2.3.5.5	Atmospheric and Vapor Geochemical Surveys	50
2.3.5.6	Rock Geochemical Surveys	51
2.3.5.7	Analytical Methods	51
2.3.5.8	Interpretation of Geochemical Surveys	53
3	**Quantitative Assessment of Mineral Potential**	**57**
3.1	Accessing the Ore	57
3.1.1	Trenching and Pitting	58
3.1.2	Drilling	58
3.1.3	Drill-Hole Geophysics	60
3.1.4	Underground Exploration	61
3.2	Sampling and Assaying of the Ore	61
3.2.1	Sampling	62
3.2.2	Sample Preparation	63
3.2.3	Chemical Assays	63
3.3	Ore Reserve Estimation	64
3.3.1	Geometric Methods of Ore Reserve Estimation	64
3.3.2	Statistical Methods of Ore Reserve Estimation	66
3.3.3	Geostatistical Methods of Ore Reserve Estimation	67
3.3.4	Classification of Reserves and Resources	68
4	**Mining and Mineral Processing**	**74**
4.1	Mining	74
4.1.1	Surface Mining	75
4.1.2	Underground Mining	77
4.1.3	Solution Mining	78
4.1.4	Marine Mining	78
4.1.5	Strategies and Trends in Mining	79
4.2	Mineral Processing	80

Part II Mineral Economics

5	**The Economic, Institutional, and Legal Framework for Mineral Development**	**85**
5.1	Mineral Development as an Economic Activity	85
5.2	Participants in Mineral Development	87
5.3	Mining Law	90

6	**Economic Evaluation of Mineral Deposits**	94
6.1	Goals and Strategies	95
6.2	Methods of Investment Analysis	96
6.2.1	Cash Flows and the Time Value of Money	97
6.2.2	Benefit-Cost Analysis	102
6.2.3	Risk Analysis	102
6.3	Evaluating Exploration Projects and Mineral Deposits	109
6.3.1	Program Design	109
6.3.2	Reconnaissance Exploration	110
6.3.3	Detailed Exploration and Initial Target Evaluation	111
6.3.3.1	Minimum Acceptable Reserves – Tonnage and Grade	111
6.3.3.2	Cutoff Grade	113
6.3.3.3	Optimal Scale of Operation	113
6.3.3.4	Cash-Flow Analysis	115
6.3.3.5	The Prefeasibility Study	117
6.3.4	Detailed Target Evaluation and the Feasibility Study	119
6.3.4.1	Legal Factors	122
6.3.4.2	Fiscal Regimes	122
6.3.4.3	Environmental Regulations	123
6.4	Financing	124
6.4.1	Exploration Projects	126
6.4.2	Mine Development	127
7	**Mineral Markets**	132
7.1	Market Structures	133
7.1.1	Determination and Alteration of Market Structures	134
7.1.1.1	The Tin Market	135
7.1.1.2	The Aluminum Market	137
7.1.1.3	The Gold Market	138
7.1.2	Market Shares of International Mining and Oil Companies	141
7.2	Market Organizations	142
7.2.1	Producer Associations	144
7.2.1.1	Petroleum Associations (OPEC, OAPEC)	145
7.2.1.2	Copper Associations (CIPEC and Others)	150
7.2.1.3	Bauxite Association (IBA)	153
7.2.1.4	Iron Ore Association (APEF)	155
7.2.1.5	Tungsten Association (PTA)	156
7.2.1.6	Uranium Association (UI)	157
7.2.1.7	Other Producer Associations	158
7.2.2	International Commodity Agreements	160
7.3	Mineral Pricing	166
7.3.1	Determination of Mineral Supply	168
7.3.1.1	Factors Determining Production Trends	168
7.3.1.2	Special Supply Factors	168
7.3.2	Determination of Mineral Demand	173

7.3.3	Competitive Prices	174
7.3.4	Producer Prices	177
7.4	Market Models and Forecasts	179
7.4.1	Types of Mineral-Market Models	180
7.4.2	Forecasting	184

Part III International Mineral Policies

8	**Policies and Cooperation Programs of International Organizations**	189
8.1	General Problems and Objectives	189
8.1.1	Objectives of Mineral Exporting Countries	192
8.1.2	Objectives of Mineral Importing Countries	193
8.1.3	Conflicting Aims and Possible Solutions	194
8.2	Mineral Policy of International Organizations	195
8.2.1	The League of Nations	195
8.2.2	The United Nations	196
8.2.2.1	UNCTAD I–VI	197
8.2.2.2	Integrated Program for Commodities	198
8.2.2.3	United Nations Conventions on the Law of the Sea	200
8.2.3	The International Energy Programme of the Organization for Economic Cooperation and Development	205
8.2.4	The European Community	207
8.2.4.1	Lomé I Convention	208
8.2.4.2	Lomé II Convention	209
8.2.4.3	Lomé III Convention	210
8.3	International Cooperation in Mineral Exploration and Exploitation	210
8.3.1	Cooperation Programs of International Institutions	212
8.3.2	The UNDP and the UN Revolving Fund	214
8.3.3	The World Bank	216
8.3.4	The European Community	217
9	**Policies in Industrialized Countries**	219
9.1	Government Programs	222
9.1.1	Programs for the Preservation and Enhancement of Domestic Mineral Production	222
9.1.2	Promoting Mineral Exploration	223
9.1.3	Promoting Mining Investments	225
9.1.4	Stockpile Programs	226
9.1.5	Promoting Mineral Trade	229
9.1.6	Promoting Mineral Research	229
9.2	Policies of Mining Companies	231
9.2.1	Investment Strategies	231
9.2.2	Improvements in Mineral Utilization	233

10	**Policies and Special Problems in Developing Countries**	234
10.1	Concession Policy	235
10.2	Participation Policy	236
10.3	Fiscal Policy	238
10.4	Marketing Policy	239
10.5	Importance of Small-Scale Mining in Developing Countries	240
10.5.1	Definition and Characteristics of Small-Scale Mining	240
10.5.2	Production of Small-Scale Mining	241
10.5.3	Advantages and Disadvantages of Small-Scale Mining	243
10.5.4	Forms of Technical Cooperation in Small-Scale Mining	245
10.6	Effects of Mining in Developing Countries	246
10.6.1	Primary Effects	247
10.6.2	Secondary Effects	247

References 253

Index 263

Abbreviations

AAC	Anglo American Corp. of South Africa Ltd.
AAS	Atomic Absorption Spectometry
ACP	Countries from Africa, the Caribbean and the Pacific
ALCOA	Aluminum Company of America
AMC	Amalgamated Metal Corporation
ASARCO	American Smelting and Refining Company
ASEAN	Association of Southeast Asian Nations
ATPC	Association of Tin-Producing Countries
bbl	barrel
BGR	Bundesanstalt für Geowissenschaften und Rohstoffe, Hannover
BHP	Broken Hill Pty Company Ltd.
BOPD	barrels oil per day
CDA	Copper Development Association
Comex	Commodity Exchange, New York
COMIBOL	Corporación Minera de Bolivia
CTS	Consolidated Tin Smelters Ltd.
Cu	Copper
DAC	Development Assistance Committee
DC	Development Committee
EC	European Communities
ECU	European Currency Unit
EDC	Canadian Export Development Corporation
EDF	European Development Fund
EM	Electromagnetic
EOSAT	Earth Observation Satellite Company
ESCAP	Economic and Social Commission for Asia and the Pacific
Ex-Im Bank	US Export-Import Bank
FE	Frequency Effect
GATT	General Agreement on Tariffs and Trade
GDP	Gross Domestic Product
GNP	Gross National Product
IBA	International Bauxite Association
IBRD	International Bank for Reconstruction and Development (World Bank)
ICP	Inductively Coupled Plasma Spectrometry
IEA	International Energy Agency
IFC	International Finance Corporation
IMF	International Monetary Fund
IP	Induced Polarization

IRR	Internal Rate of Return
ITA	International Tin Agreement
ITC	International Tin Council
JCI	Johannesburg Consolidated Investment Co., Ltd.
KLTM	Kuala Lumpur Tin Market
LDC	Least-Developed Countries
LME	London Metal Exchange
mGal	Milligal
MSAC	Most Seriously Affected Countries
MSS	Multispectral Scanning
n.a.	not available
NIC	Newly Industrializing Countries
NPV	Net Present Value
OAPEC	Organisation of Arab Petroleum Exporting Countries
ODA	Official Development Assistance
OECD	Organization for Economic Cooperation and Development
OPEC	Organization of Petroleum Exporting Countries
OPIC	US Overseas Private Investment Corporation
Pb	Lead
PTA	Primary Tungsten Association
SLAR	Sidelooking Airborne Radar
TC	Total Count
TM	Thematic Mapper
tpa	tons per annum
UN(O)	United Nations (Organization)
UNCLOS	United Nations Conference on the Law of the Sea
UNCTAD	United Nations Conference on Trade and Development
UNDP	United Nations Development Programme
UNIDO	United Nations Industrial Development Organization
US	United States
XRF	X-Ray Fluorescence
Zn	Zinc

Introduction

Since 1974, profound changes have taken place in world mineral markets, the full extent and effects of which can only now be assessed. The mineral price increases that followed the oil price rises of 1973/74 and, particularly, 1979/80, brought about a mood of euphoria among mineral exporting countries and mineral politicians at the U.N. Conference on Trade and Development. Price rises brought about extensive exploration programs for many mineral resources in the late 1970s. Simultaneously, a new awareness of the importance of mineral resources led to more rational use, to more widespread recycling of metals, and to the development of new plastic, ceramic and glass materials which substitute for some of the traditional mineral resources.

The contradictory trends of increased mine development and production on the one hand, and stagnation of consumption on the other, resulted in oversupply, which caused prices to fall to such an extent that mines were closed, mining companies went out of business, and mineral exporting countries lost an important source of foreign exchange. Such cyclical alternations between shortage and oversupply are, in fact, typical for the mining industries and world trade in minerals. The intensity, duration, and impact of the hectic developments since 1974 are, however, historically unprecedented. For world mineral markets and policy, the analysis of these changes has far-reaching implications for the future.

We do not foresee a physical exhaustion of our mineral resources in the foreseeable future, but rather limitations to their availability by economic, technical and ecologic factors involved in their production and use. Analysis of the economic and technical factors affecting mineral resources is the topic of mineral economics, which ranges from the geologic, technical and economic evaluation of mineral resources to market and mineral policy analysis of mineral commodities on a national and international level. It is the task of the economic geologist, mining engineer, and mineral economist to assure a sufficient, secure and affordable supply of mineral resources.

This book "International Mineral Economics" was conceived with these scientific, technological and economic considerations in mind. The geologic and technical chapters provide the initial framework, dealing with fundamental concepts in the make-up and genesis of mineral deposits, which serve as a basis for the discussion of the most relevant exploration strategies and methods. This section concludes with a brief introduction to mining and mineral dressing technology. The economic evaluation of mineral deposits and the approaches to market analyses are covered in the central section of the book. National and international mineral policies, which to a large extent determine world trade between industrialized and developing countries, are thoroughly discussed in the final section.

Since the subject of mineral economics involves a wide range of specialists, from geologists to mining engineers and economists, several potentially misleading professional designations must be defined beforehand:

- *Economic geology, exploration geology, mining geology:* The branch of geology that deals with the geologic factors, exploration, and exploitation of mineral resources. The three expressions overlap in meaning to a certain extent, the term "economic geology" emphasizing the research aspects, and "exploration" and "mining geology" stressing the applied aspects.
- *Mining engineering and mineral dressing:* The fields of engineering which are concerned with the mining and processing of mineral resources.
- *Mineral economics:* The field of economics that is involved in the economic evaluation of mineral prospects, and the analysis of mineral markets and international trade of mineral commodities.

There is considerable overlap between these fields, and a close cooperation between specialized professionals is essential to achieve the common goal of discovering, developing, and distributing mineral resources for use by society.

Part I
Economic Geology, Mineral Exploration, and Mineral Development

Part 1

Chapter 1 Mineral Deposits and Metallogenic Concepts

A geologist defines a mineral as a naturally occurring, crystalline solid of specific chemical composition, structural arrangement of component atoms, and physical properties, e.g., quartz, pyrite, diamond. Minerals combine to form rocks, defined as naturally occurring accumulations or mixtures of minerals formed by geologic processes, e.g., granite, limestone, oil shale. In the context of mineral economics, the term "mineral resources" is commonly used to denote all solid, liquid, or gaseous geologic materials exploitable for use.

In geologic and economic practice, mineral resources are generally classified, by use, into building materials, industrial minerals, metallic minerals, and energy resources, all of which are needed for construction, agriculture, industry, housing, or transport. Good summary descriptions of the types of resources and of their occurrence and use are given by Skinner (1986) and Wolfe (1984). The value of world mineral production and an overview over the global distribution of selected mineral resources are given in Tables 1.1 and 1.2.

Building Materials are bulk commodities, for example stone, crushed rock, gravel, sand, clay-rich rock for the manufacture of bricks, and limestone as raw material for cement. The building materials are geologically common, require little processing, and are used in large quantities for all kinds of construction, from dams and highways to housing. The price per unit is low; exploration, mining, and processing are inexpensive; and high transportation costs require closeness to market. Supply and reserves are large and substitution between the commodities is easy. Because of the market limitations imposed by high transportation costs, most individual mining operations are relatively small, even though the overall value of the commodity mined countrywide is large.

Industrial Minerals, often called nonmetallic mineral resources, are comprised of rocks and minerals which are used as fertilizers, raw materials in the chemical industry, abrasives, and fillers. The industrial minerals are geologically abundant, can often be used as they occur in nature without further processing, and are produced and consumed in large quantities.

Metallic Mineral Resources are used to extract metals. A rock that contains a high enough concentration of a metal or of a metal-bearing mineral to be exploitable under current economic conditions is called an *ore*. The metallic resources are much less abundant and much more irregularly distributed in the earth's crust than the building materials or the nonmetallic minerals. Metal-bearing minerals require processing for the extraction of the metals, and the cost of mining and processing is generally very large in comparison with the cost of transportation, making proximity of the mineral deposit to market a secondary concern. Because of the cost and supply factors inherent in the production of the metallic resources, recycling contributes sig-

Table 1.1. Value of world mineral production (in $ billion)

	1950	1960	1970	1980
Global GNP	3932.00	5805.00	9363.00	13108.00
Energy fuels				
Petroleum and gas	24.30	37.00	50.30	704.70
Coal	80.60	86.40	104.60	169.90
Others	6.20	13.00	28.80	104.90
Total	111.10	136.40	191.60	979.50
In % of GNP	2.82	2.35	2.05	7.47
Metallic resources				
Iron and iron alloys	20.80	33.50	48.40	47.60
Precious metals	4.40	5.40	7.60	39.10
Base metals	12.20	25.90	54.70	68.80
Others	0.40	0.60	1.80	1.30
Total	37.80	65.40	112.50	156.80
In % of GNP	0.96	1.13	1.20	1.20
Non-metallic resources				
Construction materials		39.90	69.30	100.60
Fertilizers	2.90	6.00	8.50	16.50
Diamonds		2.00	4.10	9.10
Chemical resources	6.30	12.10	18.00	32.40
Total	40.70	60.00	99.90	158.60
In % of GNP	1.04	1.03	1.07	1.21
Total mineral resources	189.60	261.80	404.00	1295.00
In % GNP	4.82	4.51	4.32	9.88

Source: Sutulov (1983).

nificantly to the market of a variety of metals. The metallic minerals include (a) iron and the steel alloys manganese, chromium, vanadium, molybdenum, tungsten, nickel, and cobalt; (b) the light metals aluminum, magnesium, and titanium; (c) the base metals copper, lead, zinc, cadmium, tin, antimony, bismuth, and mercury; (d) the precious metals gold, platinum, and silver; and (e) the fuel metals uranium and thorium. This array of metals of diverse geologic occurrence requires a variety of exploration strategies, exploitation approaches, and processing methods. Part 1 of *International Mineral Economics* deals mainly with these metallic resources, their geologic setting, and the exploration, exploitation and processing methods used for their discovery and production.

Mineral Fuels include the fossil fuels petroleum, natural gas, tar sand, oil shale, and coal on one hand, and the nuclear fuels uranium and thorium, which are similar in geologic occurrence to the metallic resources, on the other. Of all mineral resources, the fossil fuels have by far the highest value (Table 1.1). They share many of the exploration, evaluation, marketing, and policy concerns of the other types of mineral resources, as will be illustrated by occasional examples drawn from the petroleum industries, and as discussed more fully in Part 3 under mineral trade and policies.

Table 1.2. The regional distribution of world reserves of selected mineral resources. (Distribution in %; reserves in million metric tons)

Resource	Western Europe	European Community	Eastern Europe	Africa	North America	Central America	South America	Asia	Oceania	Total (million tons)
Al	5	3	2	36	0	10	18	8	21	5200
Cu	–	–	17	14	25	6	28	5	5	505
Pb	13	5	18	8	30	5	5	7	14	165
Sn	3	3	10	7	1	–	14	61	4	10
Zn	16	8	9	9	28	2	11	15	10	240
Fe	5	2	32	4	16	–	20	11	12	98000
Mn	0	0	26	59	–	0	3	3	9	1361
Co	1	–	9	68	1	8	–	7	6	3
Cr	1	–	1	98	–	–	–	1	–	1007
Mo	–	–	7	–	60	1	28	4	–	10
Ni	3	1	13	4	15	8	2	22	33	54
V	1	1	46	49	1	–	1	1	1	19
W	7	1	8	–	15	1	2	63	4	3
Hg	52	8	11	8	11	6	1	11	–	0.15
Sb	8	3	7	7	4	5	10	56	3	5
Ti	15	1	2	13	22	–	20	20	8	273

Source: Office of Statistics of the European Community: EC Resource Inventories, Luxemburg (1982).

The resource industries have traditionally been specialized in one or the other of the groups of mineral resources given above. As a more recent development, considerable diversification has taken place, blurring the once well-defined boundaries.

A few geologic/economic terms need to be defined for the ensuing discussion of mineral resources:

- An *ore* is a rock with a metal content that is high enough for profitable exploitation under current technologic and economic conditions.
- A *mineral deposit, ore deposit,* or *mine* is a geologically delineated, local accumulation of a specific mineral resource which can be exploited under present-day economic conditions. These terms usually refer to deposits of metals and industrial minerals. The term "mine" may be applied to a mineral deposit that is no longer exploited.
- The *grade,* or *ore grade* is the concentration of a usable mineral or metal in a rock or a mineral deposit.
- *Reserves* are measured quantities of minerals or metals which can be exploited profitably with currently available technology and under present economic conditions.
- *Resources* are potentially usable quantities of minerals or metals which may or may not be exploitable under current conditions.

The significance and limitations of these terms are further discussed in section 3.3.4.

1.1 Composition of the Earth and Types of Mineral Resources

From the center outward, the earth is made up of a core, a mantle, and a crust. The core is composed mainly of iron-nickel, and the mantle mainly of the rock-forming elements oxygen, silicon, aluminum, iron, magnesium, and calcium. Together, these two inaccessible parts of the earth account for over 98% of its radius and for more than 99.6% of its mass. The earth's crust is a thin outer shell about 10–100 km thick, which comprises about 0.4% of the earth's mass. It is composed of the rock-forming elements mentioned above, but is richer in silicon, calcium, potassium, and sodium.

The oceans are underlain by a relatively thin (10 km), homogeneous crust which covers about 70% of the earth's surface. The continents have a much thicker crust (30–100 km) which is more heterogeneous because magmatic, sedimentary, and metamorphic processes have led to segregation and local concentration of elements. The continental crust, exposed over approximately 30% of the earth's surface, is the main domain of the exploration and mining geologist, a domain which is limited at depth by the geothermal gradient (on average 30°C/km), which effectively bars any nonhydrocarbon resources located deeper than about 5 km from exploitation. Seafloor exploration and mining are relatively new pursuits for the minerals industries which have shown vast potential for the production of hydrocarbons, and promise for the production of a variety of industrial minerals and metals.

Any rock in the earth's crust contains small amounts of metallic and non-metallic resources, and the amount of these elements in a cubic km of average crustal rock is impressive (Table 1.3). However, the cost of exploitation in terms of energy, capital, and labor vastly exceeds the value of the contained resource, and we have to rely on

Table 1.3. Metal concentrations, quantities, ore grades, and ore enrichment factors for selected elements in the earth's crust

Element	Chemical symbol	Average concentration % (1)	Quantity/km³ (000 mt)	Typical ore grades % (2)	Enrichment factor
Aluminum	Al	8.1	250000	30	4
Iron	Fe	5.4	150000	53	10
Titanium	Ti	0.5	15000	0.7–15	2–40
Manganese	Mn	0.10	3000	31	310
Chromium	Cr	0.01	300	30	3000
Nickel	Ni	0.008	200	1	130
Zinc	Zn	0.007	190	4	570
Copper	Cu	0.005	135	0.5–4	100–800
Cobalt	Co	0.002	60	0.4–2	200–1000
Lead	Pb	0.001	35	5	3850
Uranium	U	0.0003	7	0.3	1100
Tin	Sn	0.0003	7	0.3	1200
Molybdenum	Mo	0.0002	4	0.2	1300
Tungsten	W	0.0001	3	0.7	5800
Silver	Ag	0.00001	0.2	0.01	1400
Gold	Au	0.0000003	0.01	0.001–0.0001	300–3000

(1) Source: Krauskopf (1979).
(2) Dependent on type of deposit. Sources Cox and Singer (1986) and Crowson (1986).

local enrichments of certain elements, formed by a variety of natural processes, for the exploitation of most mineral resources. Typical enrichment factors of some elements, required to achieve the concentration or "ore grade" needed for mining, are included in Table 1.3.

The average concentration of elements in the earth's crust controls their geologic behavior and their occurrence in mineral deposits. Nine elements (the major elements) make up over 99.5% of the continental crust, while the remaining 79 naturally occurring elements (the minor elements) account for less than 0.5% (Skinner 1979). The major elements, among them iron and aluminum, are abundant enough to form minerals and rocks readily, while the minor elements, including most of the metals, are generally not abundant enough to form minerals of their own. Instead, they substitute for the major elements in the common rock-forming minerals in subeconomic concentrations. They are locally enriched to form exploitable mineral deposits only under exceptional geologic conditions, combining with oxygen, sulfur, and carbon to form the oxide, sulfide, sulfate, and carbonate ore minerals we mine. The relationship between crustal abundance, ore formation, and resource assessment are explained in more detail in Skinner (1979), and in Harris and Skinner (1982).

A homogeneous earth crust ist devoid of the high mineral concentrations needed for profitable exploitation. The formation of exploitable mineral deposits is closely tied to the overall evolution of the earth's crust and to local geologic processes. The main task of the exploration and mining geologist is to understand the ore-forming geologic processes, and to recognize the geologic environments which are favorable to the formation of mineral deposits.

1.2 Geologic Cycles and the Formation of Mineral Deposits

According to our present interpretations, the earth evolved from a planet with a molten, homogeneous crust some 4.5 billion years ago to its present configuration of a cool, solid crust overlying a hot, partially molten interior. The processes involved in the cooling, crystallization, and re-working of the primordial earth's materials are the same which we see in action today. Under exceptional geologic circumstances, these processes from mineral deposits now, but at a rate so slow that only accumulations formed over long geologic time spans are large or rich enough to be exploitable. Mineral deposits are, therefore, largely nonrenewable in relation to the human time scale.

Ore-forming processes and ore deposits are closely tied to rock associations, geologic settings, and geologic cycles which are fairly well understood. Deep in the earth's crust, molten rock or *magma* rises and crystallizes as it cools on the way to the surface. A surface expression of this magmatism is the intense volcanic activity which we can observe, for example, around the Pacific margin. Mountain ranges are weathered and eroded down, and water carries the clastic and chemical constituents to depositional basins in which sedimentary deposits of gravel, sand, clays, limestones and other chemical precipitates form. Downwarping of the earth's crust under these basins leads to burial of the sediments to depths at which they are metamorphosed, i.e., recrystallized under heat and pressure to the point of melting to form new magma. Structural movements cause local fractures in the earth's crust which often provide channelways for ore-forming solutions, and in which ore minerals may precipitate to form veins.

The rock-forming and structural developments can be interpreted in terms of the plate tectonic model of crustal evolution (Fig. 1.1). This model explains how new crust is formed in *rift zones,* mostly at mid-oceanic ridges, by the addition of basaltic magma from depth. This process forms a homogeneous oceanic crust which has undergone few of the processes that are essential for the segregation of metals into ore deposits. Exceptions are local segregations of chromium and nickel in the deepest parts of the oceanic crust, and precipitation of massive sulfides of copper and iron at

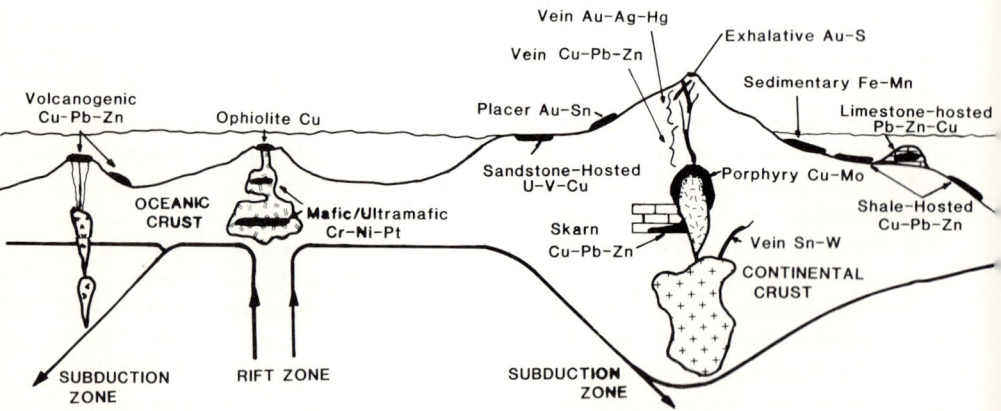

Fig. 1.1. Schematic diagram illustrating the geologic settings of mineral deposits, and their relationship to plate tectonic processes

places where hot, metal-bearing brines enter the ocean through the rift zone, as illustrated by the black smokers discovered on ocean ridges. Oceanic crust is consumed in subduction zones at places of plate collision. This process is accompanied by intense earthquake and volcanic activity, and leads to magmatic differentiation processes. The segretation of granitic magmas and the formation of a large variety of magmatic and magmatic-hydrothermal mineral deposits is related to processes of subduction. Collision and subduction build large mountain ranges such as the Andes, in which mineral deposits form by magmatic differentiation.

In the last decades, intensive research has led to a vastly improved understanding of many of the factors, processes, and geologic environments which are essential to the formation of mineral deposits, and to the recognition of criteria by which the special, ore-forming environments can be recognized. The strategies of modern exploration programs are based in large measure on models which describe these geologic environments and provide criteria for their recognition.

1.2.1 Mineral Deposits Associated with Magmatic Processes

Magma derived from the mantle and brought up in oceanic rift zones cools rapidly and crystallizes to form the basaltic rocks which constitute the oceanic crust. If, on the other hand, the magma is emplaced deep in the continental crust, it cools more slowly and crystallizes gradually, producing typical assemblages of rocks and ores of different composition at different stages in the cooling process. Early crystallization forms rocks rich in iron, and magnesium, silicates which are generally dark green to black. Ore minerals may segregate into separate layers at this stage, for example the extensive layers of chromite, or of nickel sulfides and platinoids, in the Bushveld Complex in South Africa or the Stillwater Complex in Montana, USA. The early crystallization depletes the magma of iron and magnesium and leads to its relative enrichment in silicon, aluminum, calcium, sodium, and potassium. This magma crystallizes to form quartz-feldspar-rich rock of the granite family which makes up a large proportion of the upper continental crust. The crystallization of the magma liberates much water which migrates upward and outward into areas of lower pressure and temperature, a flow which is enhanced by convective flow of water from the surrounding rocks. The hot water or *hydrothermal solution* often contains metals which are precipitated in the uppermost kilometers of the earth's crust. Depending on the depth and temperature of precipitation, different mineral and element associations prevail, for example oxides of tin and tungsten in the deep, high-temperature zones; sulfides of copper, molybdenum, lead, and zinc in the intermediate zones; and sulfides or sulfosalts of silver and native gold in the near-surface, low-temperature zones (Fig. 1.1). The minerals may be finely disseminated between silicates or concentrated in fine fractures in the igneous rock, for example in the Bingham *porphyry copper deposit* in Utah (Figs. 1.2a, 1.3; Table 1.4). Limestones surrounding the intrusive react with the hydrothermal solutions and may be partly replaced by minerals of tungsten, copper, lead and zinc, in contact metasomatic or *skarn deposits*. The Carr Fork deposit which borders the Bingham porphyry copper mine is a good example. If the solutions pass through open fractures and the metals precipitate in them, *vein deposits* of copper, lead, zinc, silver, and gold form (Fig. 1.2b; Table 1.5). Such base- and pre-

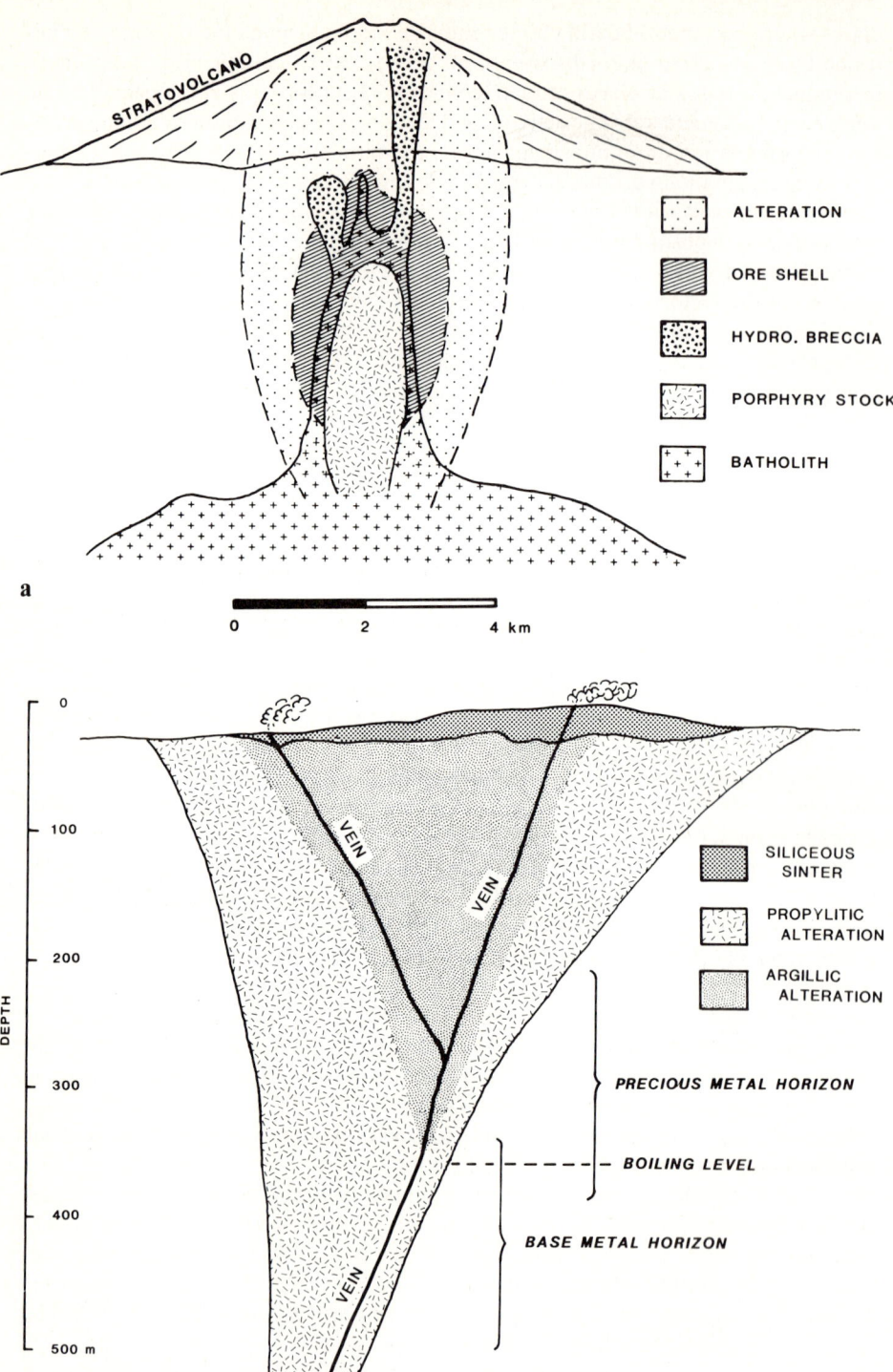

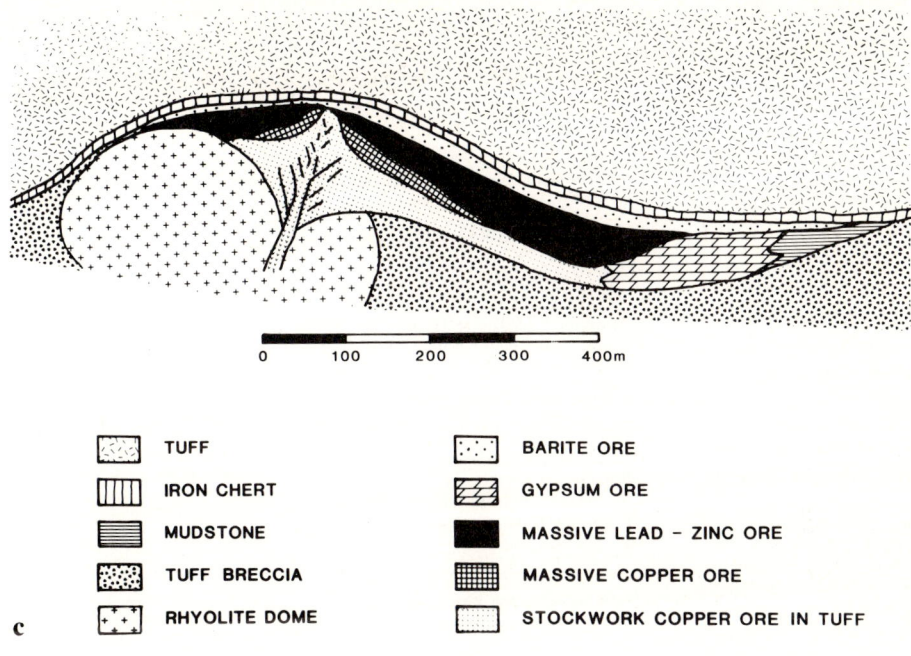

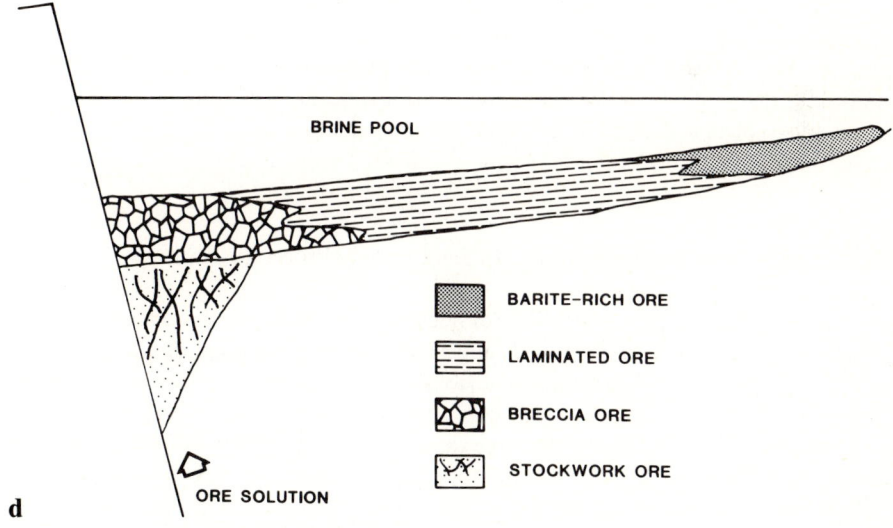

Fig. 1.2a–d. Geologic models of some major types of ore deposits. **a** Geologic model of a typical South American porphyry copper deposit. (After Sillitoe 1973). **b** Geologic model of a precious-metal vein deposit. (After Buchanan 1981). **c** Geologic model of a volcanogenic copper-lead-zinc deposit. (After Horikoshi and Sato 1970 and Sato 1974). **d** Geologic model of a sedimentary-exhalative lead-zinc deposit. After Lydon 1983)

Table 1.4. Geologic model of a porphyry copper deposit rich in molybdenum

General geology

Rock types: Quartz monzonite to tonalite intrusives into igneous, volcanic, or sedimentary rocks.
Textures: Intrusives associated with ore are porphyries with fine to medium-grained groundmass.
Age: Most are Mesozoic to Tertiary, but can be of almost any age.
Tectonic setting: Many faults.
Associated deposit types: Skarns containing copper, zinc, or gold; veins with base-metal sulfosalts and gold; gold placers.
Metal concentrations: Copper, molybdenum, lead, zinc, tungsten, gold, silver.

Deposit description

Ore minerals: chalcopyrite, pyrite, molybdenite; peripheral vein and replacement deposits with chalcopyrite, sphalerite, galena, perhaps gold; outer zone sometimes with gold and sulfides of copper, silver, antimony.
Texture/structure: Veinlets, disseminations, massive replacement of country rock.
Alteration: Approximately concentric alteration rings of clay, mica, feldspar, other minerals often several kilometers away from deposit.
Geochemical signature: Central zone of copper, molybdenum, tungsten; outer zone of lead, zinc, gold, silver, arsenic, antimony, tellurium, manganese, rubidium.
Examples: El Salvador, Chile; Silver Bell, Arizona (USA); Highland Valley, British Columbia (Canada).

Source: Cox DP (1983).

cious-metal veins account for much of Mexico's and Peru's mineral wealth. The metal-bearing hydrothermal solutions may also migrate laterally into permeable or chemically reactive rocks to form blanket-shaped sulfide deposits, or they may reach the surface and precipitate gold, silver, and mercury in siliceous or calcareous hot-spring sinters, as illustrated by the high gold contents in some vents of active geothermal fields in New Zealand. Where metal-bearing volcanic solutions enter marine environments, volcanic-sedimentary accumulations of copper-lead-zinc form, as discussed in the next section.

1.2.2 Mineral Deposits Associated with Sedimentary Processes

The erosion of continents and filling of sedimentary basins in the ocean sets up a geologic and chemical cycle that is closely related to the formation of a large variety of mineral deposits. During weathering, rocks are disassembled into insoluble, fragmental constituents, for example quartz or, occasionally, gold or heavy minerals, and into chemically dissolved elements, for example calcium, sodium, or potentially ore-forming metallic elements like iron, copper, lead, and zinc. The fragmental constituents are transported by surface waters to be deposited as clastic sedimentary rocks, for example conglomerates and sandstones, where the velocity and carrying capacity of the transporting water decreases. Clastic sediments on the continents and in near-shore environments tend to be coarse-grained and may contain local enrichments of valuable minerals which have been transported with the clastic fraction, for example placer concentrations of gold in the Witwatersrand deposits in South Africa, or placers of tin in Southeast Asia.

Table 1.5. Geologic model of an epithermal gold-quartz-alunite vein

General geology

Rock types: Volcanic dacites, quartz latites, rhyodacites, rhyolites.
Textures: Porphyritic.
Age: Usually Tertiary, but deposits exist for a wide range of ages.
Depositional environment: Within volcanic edifices, ring fracture zones of calderas, areas of igneous activity with host rock of sedimentary evaporites.
Tectonic setting: Extensive fracture systems.
Associated deposit types: Porphyry copper, acid-sulfate hot springs, hydrothermal clay.
Metal concentrations: Copper, arsenic, antimony.

Deposit description

Ore minerals: Native gold, enargite, pyrite, silver-bearing sulfosalts, with associated chalcopyrite, bornite, tellurides, galena, sphalerite, huebnerite.
Texture/structure: Veins, breccia pipes, pods, dikes.
Alteration: Quartz, alunite, pyrophyllite; often surrounded by quartz, alunite, kaolinite, montmorillonite.
Ore controls: Fractures, intrusive activity.
Weathering: Yellow limonites, jarosite, goethite, white argillization with kaolinite, hematite.
Examples: Goldfield, Nevada (USA); Guanajuato, Mexico; El Indio, Chile.

Source: Cox DP (1983).

Chemically dissolved constituents are precipitated in favorable sedimentary environments, forming chemical sediments. The products of chemical weathering may be concentrated on the continents, for example in evaporite basins which yield large amounts of sodium and boron. Chemical precipitation on a much larger scale occurs in the oceans, particularly in near-shore environments, where evaporites may form under arid conditions if the depositional basin is protected from sea water circulation. Since geologically abundant elements are mostly involved in these depositional processes, very large deposits may form. An example are the banded iron formations, our main source of iron ore, which precipitated in near-shore environments about 2 billion years ago. Many large, stratiform deposits of copper, lead, and zinc have similarly formed in marine, sedimentary basins where an adequate source of metal and sulfur and a specific sedimentary environment favorable for the precipitation of metal sulfides existed. The details of source, precipitation, and environment of deposition of many of these sulfides remains debated. Volcanogenic massive sulfides are related to submarine volcanism (Fig. 1.2c; Table 1.6), for example the Kuroko copper-lead-zinc deposits in Japan, and many of the major base-metal deposits in Canada. Frequently, the formation of stratiform sulfide deposits does not appear to be related to any magmatism or volcanism, but rather to the circulation of hydrothermal solutions from other sources, for example from the dewatering of deep sedimentary basins. The resulting deposits are very similar in many respects to those of volcanogenic origin because the trapping mechanisms are similar (Fig. 1.2d; Table 1.7). In both cases, the ore minerals may precipitate on the ocean floor, as shown for example by the present-day precipitation of sulfides in deep basins in the Red Sea, or they may precipitate below the water-rock interface within the sediment, by migration of the metal-bearing solutions to the site of deposition through unconsolidated, porous sediments.

Table 1.6 Geologic models of Kuroko-type volcanogenic massive sulfides

General geology

Rock types: Felsic-intermediate marine volcanics, associated sediments.
Textures: Flows, tuffs, pyroclasts, breccias, other volcanic textures.
Age: Archean to Cenozoic.
Tectonic setting: Local extensional faulting or fracturing.
Associated deposit types: Quartz veins with gold; bedded barite.
Metal concentrations: Barium, gold.

Deposit description

Ore minerals: Upper zone of pyrite, sphalerite, chalcopyrite, pyrrhotite, galena, barite; lower zone of pyrite, chalcopyrite, sphalerite, pyrrhotite, magnetite; stockwork zone of pyrite, chalcopyrite, gold, silver.
Texture/structure: Greater than 60% sulfides; sometimes an underlying zone of disseminated or stockwork sulfides.
Alteration: Covering and adjacent to some deposits are zeolites, montmorillonite; also silica, chlorite, sericite.
Ore controls: Located near felsic top of volcanic or volcanic-sedimentary rocks; near center of volcanism; sometimes brecciated; sometimes near felsic dome.
Weathering: Yellow, red, and brown gossans.
Examples: Kidd Creek, Canada; Hanaoka, Japan; Macuchi, Equador.

Source: Cox DP (1983).

Just as the sulfide minerals may be precipitated at the sediment-water interface or within the unconsolidated rock, the timing of ore formation may correspond to the time of deposition of the sediment, to the time of its compaction and consolidation, or to a much later time when the sediments are fully indurated and may become mineralized by solutions travelling through porous rock or geologic structures. The Mississippi Valley lead-zinc deposits may be good examples of this process.

Sedimentary processes also form accumulations of the fossil fuels coal, petroleum, and natural gas. To form coal, peat is compacted and heated by subsidence and burial. Likewise, petroleum and gas form by maturation of organic constituents in sedimentary rocks by increased temperature and pressure. Petroleum and gas may migrate through porous rocks to form large reservoirs in favorable structures, or they may remain in the source rock to form oil shales.

1.2.3 Mineral Deposits Associated with Metamorphic Processes

Metamorphism, i.e., the recrystallization and final melting of igneous or sedimentary rocks, may be caused by intrusion of new magma or by deep burial. Contact-metasomatic hydrothermal deposits form around intruding magmas, as described above. Deep burial metamorphism may overprint pre-existing mineral accumulations, for example the large, sediment-hosted lead-zinc deposits at Broken Hill, Australia. Burial metamorphism also liberates large amounts of hydrothermal solutions which can dissolve metals from the country rock, to be precipitated when the solution encounters an environment with the proper temperature, pressure, and chemical conditions for ore formation. The formation of gold deposits in many of the Precambrian metamorphic belts is attributed to the transport of gold by metamorphic waters into

Mineral Deposits and Metallogenic Concepts

Table 1.7. Geologic model of sediment-hosted, submarine exhalative lead-zinc deposits

General geology

Rock types: Euxinitic sedimentary rocks (black shale, siltstone, sandstone, chert, micritic limestone).
Textures: Sedimentary thicknesses and facies change hinge zones; slump breccias.
Age: Middle Proterozoic; Ordovician to Mississippian.
Depositional environments: Epicratonic marine basins.
Associated deposit types: Stratiform barite deposits.
Metal concentrations: Maximum expected background in black shales of 500 ppm lead, 1300 ppm zinc, 750 ppm copper, 1300 ppm barium.

Deposit description

Ore minerals: Pyrite, pyrrhotite, sphalerite, galena, barite, chalcopyrite and minor amounts of many others.
Texture/structure: Finely crystalline, disseminated.
Alteration: Silicification, tourmalization, carbonate depletion, albitization, chloritization, dolomitization.
Weathering: Gossans with carbonates, sulfates, and silicates of lead, zinc, copper.
Geochemical signature: Lateral sequence of copper-lead-zinc-barium extending outward; vertical sequence of copper-zinc-lead-barium extending inward.
Example: Sullivan, Canada.

Source: Cox DP (1983).

gold-bearing quartz veins. Excepting this type of deposit, however, regional metamorphism does not appear to be a major factor in the formation of metallic ore deposits.

1.3 Geologic Models of Mineral Deposits

The last decades have brought a vast increase in our understanding of the geologic setting, characteristics, and mode of formation of mineral deposits. At the same time, exploration has shifted from the search of deposits exposed on the surface to deposits concealed under soil or overburden. The depth penetration needed to discover concealed mineral deposits is achieved by the observation and interpretation of indirect geologic evidence, either through analogy with ore deposits which are known from similar geologic settings, or through a genetic interpretation of the processes that are known to lead to the formation of specific types of deposits. The criteria which help in the characterization, interpretation, and discovery of ore deposits are summarized in *geologic models*. Models for four of the most economically relevant and thoroughly studied types of ore deposits – porphyry copper, precious metal vein, volcanogenic massive sulfide, and sedimentary-exhalative – are discussed above and are presented in Figs. 1.2a, b, c, d, and Tables 1.4–1.7. Advanced geophysical and geochemical exploration methods complement geologic inference in the search for concealed deposits of these and other types.

Geologic models take advantage of the fact that particular types of mineral deposits are found in geologic provinces characterized by distinct suites of rocks and structures which reflect the age and geologic history of the province. Mineral deposits are formed as part of the geologic processes which have taken place and are associated with the corresponding rocks and structures.

Fig. 1.3. The Bingham porphyry copper mine, Utah

The statistical association of characteristic features observed in mineral deposits is used to develop *empirical ore deposit models*. For example, the association of rusty-brown oxidation zones with many sulfide deposits is one of the traditional guides in empirical exploration. The application of advanced methods of research, including improved analytical methods, experimental and theoretical geochemistry, fluid inclusion determination of temperature and composition of the fluids which formed a mineral deposit, and isotope geochemistry, have given new evidence for the genetic interpretation of mineral deposits. These methods, now often used in exploration programs, provide criteria which help the exploration geologist to predict the favorability of geologic environments for the formation of mineral deposits. The most recent example of such application of research tools to exploration may be the routine analysis of fluid inclusions to guide exploration for precious metals. Progress in the genetic interpretation of mineral deposits leads to progress toward *genetic ore deposit models*, which provide a coherent framework for the organization and evaluatation of empirical data, and for the choice of the most relevant exploration approaches.

From both the empirical and the genetic models, the exploration geologist assembles an *exploration model,* i.e., a set of recognition criteria for exploration. Some of these criteria are diagnostic for the presence or absence of an ore deposit while others are permissive. The criteria chosen should be as diagnostic as possible, and should be both cost- and time-effective. Excellent summaries of mineral deposit models are given by Cox and Singer (1986) and Eckstrand (1984); the former includes ore grade and tonnage statistics for a large variety of deposits.

The ability to recognize areas of mineral potential on the basis of geologic characteristics and genetic interpretations is essentially a function of the training and experience of the exploration geologist. From experience and by analogy, geologists infer

Mineral Deposits and Metallogenic Concepts

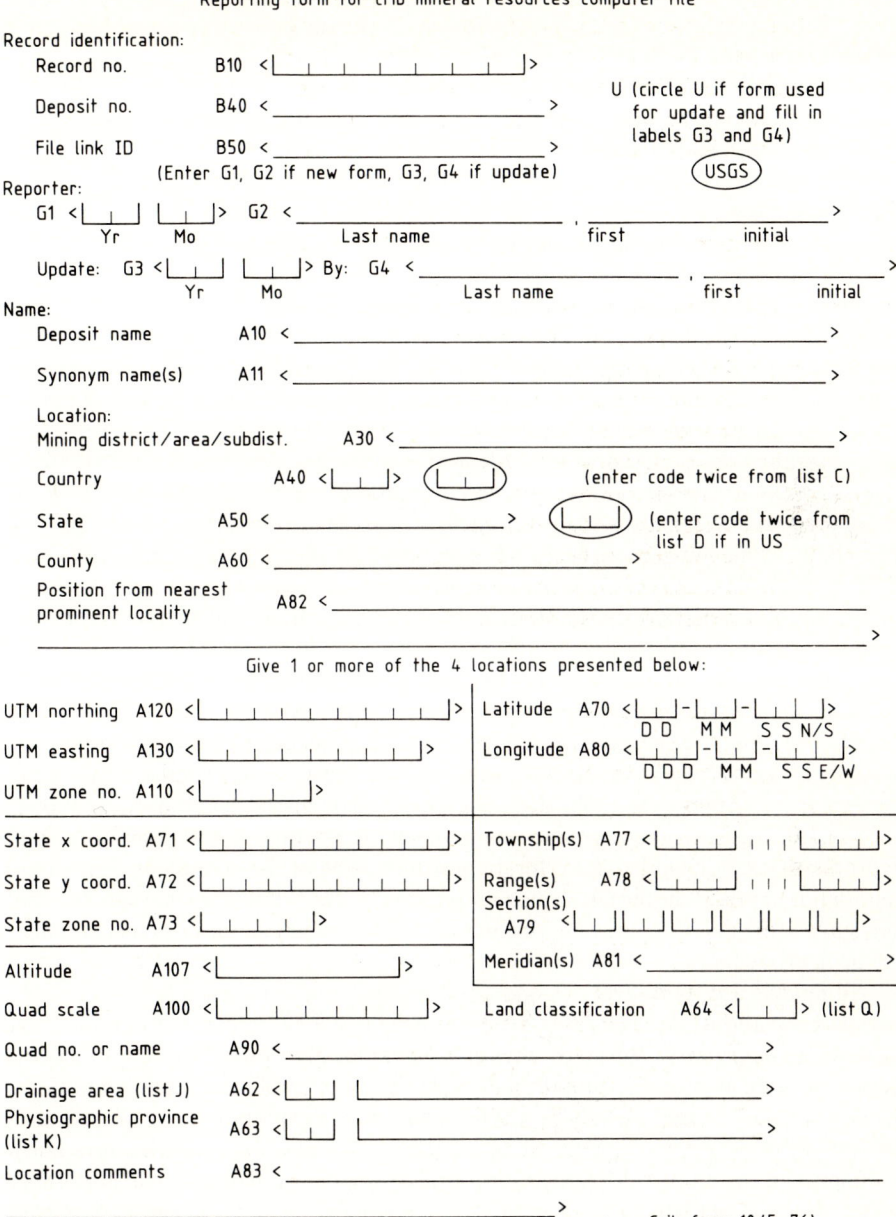

Fig. 1.4. Standard reporting sheet for the CRIB mineral resources computer file. Note that this is just one of a number of reporting sheets covering different aspects of mine geology, exploration, and valuation. (Dalkins et al. 1977)

which types of mineral deposits may be expected in a particular geologic environment. The importance of geologic experience in the ability to recognize these environments, and the need for exposure of exploration geologists to a large variety of types of ore deposits cannot be overemphasized. To help the geologist, a wealth of data about mineral deposits has been assembled in computer data banks, and programs have been elaborated which compare geologic features of a mineral prospect with the characteristics of ore deposit models. The Prospector System, developed by the Stanford Research Institute, is one example of such programs, and a typical data sheet for computer storage developed by the USGS is shown in Fig. 1.4.

The usefulness of models in exploration can be illustrated, for example, by the porphyry copper models presented by Lowell and Guilbert (1970) and by Sillitoe (1973) (Fig. 1.2a; Table 1.4). The association of the porphyry copper deposit with subvolcanic rocks, the configuration of the alteration zones and metal zones outward from its center, and the red-brown iron oxides of the weathering zone are all part of exploration models which led to the discovery of a large number of copper deposits in the southwestern US, northern Mexico, and South America. Instructive case studies on the use of these models in exploration are given by Lowell (1968) for the discovery of the Kalamazoo deposit in the San Manuel district, Arizona, and by Wallace (1975, 1978) for that of the Henderson porphyry molybdenum deposit in Colorado. In these two successful exploration programs, drilling was guided by the observation of characteristics of alteration, mineral composition, and metal distribution encountered in the drill holes, which were compared to the genetic model for this type of deposit. It took courage on the part of the geologists and management to propose and carry through the expensive drilling program on the basis of circumstantial evidence, but the success justified the risks. The Buchanan (1981) model for precious-metal veins (Fig. 1.2b; Table 1.5) has been extensively and successfully used in the exploration of precious metal deposits. This model illustrates how a study of alteration assemblages, ore and gangue minerals, and temperatures and compositions of hydrothermal fluids can guide exploration by indicating the level of the hydrothermal vein that is exposed on the surface or underground, and whether or not any ore that may have formed at the time of hydrothermal activity might be lower in the system or has been eroded away.

All models are simplifications and abstractions based on a large number of individual observations. As such, they need refinement as new evidence is obtained and have to be adjusted to the local situation as exploration proceeds. These models should be used with much caution and with an understanding of their limitations. While empirical models now form a solid basis for the exploration of most types of mineral deposits, genetic models are less reliable. We may, for example, have much confidence in the models developed for porphyry copper or hydrothermal gold-silver veins; however, our genetic interpretations of many sedimentary deposits are not yet as reliable. Even within their limitations, the models that have been developed have been very successful in predicting geologic environments which are favorable for the formation of ore deposits, and in the recognition of diagnostic indicators of mineral potential.

Notes on the Literature

Books and scientific articles by Brian J. Skinner provided inspiration for Chapter 1. Earth Resources (Skinner 1986) summarizes the geologic occurrence, uses, and consumption patterns of resources for the non-specialist, briefly and succinctly. In two benchmark papers (Skinner 1979; Harris and Skinner 1982), the formation of mineral deposits is related to the distribution of elements in the earth's crust, and the evolution of exploration, exploitation, and future availability of resources is analyzed on this basis. In the 75th Anniversary Volume of Economic Geology, edited by Skinner (1981), comprehensive descriptions of types of metallic-mineral deposits and exploration strategies, exploration methods, and resource assessments are presented, each topic by a specialist in the field.

Information about the uses, production, geologic setting, and future availability of metallic resources are assembled in Brobst and Pratt (eds.) (1973), and of non-metallic resources by Lefond (ed.) (1983). Crowson (1986) provides an instructive yearly update of mineral statistics – uses, demand/supply, trade, prices, and trends.

The geologic aspects of metallic mineral deposits are summarized in very accessible form in a textbook by Edwards and Atkinson (1986). For a more comprehensive treatment, the reader is referred to Guilbert and Park (1986), and for a clear and readable account of the relationships between plate tectonics and mineral deposits to Sawkins (1984). Peters (1986) combines a discussion of the geology of ore deposits with a comprehensive treatment of engineering, exploration, mine valuation, and mining aspects. His very comprehensive treatise stands out by its clarity and lively style. Geologic models for ore deposits are presented descriptively in Eckstrand (1984), and very succinctly, including grade and tonnage statistics, by Cox and Singer (1986). An atlas illustrating a variety of types of ore deposits (Dixon 1979) and another relating global geologic provinces to types of ore deposits (Derry 1980) complement the books on geologic models.

Chapter 2 Exploration Methods

The methods used to find and evaluate mineral resources are resource-specific. This is very apparent, for example, in the differences of exploration approaches and costs between the search for metallic ores and for petroleum, as illustrated in Figs. 2.1, 2.2, and 2.3. Likewise, the exploration for different metallic resources, for instance volcanogenic massive sulfide deposits of copper, lead, and zinc on the one hand, and vein deposits of gold and silver on the other, require fundamentally different methods of exploration.

Most mineral deposits located at or near the earth's surface have probably been discovered. The trend toward exploration of mineral deposits that are less exposed or not exposed on the surface has led to the increased application of a broad spectrum of new exploration tools which allow both a deeper penetration into the subsurface and a higher resolution of geologic, geochemical, and geophysical signals. The sophistication of the exploration tools applied, the large size required of mineral deposits for economic exploitation of most commodities, and the large expenditures involved in the development of these deposits, limits their exploration and development to well-organized mining or petroleum companies, or to government organizations which can support the large instrumental and personnel expense. Small com-

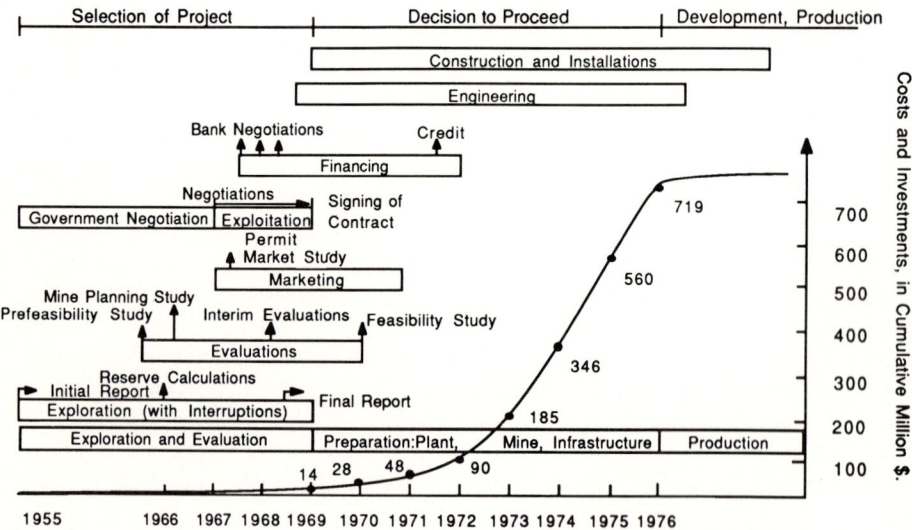

Fig. 2.1. Stages and costs of exploration and development of the Cuajone copper mine, Peru (Southern Peru Copper Corp)

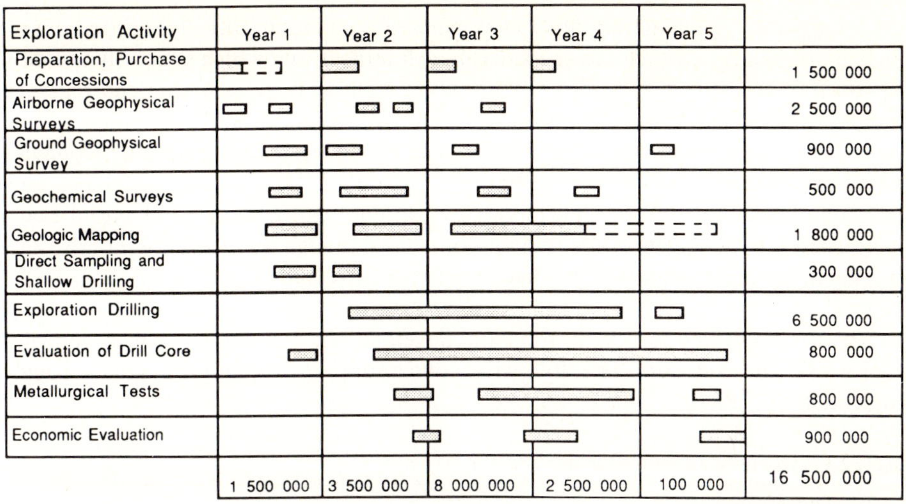

Exploration Activity	Year 1	Year 2	Year 3	Year 4	Year 5	
Preparation, Purchase of Concessions	⊐ ː ⊐	⊐	⊐	⊐		1 500 000
Airborne Geophysical Surveys	▫ ▫	▫ ▫	▫			2 500 000
Ground Geophysical Survey	⊏⊐	⊏⊐	⊏⊐		⊏⊐	900 000
Geochemical Surveys	⊏⊐	⊏⊐	⊏⊐	⊏⊐		500 000
Geologic Mapping	⊏⊐	⊏══⊐	⊏════⊐	⊏═ ═ ═ ═ ⊐		1 800 000
Direct Sampling and Shallow Drilling	⊏⊐	⊏⊐				300 000
Exploration Drilling			⊏══════════⊐		⊏⊐	6 500 000
Evaluation of Drill Core		⊏⊐	⊏══════════════════⊐			800 000
Metallurgical Tests			⊏═⊐	⊏══════⊐	⊏⊐	800 000
Economic Evaluation			⊏═⊐	⊏══⊐	⊏══⊐	900 000
	1 500 000	3 500 000	8 000 000	2 500 000	100 000	16 500 000

Fig. 2.2. Exploration methods and costs involved in the discovery and development of an intermediate-size Cu-Ni mine (cost estimate 1982)

panies and individual prospectors continue to be very successful in the exploration and mining of small gold deposits which may be very profitable because of the high unit value of gold.

2.1 Development Phases of Exploration and Mining Projects

Mining projects begin with the search for a mineral target that meets a set of economic, geologic, and technical requirements established by the mining company or organization. The motivation and extent to which the search is carried out will vary and depends on the organization that carries out the exploration. Private organizations, be they large holding companies or oil and mining companies, seek mineral deposits that yield an attractive profit or provide secure mineral supplies for company-owned manufacturing units. Government agencies may be motivated by political and societal rather than by purely economic aims, for instance the development of underdeveloped regions, the creation of employment opportunities in areas of high unemployment, the assurance of an adequate supply of strategic minerals, or the earning of foreign exchange. While private companies and government organizations use similar approaches to explore and develop mineral potential, government agencies are predominantly involved in regional assessments of mineral potential, either for land use planning or as a first step in exploration, often leaving further development of any discovered targets to private business. International organizations, for example the United Nations, and the international branches of some geologic surveys, for example the US Geological Survey, the French Bureau de Recherches Géologiques et Minières, and the German Bundesanstalt für Geowissenschaften und Rohstoffe, are involved in large-scale, regional exploration in many developing countries.

The discovery, evaluation, and development of a mineral project is carried out in a sequence of stages which are specific for each project, but which generally fall into a pattern of activities and decisions that evolve from strategic planning and regional geologic assessment to specific and local geologic, technical, and economic evaluation. The search evolves from an initial design of a program, to geologic reconnaissance exploration, delineation of targets for detailed follow-up, technical and economic feasibility studies, development of a mine, and mineral production. Each stage builds on the results of the previous one, even though there is considerable time overlap in the exploration and evaluation activities involved (Figs. 2.1, 2.2, 2.3; Table 2.1).

The time required for exploration and development of a mining project depends on its size and location. The following time requirements may provide a rough guideline:

Table 2.1 Typical exploration sequence and some costs

1. Program design		
Literature and map study	General personnel cost	
Geologic studies	General personnel cost	
Rock-ore associations	General personnel cost	
Structures – tectonic settings	General personnel cost	
Past exploration	General personnel cost	
→ Recommendations for regional reconnaissance		
2. Reconnaissance exploration		
Remote sensing	$100 – 150	per 1000 square km
Photogeology	$ 10 – 40	per square km
Reconnaissance mapping	$ 50 – 200	per square km
Airborne geophysics	$ 40 – 70	per square km
Stream sediment geochemistry	$ 20 – 100	per square km
→ Areas for follow-up examination		
3. Follow-up examination		
Detailed mapping	$400 – 1000	per square km
Ground geophysics		
Magnetic surveys	$100 – 250	per line km
Electric surveys	$800 – 1000	per line km
Electromagnetic surveys	$250 – 600	per line km
Radiometric surveys	$ 80 – 160	per line km
Gravimetric surveys	$300 – 600	per line km
Seismic surveys	$500 – 800	per line km
Stream sediment sampling	$ 50 – 100	per square km
Soil sampling	$300 – 1500	per square km
Rock chip sampling	$400 – 1500	per square km
→ Recommendation for prospect evaluation		
4. Prospect evaluation		
Drilling	$ 50 – 100	per m
Bulk sampling (adits, shafts)	$ 10 – 70	per cubic m
Reserve and grade estimation	Personnel cost	
Economic evaluation	Personnel cost	
→ Feasibility studies	Personnel cost	
→ Recommendation for development		

Cost figures after Peters (1987).

- Small deposits from 2 to 3 years exploration; from 1 to 2 years development of the mine and processing facilities.
- Medium-sized deposits, from 3 to 4 years exploration; from 2 to 4 years preparation of infrastructure, mine, and processing facilities.
- Large deposits, from 5 to 10 years or more exploration; from 5 to 8 years or more preparation of infrastructure, mining, and mineral processing facilities.

As these estimates show, the discovery and development of a mineral project is a long-rang proposition, lasting normally from at least 3 years for small projects to over 20 years for large projects. A longer time must generally be budgeted for projects carried out in remote areas, particularly in case of a lack of infrastructure, lack of qualified personnel, or extreme climatic conditions. The time frame, stages, and costs of mineral exploration and development involved in the development of the large Cuajone copper deposit in Peru from 1955 to 1976 are shown in Fig. 2.1, and the time and cost involved in the exploration of a medium-sized copper-nickel deposit in Canada is illustrated in Fig. 2.2. The development of large, offshore petroleum projects (Fig. 2.3) typically requires less time and a much shorter investment period.

The following terms need definition for a discussion of exploration activities:

- *Prospecting,* in American usage, is defined as the search for mineral resources by individuals on the basis of direct field observation of diagnostic rock or mineral assemblages. Historically important as a first step in finding mineral prospects and mines, this step has now been largely replaced by more systematic exploration approaches. In many countries, the term prospecting is used as a synonym for reconnaissance exploration.
- *Exploration* is the systematic search for mineral resources by mining companies or government agencies on the basis of direct and indirect geologic, geophysical, and

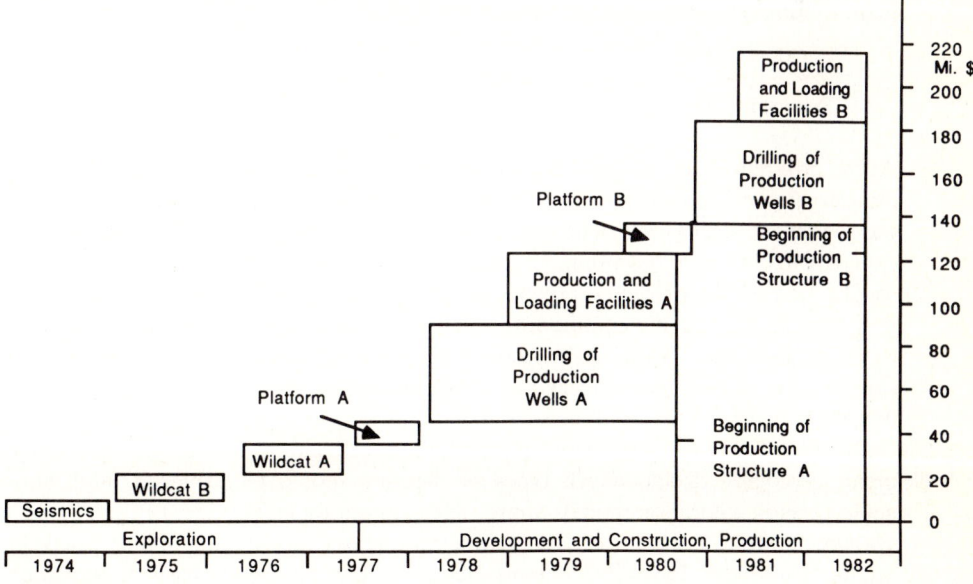

Fig. 2.3. Development costs of an off-shore petroleum field. (After Deutsche Schachtbau, Lingen)

geochemical evidence, and with the help of empirical and genetic models of mineral occurrence and formation. Exploration activities range from regional assessment to pre-development geologic evaluations of mining prospects.
- *Reconnaissance Exploration* is a preliminary exploration of large areas of unknown or little known geologic characteristics and mineral potential. Also called *grass roots exploration,* its aim is the discovery of target areas for detailed follow-up exploration.
- An *Exploration* or *Mining Lease* is the exclusive right, generally granted by the government, for the exploration or exploitation of mineral commodities in a specified area over a given time period.
- The *Pre-feasibility Study* involves an assessment of the profitability of a mineral prospect on the basis of initial exploration data and by use of economic parameters from comparable projects. The results of the prefeasibility study determine whether the increasingly large expense of a full geologic, technical, and economic evaluation of a prospect are justified.
- The *Feasibility Study* is the final evaluation of profitability of a mining venture in light of the results of exhaustive geologic exploration, assessment of mining and processing costs, environmental factors, and market analysis. This study usually forms the basis for the "go/no go" decision on developing a deposit.

2.2 Exploration Strategies, Procedures, and Stages

Geologic exploration develops in a sequence of strategies, procedures, and stages which go from large-scale to small-scale, inexpensive to expensive, and predictive to factual. Geologic, geophysical, and geochemical strategies and methods of exploration are discussed in the remainder of this chapter; the quantitative geologic assessment of mineral potential is the subject of Chapter 3.

2.2.1 Program Design

At the program design stage, management defines the economic parameters for mineral targets and geologists design the exploration program that promises success in the search for such target. The economic parameters vary widely depending on the expected exploration, development, and operating cost of the type of mineral deposit sought, and on the predicted revenue, payback period, and mine life. Management may also choose specific commodities, depending on the needs and expertise of the company involved. While specialization by commodities, for example petroleum, copper, or lead-zinc, characterized the 1960s and 1970s, most large mining companies now seek diversification in mineral and metal production.

In their search for deposits which fulfill the specifications established by management, geologists decide which types of deposits should be explored, and which geologic and exploration models apply. The exploration models identify geologic environments, rock associations, and geologic structures in which the mineral deposits are likely to occur. Favorable regions are selected, either on the basis of known potential as expressed by existing mines or mineral occurrences, or on the basis of a general knowledge of their geologic makeup. Exploration procedures are recommended to management by the geologic staff, and a time schedule and budget can be set.

2.2.2 Reconnaissance Exploration

Reconnaissance exploration is the geologic survey of large areas to assess their overall mineral potential, to define targets for detailed follow-up, and to reject areas of little apparent promise. In preparation for a reconnaissance program, a geologic staff is assembled, logistics are organized and, depending on the legal requirements of the state or country involved, exploration permits or exploration concessions are secured. Reconnaissance exploration evolves from regional geologic, geophysical, and geochemical surveys to local follow-up, with overlap in the sequence and intensity of activities as a differentiated picture of areas of varying potential emerges. In a comprehensive exploration program of an unknown region, a preliminary literature study of the geology and known mineral occurrences is typically followed by an overall geologic interpretation of remote sensing images obtained by satellites, or, more commonly, of aerial photographs. These investigations give a first impression of rock types, geologic structures, and other features which may be diagnostic of mineral accumulations. They are followed by airborne visual and geophysical surveys and surface geologic, geophysical, and geochemical work as required to recognize both the overall geologic framework and the specific mineral potential of the area. Geophysical surveys such as magnetic, electric, or gravity surveys, take advantage of physical properties of rocks and minerals to indicate the presence and distribution of rock types, geologic structures, and mineral deposits. These surveys cover large areas rapidly and allow geologic interpretations through cover by soil or vegetation which may make direct geologic observation impossible. Regional geochemical surveys outline areas of anomalous element concentrations in streams, soils, vegetation, and rocks, reflecting the presence of ore deposits. An integral part of these approaches is old-fashioned prospecting, that is, direct observation of telltale diagnostic features which indicate proximity or presence of mineral concentrations, for example the bleached appearance of alteration zones over porphyry copper or vein deposits, or the deep red or brown oxidation zones over weathered sulfides.

Most assessments of mineral potential for land use by government agencies, and most programs undertaken by international organizations in developing countries conclude at this point and leave further exploration to mining companies, often on a contractual basis. For the mining company, reconnaissance exploration ends with the selection of targets for detailed follow-up and rejection of areas of little apparent potential.

2.2.3 Detailed Follow-up Exploration

The aim of detailed exploration is to provide an exhaustive geologic evaluation of the potential of targets defined by reconnaissance exploration. The geologic follow-up determines the geologic setting, depth, geometry, grade, tonnage, and quality of the ore in mineral prospects.

Detailed exploration is restricted to relatively small areas, is intensive, and is expensive. Accordingly, at this stage it is essential to protect the investment and potential revenue from the prospect by obtaining exclusive exploration or mining rights, and to enter into negotiations with owners of surface property in preparation for later mine development.

The initial procedures of detailed exploration are similar to those of regional reconnaissance, but they are carried out in much more detail over a much smaller area. Detailed geologic mapping leads to an understanding of the arrangement and geometry of rock bodies and geologic structures which may host ore, and identifies the presence or absence of diagnostic and essential features of prospective ore deposits as given by the exploration model. Low-level airborne magnetic, electromagnetic, and radioactivity surveys provide information about the presence or absence of specific mineral commodities and, if present, lead to conclusions as to their geometry, arrangement, magnitude, and depth under overburden. Ground-based surveys give more detailed information, and close-grid geochemical sampling determines element dispersion patterns which may indicate unexposed targets. Thus, geologic inference, geophysical surveys, and geochemical sampling point to subsurface mineral potential that cannot be seen on the surface.

These procedures narrow our attention to a final target. If exposed at or near the surface, sampling of rock outcrop or of trenches will give a first indication of the type of ore and of its grade. If the ore does not crop out, which is the case in all but a few discoveries made in the last decades, drilling is needed to investigate the target. As a final step in exploration, the setting, geometry, grade, and tonnage of the mineral oc-

Table 2.2. Discoveries of non-ferrous metal mines in the US and Canada from 1951 to 1975 by principal exploration method

Discoveries of US metal mines 1951–1970					
Year of Discovery	Concentional Prospecting	Geologic Inference	Geophysical Anomaly	Geochemical Anomaly	Total Number
1951–1955	1	9	2	–	12
1956–1960	2	10	2	1	15
1961–1965	–	13	2	–	15
1966–1970	–	15	2	2	19
1951–1970	3	47	8	3	61

Discoveries of Canadian metal mines 1951–1975					
Year of Discovery	Conventional Prospecting	Geologic Inference	Geophysical Anomaly	Geochemical Anomaly	Total Number
1951–1955	16	14	5	–	35
1956–1960	6	4	14	–	24
1961–1965	4	4	5	2	15
1966–1970	2	4	13	1(?)	20
1971–1975	1	5	15	3	24
1951–1975	29	31	52	6	118

Source: Skinner (1979), from data gathered by Derry and Booth, quoted by Charles River Associates (1978).

Exploration Methods

Table 2.3 1985 mineral exploration statistics reported by US companies

	US	Canada	Al other countries
Total exploration expenditures by country (Thousands of US dollars)	$98,384	$7,906	$26,852
Expenditures by commodity group			
Base and precious metals	$88,314	$7,776	$24,138
Other metals	540	—	1,327
Uranium	6,016	13	50
Industrial minerals	3,514	117	1,337
Total	98,384	7,906	26,852
Expenditures by activity			
Drilling	$24,402	$2,198	$4,158
Airborne geophysics	197	271	595
Ground geophysics	1,630	221	581
Geochemistry and prospecting	13,336	3,243	7,776
Land acquisition	7,459	294	818
Other (geologic studies, etc.)	25,520	634	5,885
General administrative	25,840	1,045	7,039
Total	93,384	7,906	26,852

The figures represent an estimated 60% of the total expenditures by all US mining companies.
Source: Data collected and compiled by the American Bureau of Metal Statistics Inc. (ABMS) and reported by McKelvey et al. (1986).

currence is determined with sufficient accuracy to serve as the basis for a prefeasibility study which, if positive, is followed by a systematic geologic, technical, and economic evaluation of the project.

Table 2.2 illustrates the changes in the success rate of different exploration methods over the last 35 years. Of particular note are the growing importance of geologic inference in the discovery of ore deposits in well-exposed and well-known terrain, for example the US, and of geophysical methods in covered and geologically less known terrain, for example a large proportion of northern and northwestern Canada. Eggert (1987) gives an analysis of the exploration of metallic mineral resources, and Table 2.3 summarizes exploration expenditures by US and Canadian companies which responded to a yearly questionnaire. An estimated 60% of total exploration expenditures are given in the table, which illustrates the distribution of funds expended in the exploration of different groups of metals, and the proportion used for different exploration approaches.

2.3 Exploration Methods

The geologic, geophysical, and geochemical methods applied at different stages of exploration are discussed in the next sections (Table 2.4). The methods are presented in order of scale and stage, from satellite-based remote sensing to drill-hole geophysics, and by methodology, from image processing to geologic investigations,

Table 2.4. The applicability of geophysical and geochemical methods in exploration

Exploration methods	Fe	Cr	Cu/Pb Zn	Au	Ag	Sn	U	Hydro-carbons
Magnetic surveys	+ +	○	○	– –	–	– –	– –	– –
Electric surveys	–	–	+ +	○	+	– –	– –	– –
Electromagnetic surveys	○	–	+ +	○	+	– –	○	○
Radiometric surveys	– –	– –	–	○	–	○	+ +	– –
Gravimetric surveys	+	+	○	○	–	–	– –	+
Seismic surveys	– –	– –	– –	○	– –	○	– –	+ +
Geochemical surveys	– –	–	+ +	+	+ +	○	+	–
Heavy Mineral surveys	+	+ +	–	+ +	– –	+ +	○	– –
Hg detectors	– –	– –	+	+	+	– –	–	– –

– – Not applicable; – Rarely applicable; ○ Applicable for indirect evidence; + Occasionally successful; + + Very successful.

geophysical surveys, and geochemical sampling programs. Only the essence of each method can be given here, and the reader is referred to the specialized literature for in-depth treatment of individual exploration methods and approaches.

2.3.1 Remote Sensing

Remote sensing is the characterization of the earth's surface by measurement of reflected or emitted electromagnetic radiation, which allows us to recognize major regional or local topographic features and geologic relationships, and helps in the discovery of areas of mineral potential. Remote sensing instruments mounted on satellites scan the earth's surface and measure reflected solar radiation, or radiation emitted from the surface (Fig. 2.4) in wavelengths of 0.3 to 3m (Fig. 2.5). These wavelengths span the range from the ultraviolet through the visible-infrared to the microwave-radar spectrum. Most measurements are made in the visible or near-visible range by passive methods, in which reflected natural radiation is measured. Measurements done in the radar frequency, on the other hand, measure the reflection of radiation emitted by the measuring craft, and these methods are referred to as the active methods of remote sensing.

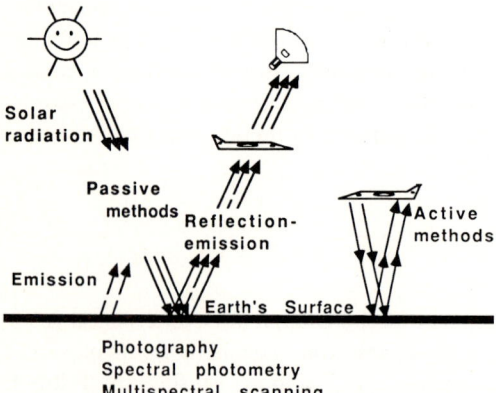

Fig. 2.4. Illustration of active and passive methods of remote sensing. (After Günther 1972)

Exploration Methods

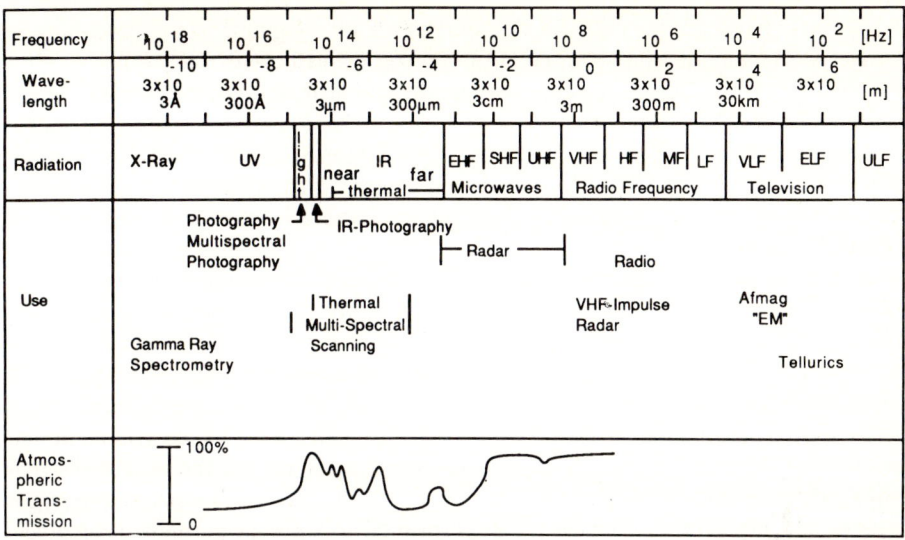

Fig. 2.5. Electromagnetic radiation spectra used in remote sensing and geophysical exploration. (After Günther 1972 and Peters 1987)

Because different rock types reflect radiation to different degrees and in different spectral ranges, remote sensing allows preliminary geologic interpretations of an area. Some of the geologic features intimately associated with ore deposits provide strong signals that can be detected by remote sensing, for example the bleached rock and red cappings associated with hydrothermal alteration and oxidation of porphyry copper deposits, or the faults and fractures in volcanic terrain which host precious metal veins. These features are often clearly recognizable, even through soil cover or vegetation (Stone et al. 1982). Vegetation itself may reflect the effects of unusual metal concentrations in the subsurface by measurable changes in types of plants, or by anomalies in the reflectance of individual plant species (Birnie and Francica 1981).

Remote sensing for commercial use began with the launching of Landsat I by NASA in 1972; the last, improved versions, Landsat IV and V, were launched in 1984. The remote sensing program has since been transferred from government to commercial control under the name of EOSAT (Earth Observation Satellite Company). A French commercial satellite, the SPOT, was launched in 1986, and several launches are projected for the late 1980s and early 1990s, including two more US Landsat and one more SPOT satellites, and one satellite each from the European Space Agency (ERS-1), Canada (Radar Sat), and Japan (MOS-1).

The multispectral scanning (MSS) devices carried by the remote sensing satellites of the Landsat series measure intensities of radiation in narrow windows or *bands* in the visible and infrared spectrum in wavelengths from 0.3 to 300 microns. Intensities of radiation are measured for each band separately and are recorded on magnetic tapes, each containing the data for an area or *scene* of 185 x 185 km. The Landsat IV and V satellites include the Thematic Mapper (TM) scanning unit, which increases the resolution of the early Landsat satellites from 80 x 80 m areas or *pixels* to the present 30 x 30 m pixels.

The electromagnetic data can be purchased on magnetic tape for computer analysis and manipulation, or as composite images which are derived from the electromagnetic record by transforming the signals into artificial, photograph-like black/white or false-color composites which may be purchased through EOSAT as negatives or as prints at scales ranging from 1:1000000 to 1:250000. Different spectral bands or combinations of spectral bands can be chosen to highlight the most important geologic features – rock types, hydrothermal alterations, or structures – in the false-color composites.

Computer-aided image processing of the original electromagnetic data enhances geologic signatures and can provide more useful information than that given in the commercially available images. A classification of rock types is achieved by choosing characteristic band intensities, or intensity ratios. The data combination or *algorithm* which best represents a rock type is assigned a color, and all pixels on a scene which correspond to the algorithm – and to the particular rock type – will be printed in that color. Once the most relevant rock types or alteration zones and their spectral signatures have been recognized for an area, the signatures can be extrapolated to larger regions and false-color composites can be printed to serve as preliminary geologic maps and for an assessment of the region's mineral potential.

Remote scanning in the infrared wavelengths records the thermal radiation from the earth's surface, outlining regions of high or low heat flow, and identifies rock types which retain or emit accumulated heat to different degrees. Radar wavelengths are used mostly for precise representation of the earth's surface because they penetrate vegetation but are sharply reflected by the land surface. The Sidelooking Airborne Radar (SLAR) system is ideally suited for regional mapping of geologic structures.

The ability to give a coherent overview of rock types and geologic structures over large areas is the major exploration strength of remote sensing. Even though progress in image processing has been very rapid, much research remains to be done before it can be used as a routine exploration tool. Excellent summaries of remote sensing methodology and geologic applications are given by Short (1982) in a tutorial workbook published by NASA. Numerous examples of the application of remote sensing to the exploration of mineral deposits are given in the special Economic Geology issue on Remote Sensing (1983), and in the Paley (1984) volume of the Joint NASA/GEOSAT Test Case Project, which includes remote sensing applications to the exploration of hydrocarbons.

2.3.2 Aerial Photography

While remote sensing and satellite imaging cover very large areas of the earth's surface, aerial photography and photogeologic interpretation provide the topographic and geologic basis for most exploration work of smaller areas of tens square kilometers or less. In contrast to the wide electromagnetic spectrum used in remote sensing, aerial photography works only in the visible and near-infrared part of the spectrum.

Some excellent photographs of large regions have been taken from manned satellite missions and the space shuttles, but unfortunately there is very little systematic coverage. Vertical photography carried out by aircraft provides the basis for most geologic work. Aerial photographs are sold by government agencies as black/white negatives or contact prints which measure about 20 x 20 cm and are offered at scales ranging from slightly over 1:100 000 to less than 1:20 000. Color prints, which are not usually available from government sources and have to be contracted commercially, are very useful for mineral resource exploration because the colors highlight important geologic subtleties.

Aerial photographs show much detail of the earth's surface which is essential for work in remote areas, and they are an extremely important supplement to topographic base maps in any exploration program because topographic or cultural features can be located precisely. The power of aerial photography lies in its precise rendering of landforms which reflect bedrock and structures, and in its greytone or color discrimination which allows the recognition, for example, of zones of rock alteration and sulfide oxidation. The detail of landforms is particularly apparent because adjacent photographs taken by the aircraft make stereoscopic viewing possible. These adjacent photographs, or stereopairs, overlap about 60% in the forward direction and about 30% laterally. Stereoscopes used for three-dimensional viewing may be pocket-type for fieldwork, or mirror- or single-prism type for office use. Because individual aerial photographs are taken vertically in central perspective, they have edge and elevation distortion which can, however, be corrected by joining and overlapping the photographs to form an uncorrected photomosaic, or by graphically correcting the distortion on transparent overlays which serve as rough base maps on which geologic features can be plotted. Fully corrected topographic maps made from aerial photographs by photogrammetric methods can be obtained from government agencies or made to order by private companies.

The combination of topographic, geologic, and cultural features recognizable on aerial photographs makes them an ideal basis for planning the logistics of exploration, geologically by pointing out areas of heightened interest; topographically by allowing the identification of access routes, camp sites, geochemical sample points, or lines for geophysical surveys; and during geologic mapping for the exact location on maps of geologic features and sampling points.

Special cameras with distortion-free and color-corrected objectives have been developed and a variety of films is available. While black-white films are by far the most common, infrared film and different types of color films may be used. Multispectral photographs are taken by combining different films and filters to identify reflections in narrow spectral bands, a method reminiscent of multispectral scanning in satellite imagery. The data can be manipulated to yield high contrast, but because of the narrow spectral ranges covered, the method has not yet been very successful.

Observation and interpretation of geologic features which are recognizable on the aerial photographs yields preliminary photogeologic maps. Through close coordination of photogeologic interpretation and geologic field work, more reliable geologic maps, which are essential ingredients in most regional exploration programs, are assembled.

2.3.3 Geologic Exploration

Geologic practice and research have shown the intimate relationship of geologic settings with particular types of ore deposits. This knowledge, both empirical and metallogenic, is summarized in the mineral deposit models described before. It is the task of the exploration geologist to recognize and interpret the features which indicate that a geologic environment may contain mineral deposits or has all the attributes which promise mineral potential. The geologic expertise is used on all scales, from global strategies and exploration planning to regional grass roots exploration and local target evaluation. At the mining stage, mining geologists apply their knowledge and experience to the discovery of the additional ore reserves needed to sustain production in a mine or mining district. Because oft the large array of geologic methods used in exploration, the exploration and mining geologist must have a thorough knowledge of many geologic subdisciplines, including petrology, geophysics, geochemistry, and statistical geology.

Geologic exploration often begins with photogeologic interpretations, which are followed by geologic mapping and sampling of rocks and potential ore materials. By geologic mapping, the geologist is best able to recognize and interpret the three-dimensional distribution of rock bodies and structures. Samples are studied microscopically to determine their mineral composition and texture, and chemically to detect unusual element concentrations and distributions. Less frequently, more advanced methods of investigation are applied in exploration, for example fluid inclusion studies to determine the temperature and composition of mineralizing solutions in precious metal deposits, or isotope studies to ascertain the source of elements and the processes responsible for the precipitation of ore minerals. These studies help the geologist to understand the geologic and geochemical processes which shaped a region or a mineral prospect, and to develop the most appropriate genetic and exploration model.

The geologic interpretation of a mineral prospect concludes, if positive, with an estimate of the geometry, tonnage, and grade of potential mineral concentrations and, if negative, with the rejection of apparently barren areas. A good cooperation between geologists and engineers, while of great value at all stages of this evaluation, is particularly relevant during the final stages of detailed exploration and mining. While the geologist's task is mostly the prediction of where the ore is and how much there is, the mining engineer is more concerned with the approaches and strategies of mining and mineral extraction.

2.3.4 Geophysical Exploration

Geophysical exploration takes advantage of measurable differences in physical properties of rocks, minerals, and ores in the search for mineral deposits (Table 2.4; Fig. 2.6). Magnetism, specific gravity, and radioactivity of rocks can be measured by passive methods while electric conductivity, electromagnetic properties, and seismicity are generally measured as responses to man-induced pulses. The geophysical signal may be directly related to mineral deposits, for example a magnetic anomaly caused by magnetite ore in an iron deposit; more commonly, however, geophysical methods

Exploration Methods

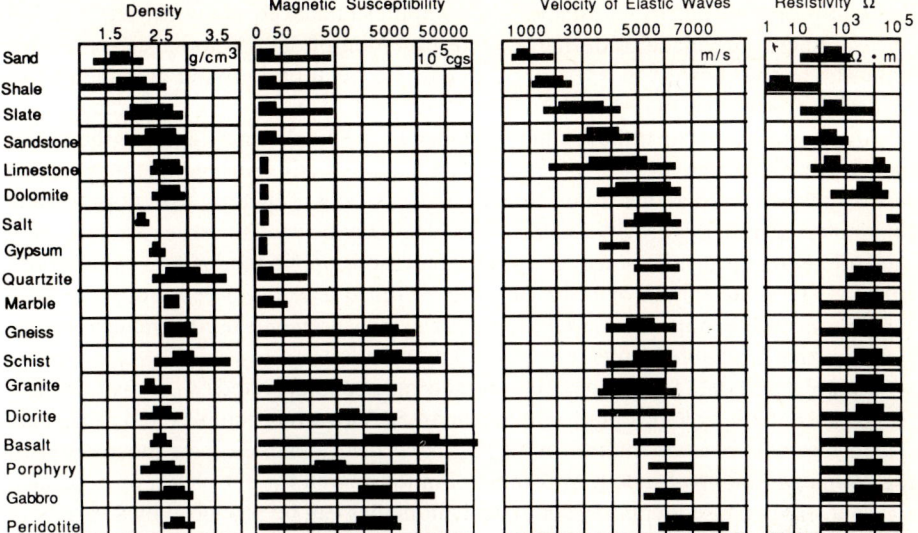

Fig. 2.6. Some physical characteristics of rock types used in geophysical exploration

provide indirect evidence which leads to interpretations of the subsurface geologic distribution of rocks but does not directly or necessarily reflect the presence of a mineral deposit. Accordingly, geophysical methods are applied both to mineral discovery and geologic mapping, and they are so useful because geophysical responses of rocks and ores can be measured through vegetation, soil cover, and extraneous overburden. In many cases, geophysical measurements provide the only means of interpreting the geologic characteristics of the subsurface short of drilling, which is much more expensive.

Geophysical and geologic investigations complement each other on all scales. Airborne geophysics, carried out either by fixed-wing aircraft or by helicopter, combines with photogeologic interpretations and regional geologic mapping. Airborne geophysical work most commonly involves magnetic, electromagnetic, and radiometric surveys which cover large areas of hundreds of square kilometers, rapidly and inexpensively per unit area (Table 2.1). The speed of the survey comes at the expense of sensitivity and resolution of detail in the data. Ground geophysical surveys combine with local geologic mapping to enable correlation of geophysical signals from the subsurface with geologic surface information. Ground surveys, while slower than airborne surveys and therefore restricted to smaller areas, allow the application of a larger variety of geophysical methods and provide a higher sensitivity and better resolution. On a still smaller scale, drill hole logging combines with geologic logging of rocks recovered from drill holes to gain three-dimensional information about rock and ore assemblages.

The signals measured by geophysical instruments may reflect noise caused by the instrument or extraneous environmental factors, the background which is typical for a particular region or location, and anomalies which reflect the presence and distribution of rocks or mineral concentrations of contrasting physical properties. Noise is generally filtered out by mechanical or computer-enhanced methods, either while measurements are made or during the interpretation of data. Anomalies are a function of: (a) the contrast in physical properties between background and anomalous material; (b) the size and shape of the geologic body which causes the anomaly; and (c) the depth from which the anomaly originates, or the distance between the site at which the measurement is taken and the anomalous body. Seismic methods, which are extensively used in the exploration for hydrocarbons, penetrate to depths of many kilometers. Similarly, gravity and magnetic data may be used to sense structures several kilometers deep. The other geophysical methods applied to mineral exploration, such as electromagnetics or resistivity, are useful to a depth of 300m or more only under unusual circumstances. Most geophysical measurements used in mineral exploration provide reliable results to depths of 100m at most.

Geophysical data from exploration targets may be interpreted by comparison with geophysical and geologic data from known areas. However, complex and computer-aided methods of mathematical modeling have been developed for more sophisticated interpretations needed, for example, in the search for petroleum or concealed ore bodies. Because identical geophysical anomalies can be produced by a variety of geologic bodies of different contrast, size, shape, and depth, unequivocal modeling of geophysical anomalies is difficult and a variety of interpretations of geophysical data is generally possible. Therefore, additional evidence is essential to constrain the interpretations and to develop a model that most closely approaches reality. The additional information comes from independent interpretations of the local geology achieved by geologic mapping, or from the application of other geophysical methods which measure different characteristic rock or mineral signatures.

Geophysical exploration methods are particularly useful to gain geologic information in areas where rocks are concealed or in the exploration of remote regions, because the surveys are largely independent of topography and accessibility. Airborne geophysical surveys have the added advantage that they are fast and inexpensive per unit area covered. Extensive cover and remoteness of large parts of Canada, for example, explain why a much larger proportion of mineral deposits has been found by geophysical methods there than in the US, where geologic inference has been more successful (Table 2.3). The table also shows that the discovery of ore deposits by geophysical prospecting has increased significantly in the last decades. This trend reflects both the fact that the proportion of mineral deposits exposed on the surface has decreased and that, at the same time, geophysical exploration methods have been refined by technological improvements and computer support of data interpretation. Success in geophysical exploration does not come easily, however, if we consider that the vast majority of geophysical anomalies discovered does not result in the discovery of a mineral deposit, but rather reflects unusual rock types or structures. The importance of geophysical methods in mineral exploration is illustrated by the high proportion of funds devoted to geophysical surveys.

Because of the common ambiguity of geophysical interpretations in the absence of constraining geologic information, a close cooperation between geophysicists and geologists is called for. By the time an area is selected for geophysical exploration, geologists have a general knowledge of the types of rocks and mineral deposits which might be expected and have a variety of geologic and exploration models in mind. While the geologic features seen on the surface can be projected into the subsurface by geologic reasoning, geophysical surveys provide direct evidence for the distribution of rock types, favorable structures, or mineral deposits at depth.

A variety of geophysical exploration methods is used, and each boasts many subsystems and instrumental configurations. Only the principles and applications of the main methods: magnetic, electric, electromagnetic, gravity, seismic, and radiometric, will be given here. The reader is referred to the geophysical literature for more detail. In the search for petroleum and natural gas, seismic and gravity surveys are by far the most relevant, while in the exploration for mineral deposits, magnetic, electric, electromagnetic, and radioactivity surveys are more important. In both cases, the geophysical methods are used as much to interpret the subsurface distribution and arrangement of rock types and structures, which give indirect evidence for mineral potential, as to detect the presence of mineral resources themselves.

2.3.4.1 Magnetic Surveys

Magnetic surveys measure the intensity of the earth's magnetic field. Local deviations from this field are caused by the presence of minerals and rocks which are magnetic or in which magnetism is induced by the earth's magnetic field. By far the most important of these minerals is magnetite, but in some surveys the presence of ilmenite, hematite, or pyrrhotite may be significant. The natural or remanent magnetism, locked in minerals at the time of their formation, is generally weaker than the magnetism induced by the earth's magnetic field. The degree of induction is measured as the magnetic susceptibility of minerals, or of rocks containing these minerals (Fig. 2.6). Since the magnetic field measured during a survey is a reflection of the underlying rock type, magnetic surveys are useful for the direct detection of, for example, magnetite-bearing or pyrrhotite-bearing iron, nickel, or copper-lead-zinc deposits, and for geologic mapping.

Airborne magnetic surveys outline the geologic features of large areas, even if deep soil or overburden covers the rocks. The presence of occasional rock outcrops is very important, however, to help in assigning magnetic signatures to specific rock types. Ground magnetic surveys are increasingly used in mineral exploration since small, portable magnetometers have been developed.

The intensity of the magnetic field is measured in gammas (1 gamma = 10^{-6} nT, i.e. Nanotesla) (Fig. 2.6). The total earth's field ranges from 20,000–50,000 gammas and the local magnitude depends on latitude and longitude; it is subject to variations of 10–30 gammas in daily background, and of 1000 gammas or more caused by magnetic storms which are related to sunspot activity. These disturbances may at times interfere with magnetic surveys in which minor contrasts have to be resolved, but most anomalies of exploration significance are of a much larger magnitude. The proton precession magnetometer, which measures the total magnetic field, and the fluxgate magnetometer, which measures either the total magnetic field or a single component (vertical, horizontal field) are most commonly used in exploration. They have a sensitivity down to 1 gamma or less.

The results of magnetic surveys, corrected for interference, are presented as contour maps of magnetic field intensity (Fig. 2.7) or as magnetic profiles. Ideally, these maps and sections are made to the scale of the geologic maps to facilitate geologic in-

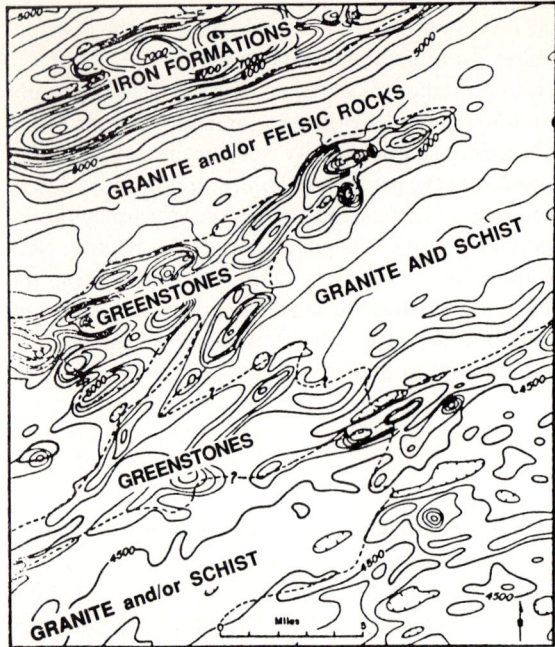

Fig. 2.7. Aeromagnetic survey over iron formations in the Precambrian shield of Wisconsin. (After Wright 1981)

terpretation of anomalies. To find the best match between the geophysical model and the real geological setting, geologic mapping or other geophysical surveys are done in concert. Both a mafic (iron-rich) igneous rock and a pyrrhotite-bearing massive sulfide ore deposit, for example, cause a strong magnetic anomaly. Their distinction is possible, however, by measurements of their electric conductivity, which is high for the massive sulfide deposit but low for the mafic rock.

Magnetic surveys cover large areas quickly and provide a preliminary assessment of the distribution of rock types, structures, and ore deposits. A good summary of the method and its application to geologic mapping and exploration is given by Wright (1981).

2.3.4.2 Electric Surveys

Electric surveys use mineral and rock conductivity or its reciprocal, resistivity, to gain geologic information about subsurface geology. The presence and distribution of sulfide minerals at depth can be determined by measuring the effect of their conductivity on the flow of a current applied to the ground. Since these electric measurements reflect the presence of sulfides directly, they are used more to detect ore minerals rather than to aid in geologic mapping.

Resistivity and induced polarization surveys are most common. To measure resistivity, a current is applied to the ground by two input or *current electrodes*. The potential produced by the current is measured as a voltage between two output or *potential electrodes*. Because only a relatively small space can be covered by the electrode arrays, electric methods are applied in local follow-up surveys of relatively small areas.

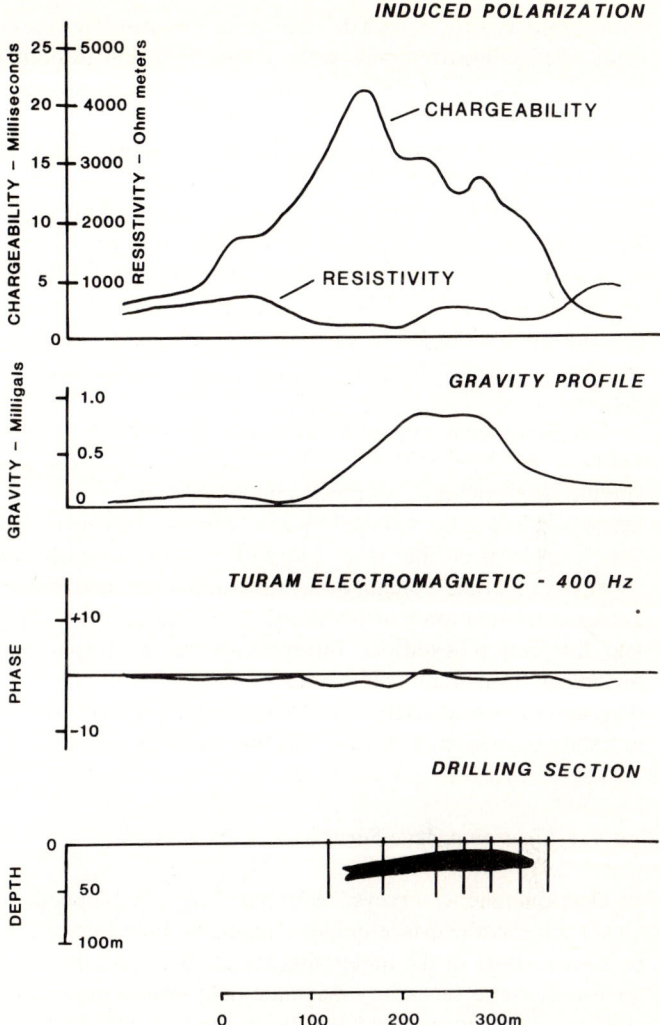

Fig. 2.8. Results of an induced polarization (IP), gravity, and electromagnetic (EM) survey at the Pyramid orebody, Northwest Territories, Canada. (After Wright 1981)

Depth penetration of the measurements is proportional to the distance between current and potential electrodes, and a variety of electrode configurations is used in different deposits. Data are presented either as resistivity profiles (Fig. 2.8) along grid lines, or as contour maps with contours joining points of equal resistivity. Resistivity surveys may indicate, for instance, the location of massive sulfide bodies, and they have been successfully used in the exploration for gold and tin placer deposits by taking advantage of the difference in conductivity between the unconsolidated placer material and the underlying bedrock.

When a current flows through electrolyte-bearing pore water in rocks, it induces polarization in conducting minerals which are in contact with the water. This polarization increases the resistivity of the rock to current flow and builds up an electric potential which decays rapidly when the current flow is interrupted. These two related processes form the basis for *induced polarization* (IP) surveys. In its practical implementation, the method is very similar to the conductivity/resistivity method described above. Current and potential electrodes are placed in the ground and an electric current is applied. The increase in resistivity caused by the polarization of disseminated sulfides is a function of the frequency of the current applied. The effect of frequency on resistivity is measured as the frequency effect (FE), which is the basis for frequency-domain IP measurements. The decay of the induced polarization when the current is cut off produces a decay curve which is characteristic for different conductors and is recorded in time-domain IP measurements. As in the case of resistivity surveys, depth of penetration of the measurements increases with the spacing of the electrodes.

The great advantage of IP surveys is that both disseminated and massive sulfides are detected. Therefore, the method can be applied to all types of sulfide deposits, including, for example, porphyry copper or vein deposits with disseminated sulfides which would not be detected by normal resistivity surveys. The output from IP surveys is given as profiles (Fig. 2.8) and contour maps which relate frequency-domain resistivity and time-domain decay measurements. Interpretation of the data, aided by geologic information if available, gives indications as to the presence, concentration, and distribution of sulfides. Interpretation of the output requires consideration of interference from graphite and other conductors, including water-rich clays or overburden, and man-made structures likes cables, pipelines, and fences. A summary of IP methods is given in Hohmann and Ward (1981).

2.3.4.3 Electromagnetic Surveys

In electromagnetic surveys, alternating currents applied to transmitting coils at or above the earth's surface induce a magnetic field in electric conductors, for example massive sulfides in the underlying rocks. The secondary current induced in the conductors causes a secondary magnetic field which, in turn, is measured in a detecting coil. The transmitting field may be applied specifically for the purpose, or it may be given by atmospheric currents (AFMAG) or very high frequency signals used for telecommunications (VHF) (Fig. 2.5).

Electromagnetic anomalies indicate the presence of conductors in the subsurface. Because the transmitting and detecting coils are easy to transport, they can be mounted on aircraft or helicopters enabling airborne surveys for a rapid coverage of large areas. Such surveys are flown at an elevation of about 50–150 m above the ground and may effectively measure the response of strong conductors to a depth of 100 to 200 m. Electromagnetic ground surveys are a part of most follow-up exploration programs and give more detail and a better resolution than airborne surveys.

A variety of configurations of transmitting and detecting coils is in use for airborne and surface surveys. Measurements of the strength of the secondary field as compared to the primary field, of the phase differences between the secondary and primary field, and of decay curves for the secondary field in pulsating on/off systems are recorded and interpreted. To save time, effort, and funds, airborne elec-

Exploration Methods

tromagnetic surveys are generally combined with magnetic and radiometric surveys to give a rapid and comprehensive geophysical characterization of the ground surface and subsurface.

2.3.4.4 Radiometric Surveys

The decay of radioactive uranium, thorium, and potassium isotopes contained in rocks emits shortwave electromagnetic radiation or *gamma radiation* (Fig. 2.5) which is measured by spectrometers. The intensity of radiation is a function of the uranium, thorium, and potassium content of the rocks, and the method is extensively used both in airborne and ground surveys to determine the presence of radioactive elements and of potassium. Radiometric surveys may be applied directly to the detection of uranium deposits and have led to the discovery, for instance, of some of the major uranium deposits in the Northern Territories of Australia and in Saskatchewan, Canada. They also help in geologic mapping of potassium-bearing rocks, for example granites, or of potassically altered rocks associated with hydrothermal ore deposits.

The gamma radiation measured by spectrometers records total radiation, or radiation from uranium, thorium, and potassium separately. The sensitivity of the instruments used for air or ground surveys is appropriate to measure contents of 1 ppm U, 1 ppm Th, and 0.1 % K in rock (Ward 1981). Because gamma radiation is strongly absorbed in rocks and overburden, most of the measured radiation is from the uppermost 10–30 cm of the surface. This fact limits the use of radiometric surveys to areas with abundant outcrop and without extraneous overburden which would not only obliterate the signal from the underlying rock, but provide a signature of its own, invalidating the survey. The radiation signal is attenuated exponentially with elevation of the surveying instrument above the ground, limiting effective flying heights to less than 100 m for most instruments.

Results of radiometric surveys may be presented as profiles or contour maps of total counts, separate counts of uranium, thorium, and potassium, and of spectral ratios between the radiation of these elements. Because of the shallow source of the anomaly and its corresponding sensitivity to surface effects, it is essential that the geology, rock types, and overburden in the survey area be carefully considered in the final interpretation of the data. Other sources of error include the presence of cosmic radiation (20–50 counts/second), other extraneous radiation, for example from the aircraft itself, and the strong dependence of the intensity of radiation on rainfall.

2.3.4.5 Gravimetric Surveys

Gravimetric surveys measure the distribution of the acceleration of gravity on the earth's surface. The acceleration, about 983 cm/s^2 (Gal) on average, is influenced by (a) latitude, because of the irregularity of the earth's shape and the variation of the centrifugal force from the equator to the poles; (b) the tides; (c) differences in elevation; and (d) the density of the subsurface rocks at the point of measurement. Gravity is measured as Milligal (mGal, 1/1000 cm/s^2) by gravimeters which operate like very sensitive balances and can measure differences down to 0.01 mGal (Milligal, 1/1000 cm/s^2). The average upper crust has a density of approximately 2.67 g/cm^3, and geologic materials range in density from less than 2.0 g/cm^3 for soils to more than 4.0 g/cm^3 for massive sulfides or iron ore deposits (Fig. 2.6). For exploration purposes, heterogeneities in the gravitational field are interpreted in terms of variations of rock densities in the subsurface, and relative rather than absolute values are important.

Gravimetric surveys are slow and the low density contrast between different geologic materials limits their use in exploration. In the exploration for petroleum and gas, gravity surveys outline low-density salt domes which often form traps for hydrocarbons. In mineral exploration, gravity surveys help to measure the thickness and distribution of alluvial cover over bedrock, for example in the delineation of placer deposits of tin, rutile or gold; or to outline the distribution of deep-seated granitic complexes for the discovery of associated hydrothermal ore deposits. On a smaller scale, gravimetric methods may help to predict the size, tonnage, and grade of massive sulfide or other dense mineral deposits whose general outline is known.

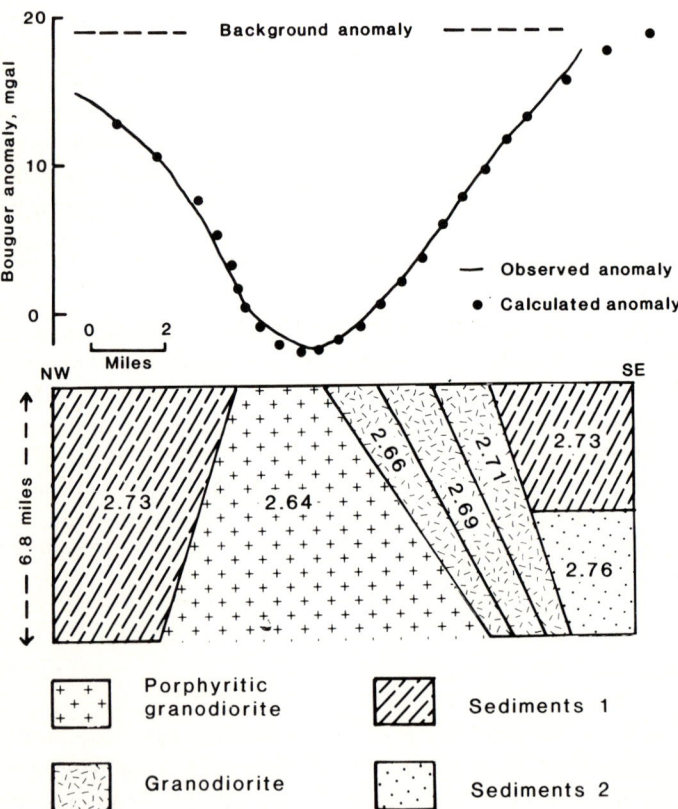

Fig. 2.9. Observed gravity anomaly, geologic model, and calculated anomaly to fit the model. Note that a variety of distributions of rock types at depth could produce the observed anomaly. (After Griffiths and King 1981)

Gravimetric measurements must be corrected for latitude and the effects of local topography. Therefore, close control of the location and elevation of measuring stations is essential. To place a gravity survey into a regional context, gravity values are referred to a "reference elevation", which may be sea level or any more local level that appears convenient. Measurements corrected for the effects of topography and for the thickness and density of the slab of rock that lies between the measuring station and the chosen reference elevation give the *Bouger anomaly* which is used in interpretations. The data are presented as gravity profiles (Figs. 2.8, 2.9) and as contour maps outlining gravity highs, for example over dense rocks or ore deposits, and gravity lows, for example over alluvial cover or salt domes.

Because gravitational anomalies depend on a variety of factors including density contrast, size and shape of the anomalous body, and depth, a multiplicity of interpretations is possible. Small to medium-sized rock bodies or ore deposits can be detected only at very shallow levels. As was discussed for some of the other geophysical methods, geologic information or results of other geophysical surveys are used to eliminate alternate interpretations. For example, a gravity anomaly caused by a massive sulfide deposit will be mirrored by an anomaly in electric conductivity, while a gravity anomaly caused by a magnetite-bearing iron deposit will not. Again, a preliminary knowledge of the geologic setting, and a close cooperation between geophysicist and geologist, go a long way toward successful interpretation. Good summaries of the application of gravity surveys to geologic problems are given in Wright (1981).

2.3.4.6 Seismic Surveys

Seismic waves, which propagate with different velocities in different rock types, are reflected and refracted at bedding or fault contacts. Seismic waves are induced at the earth's surface by a hammer, a gun, or dynamite for local, near-surface applications, and by large, truck-mounted vibrators for deep surveys. Geophone arrays on the surface measure the reflected and refracted vibration in multichannel recorders. After complex filtering of interfering noise and computer processing, seismic profiles of distance vs. arrival time of seismic waves are plotted. These profiles can be transformed to actual depth and configuration of geologic structures if the velocities of wave propagation in the underlying rock units are known. Seismic reflection and refraction surveys, the most important geophysical tools used in the exploration for petroleum and natural gas, help to detect hydrocarbon trapping structures such as anticlines, faults, or salt domes. These methods have found much less use in mineral exploration because of a strong interference between the emitted and reflected/refracted seismic waves at shallow depths, and because the very high resolution needed to successfully reflect the structural complexities associated with most ore deposits is difficult to achieve with presently available instrumentation. Research into higher-frequency seismographs, closer geophone arrays, and filtering of direct waves from the source is needed to make seismic surveys more useful for mineral exploration. A good summary of the potential applications of seismic methods to mineral exploration is given by Wright (1981).

2.3.4.7 Airborne and Ground Geophysical Surveys

A short summary will help to clarify how airborne and ground geophysical surveys, using the methods described above, work hand in hand in narrowing regional reconnaissance exploration to local target evaluation.

Airborne surveys are used for reconnaissance exploration because they enable coverage of large areas quickly and inexpensively per unit area. Airborne surveys measure physical properties of rocks and ores through dense vegetation, glacial overburden, swamps, lakes, and soil, unencumbered by the common lack of access in these landscapes. Fixed-wing aircraft are used for rapid coverage of large areas; helicopters are slower and more expensive to operate, but they achieve better resolution because the craft can stay closer to the surface and has better elevation control. Airborne surveys are generally carried out at elevations above ground of less than 150 m, and achieve the best compromise between rapid coverage and resolution at elevations between 100 m and 50 m. Instruments for magnetic, electromagnetic, and radiometric surveys can be installed on the same carrier to provide, simultaneously, a spectrum of geophysical measurements.

Airborne magnetic surveys generally use proton-precession magnetometers of about 1 gamma sensitivity. Electromagnetic surveys use a variety of configurations of transmitting and receiving coils in which the transmitting coil may be mounted on the plane and the receiver dragged on a cable at a distance, or the transmitting and receiving coils may be housed together on the plane or in a separate sonde. Airborne radiometric surveys use highly sensitive gamma-ray spectrometers to measure gamma radiation as TC (total count) and separately for potassium, uranium, and thorium. The spectrometer is flown at low ground clearance and must be shielded from cosmic radiation.

While airborne surveys serve to discover anomalies over large areas during reconnaissance exploration, geophysical ground surveys are used to define the physical characteristics and geometry of local mineral targets. Ground surveys are generally combined with geologic mapping and with geochemical surveys. Ground surveys have a higher sensitivity and better resolution than airborne surveys because measurements are taken closer to the target and because a closer sampling density is possible. Electric, gravity, and seismic surveys are possible only on the surface because direct contact of the measuring devices with the ground is needed. Drawbacks inherent in ground surveys are that they require accessibility and that progress is comparatively slow, at a few line-kilometers per team per day, unless field vehicles can be used.

The application of magnetic and electromagnetic systems in ground surveys is similar to that described for airborne systems, except that a larger variety of instruments is available. Resistivity, induced polarization (IP), gravimetric, and seismic surveys are specific to ground surveys. The measurement of self-potential, an electrochemical potential established by surface oxidation of massive sulfides, has lost importance in ground surveys but continues to be useful in drill logging. Ground radiometric surveys are useful mainly for the detection of uranium with the help of geiger counters. Instead of gamma radiation, alpha radiation, which is produced by the decay of radon, may be measured. Because radon gas migrates to the earth's surface through porous rocks or fractures, it is a good indicator of the presence of radioactive minerals at depth.

2.3.5 Geochemical Exploration

Geochemical exploration has become an essential tool in the search for ore deposits and is finding applications in the search for hydrocarbons as well. In geochemical exploration, anomalous surface enrichments of elements that point to potential ore deposits in the subsurface are sought.

The geologic processes which form ore deposits lead to the enrichment of the ore-forming elements in *primary dispersion halos,* which are larger than the ore target itself. Weathering of ore deposits leads to a redistribution of elements in *secondary dispersion halos* in weathered rock, soil, vegetation, and drainage systems, vastly increasing the area in which evidence for the presence of a mineral target can be detected (Fig. 2.10). First applied in Europe in the 1930s, geochemical exploration has found universal acceptance with the development of rapid, sensitive, and accurate analytical methods.

Geochemical surveys are conducted on all scales, from regional reconnaissance to local follow-up. In surveys which cover large concessions in developing countries and in areas where little if any geologic information is available, stream sediments are collected at very low sampling densities to detect areas of exploration potential. On a more detailed scale, geochemical surveys follow geologic or geophysical work, covering smaller areas at higher sampling densities. Geochemical anomalies, or enrichments of elements with respect to crustal or regional averages, guide exploration to potential mineral deposits. The absence of such anomalies helps to eliminate areas from further consideration, but much care has to be exercised because the failure to detect a geochemical anomaly does not in itself preclude the presence of an ore deposit.

Primary dispersion halos of the elements sought for exploitation, or of other elements involved in the ore-forming process which may be used as pathfinder elements (Table 2.5), are developed in the host rocks at the time of ore formation. The composition and distribution of primary halos depends on the type of deposit. Porphyry

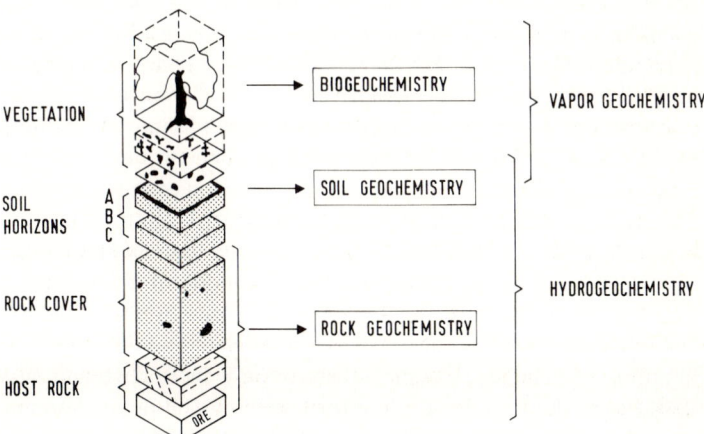

Fig. 2.10. Methods of geochemical exploration, and geologic materials sampled to detect primary and secondary dispersion halos

Table 2.5 Major and pathfinder elements used in geological exploration

Metal	Type deposit	Major and pathfinder elements
Ag	Magmatic hydrothermal Volc. massive sulfide	Ag, Pb, Zn, Cu; Au; As, Sb, Mg; Mn
Au	Magmatic hydrothermal Placer	Au, As, Sb, Hg, Tl, Te; Cu; Ag, Pb, Zn Heavy metals: gold
Cr	Magmatic	Heavy metals: chromite, ilmenite, magnetite
Cu	Magmatic-hydrothermal Volcanogenic Sedimentary	Cu, Pb, Zn; Au, Ag; Mo
Mo	Magmatic-hydrothermal	Mo, Cu; Sn, W; F
Ni	Magmatic	Ni, Cu; Pt
Pb	Magmatic-hydrothermal Volcanogenic Sediment-hosted	Pb, Zn, Cu; Ag
Pt	Magmatic	Pt, Cr, Pd, Ni; Cu
Sn	Magmatic-hydrothermal Placer	U, Be, B, F Heavy minerals: cassiterite, ilmenite, rutile
U	Sandstone-hosted Vein	Se, V, Mo, Cu Cu, Pb, As, Co, Mo, Ni
Zn	Magmatic-hydrothermal Volcanogenic Sediment-hosted	Zn, Cu, Pb, Ag Cu, Pb, Ag Pb, Cu

copper deposits, for example, may have chemical halos that measure hundreds of meters horizontally and vertically, while the stratigraphic horizons which host sedimentary sulfide deposits may have large zones of anomalous metal concentrations along the horizon, but little if any at right angles to it.

Secondary dispersion halos form by mechanical breakdown and chemical dissolution of rocks and ores. Mechanical breakdown and consequent transport in surface runoff concentrates the resistant minerals, for example cassiterite (tin), rutile (titanium), monazite (thorium), diamonds, and gold in beach sands and river conglomerates. Anomalies are detected and traced back to a primary source by heavy-mineral panning of stream sediments or soils. Chemical dissolution of rocks and ores in the presence of water causes dispersal of metals into overburden, soil, and streams. During dispersal, the elements may be re-concentrated in vegetation or, by adsorption, on clay minerals, organic matter, and iron-manganese oxides, all of which are attractive sampling media in geochemical exploration. Chemical dissolution effectively disperses the sulfides of all the common base and precious metals.

Climate and topography control the mobility of elements in the secondary environment. In a cold climate, for example, large, well-defined anomalies do not develop because chemical dissolution is inefficient and drainages are poorly developed; in a dry, arid climate, chemical dissolution is ineffective and dispersal by occasional flash-floods does not lead to the formation of well-defined anomalies; in tropical climates, on the other hand, decomposition and leaching of the ore-forming elements may be so complete that no traces of the metals remain in weathered rocks or soils. The best environment for the application of geochemical exploration methods is given in a temperate climate in areas of gentle topography, in which abundance of water and warm temperatures lead to effective dissolution of ore minerals, and the gentle topography fosters both chemical dissolution and the development of good secondary dispersion halos. Both the primary and secondary halos are sampled in geochemical exploration: rock chips from primary dispersion halos; partly weathered rock from primary and secondary halos; and soil, vegetation, stream water, stream sediment, and heavy-mineral concentrate, from secondary halos.

Geochemical exploration and sampling evolve from regional to local, from helicopter-supported to pedestrian, and from low-density to high-density. Most programs begin with regional stream sediment sampling, followed by soil and finally rock sampling. Rock outcrops and transported rock fragments are observed, described, and plotted at geochemical sampling locations during the sampling campaign to provide a first impression of the geologic framework of a region. Geologic mapping and geophysical surveys are generally carried out concurrently with geochemical exploration in follow-up programs.

The sampling medium, sampling density, and ore- and pathfinder elements are chosen with reference to the most applicable geologic ore deposit model or models (Table 2.5). In unknown areas, orientation surveys identify the most suitable target and pathfinder elements and determine the most effective sampling density. As the program evolves and more information is gained, adjustments are made to both the geologic model and the sampling methods because the failure to detect a coherent anomaly is not necessarily a sign that no mineral deposits are present; in fact, many potential deposits must have remained undiscovered because of inadequate interpretation of geochemical data. The geochemical exploration that helped in the discovery of the Mt. Pleasant tungsten-molybdenum-tin deposit in New Brunswick, summarized by Peters (1987), is an example of an effective program which passed through several stages of modeling and adjustment until an ore deposit was finally found.

2.3.5.1 Stream Sediment Geochemical Surveys

Stream sediment surveys are the mainstay of geochemical reconnaissance exploration. Stream sediments are taken from active stream channels and analyzed to detect anomalous element concentrations. Small streams give maximum resolution and sharpest contrast, as opposed to large streams in which any anomaly from a mineralized zone will be diluted by large amounts of stream sediment from barren areas. In regional surveys, sampling density is a compromise between the minimum density needed to detect a significant anomaly, and the maximum density affordable in terms of time and cost. Sampling densities range from one sample for over $100\,km^2$ in regional reconnaissance programs to several samples per km^2 in local follow-up.

To explore metals which form chemically resistant minerals (gold, platinum, tin, titanium, and thorium), heavy-mineral concentrates are panned from stream channels (Guigues and Devismes 1969; Zantop and Nespereira 1978). In the exploration for soluble base metals (copper, lead, zinc, molybdenum, nickel, and cobalt), stream sediments are collected where tributaries enter main streams and at regular intervals in stream channels. Samples are sieved to -80 mesh (0.157mm) and the fine fraction, which reflects metal anomalies best, is analyzed. Samples may be analyzed on the spot by simple field methods or, more commonly, in large, specialized laboratories. The values for background and anomalous element concentrations are determined statistically, and metal and pathfinder element distributions are plotted on drainage, topographic, or geologic maps. Anomalous zones discovered in stream sediment surveys are re-sampled at a denser grid and further investigated by soil sampling.

Care must be taken to avoid contamination from cultural wastes, old mines and mine tailings, railroad tracks, highways, and bridges, all of which may cause intense anomalies. False anomalies may also result from inadvertent sampling of sediments with a high organic content, because organic material preferentially enriches metals by adsorption or by formation of organo-metallic complexes.

2.3.5.2 Hydrogeochemical Surveys

Surface water and groundwater in streams, springs, and wells may be analyzed in reconnaissance hydrogeochemical surveys. The element concentration in these waters is a function of a variety of factors, including the solubility of target and pathfinder elements, the season-dependent quantity of runoff, the amount of suspended clay minerals or organic matter, and chemical parameters like pH, oxidation potential, and the presence of ligands in the water. To compound these complexities, the concentration of elements of geochemical interest is very low compared to that in stream sediments and analyses are therefore not often reliable. However, while stream sediment surveys are restricted to areas of little cultural activity, well sampling can be successfully done even in moderately cultivated areas without excessive fear of contamination. Because of the uncertainties mentioned above, hydrogeochemistry is not a widely used exploration tool.

2.3.5.3 Soil Geochemical Surveys

Soil sampling is a very powerful tool in the follow-up exploration of anomalies found by stream sediment or geophysical surveys. In areas which do not have well-developed drainages to allow stream sediment surveys, as is true for large parts of the northeastern US, soil geochemical surveys may be done on a regional basis. The dispersion halos of elements in soils are much smaller than those in stream sediments, but still considerably larger than those in primary halos in rocks (Fig. 2.11). As a first step in the follow-up of stream sediment anomalies, irregular soil samples may be taken along stream banks or along adjacent ridges and spurs to determine the source area of the stream sediment anomaly. Systematic soil sampling of the anomalous areas in grid patterns points to the source of the anomaly and identifies drilling targets.

The concentration of soluble elements varies between soil horizons. The C-horizon, which is closest to the rock, generally shows little dispersion of the target elements. The A-horizon, the uppermost soil horizon, may show the largest dispersion, but a variable content of organic matter leads to irregular element distributions. The B-horizon, which is intermediate between the two, is the most homogeneous horizon

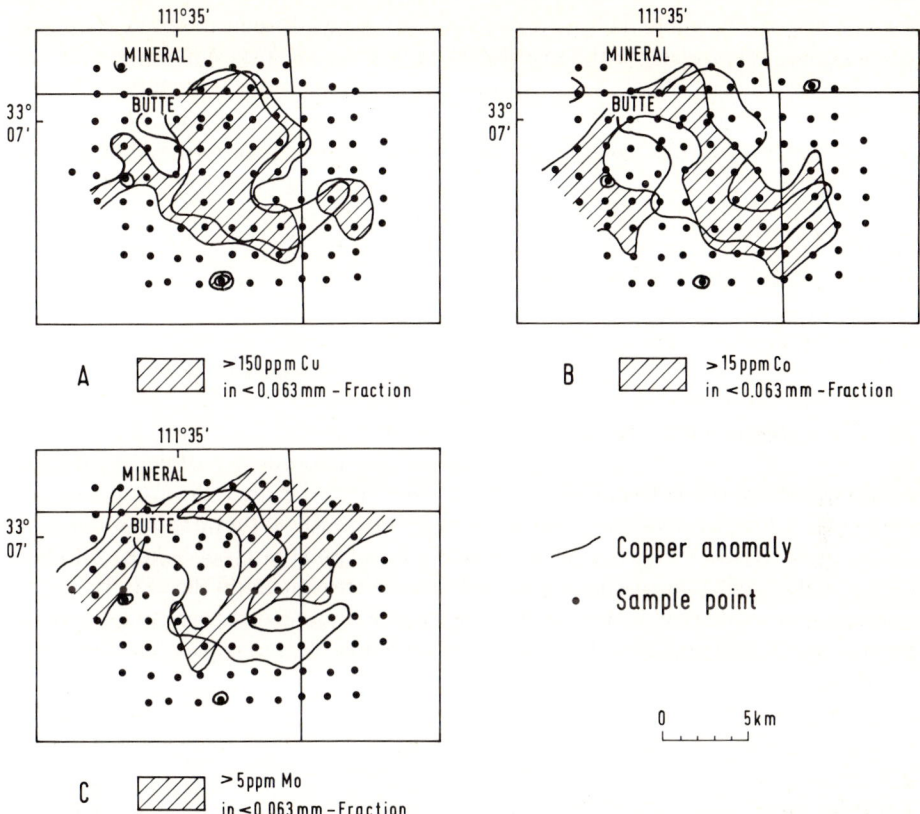

Fig. 2.11 A–C. Soil anomalies for copper, cobalt, and molybdenum surrounding Mineral Butte, Arizona

and provides the best sampling medium. Samples are taken by digging or shallow drilling with augers, and the -80 mesh (0.157 mm) fraction is analyzed. Sampling density varies depending on the stage of exploration and the orientation of the anomaly: a square grid with points every few meters to tens of meters is sampled for equidimensional anomalies, and an oriented grid is sampled with close spacing of samples (10–20 m) at right angles and wide spacing (100–200 m) parallel to the elongation of an anomaly. Results from soil surveys are interpreted statistically and presented as contour and color maps of elements or element combinations.

2.3.5.4 Biogeochemical Surveys

Plants incorporate elements from soil and groundwater into their branches and leaves; the uptake of trace elements is specific to plant species, plant organs, growth stage, growing season, and soil chemistry. These complex and poorly understood relationships make the unambiguous interpretation of biogeochemical anomalies difficult, but biogeochemistry has nonetheless been used successfully in the exploration for a variety of elements, including copper, zinc, molybdenum, uranium, mercury, and gold. It is particularly useful in arid terrain where poor surface drainage and lack of soil cover preclude stream sediment or soil geochemical surveys. In the southwestern US and Mexico, for example, mesquite trees, whose roots often extend more than 20m into the ground to reach groundwater level, have successfully been used to define geochemical anomalies.

2.3.5.5 Atmospheric and Vapor Geochemical Surveys

In atmospheric and vapor geochemistry, the concentrations of volatile elements and gases released from oxidizing ore bodies, or from accumulations of petroleum and natural gas, are measured. Mercury and sulfur vapors guide to sulfide deposits; radon gas reflects uranium deposits; helium anomalies characterize a variety of ore environments; and gaseous hydrocarbons give evidence for the presence of petroleum and natural gas in the subsurface. The vapors may be detected in air at the ground surface at extremely low concentrations, or in soils and water, where they are enriched.

Mercury is measured to concentrations as low as 1 ppb (parts per billion) with Hg-spectrometers and is increasingly used as a pathfinder element in the search for sulfide deposits. Radon anomalies on the surface or in groundwater are determined by measuring the alpha particles emitted during its radioactive decay. Because radon gas

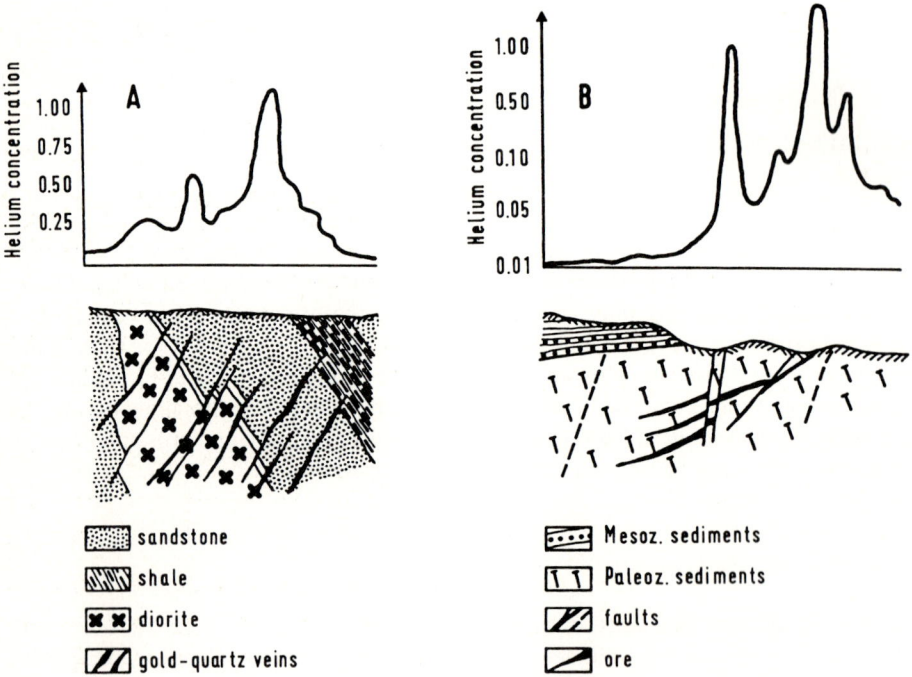

Fig. 2.12 A, B. Helium exploration in the USSR. **A** Goldquartz veins, N-Kasachstan. **B** Copper deposit, Kaukasus

migrates upward in waters circulating through faults or porous rock, unusual concentrations in surface waters help in the localization of uranium deposits at depth. The presence of helium in groundwater may reflect the presence of uranium deposits, sulfide deposits, and deep crustal fractures, as shown by impressive results obtained, for example, by helium surveys done in the Kaukasus (USSR) for carbonatites, gold-quartz veins, and lead-zinc-copper occurrences (Fig. 2.12).

2.3.5.6 Rock Geochemical Surveys

Rock geochemical surveys seek the primary dispersion halo around mineral deposits. Because primary dispersion halos are restricted to a small area immediately surrounding any prospective mineral deposit, rock surveys are applied to evaluate specific targets outlined by regional surveys. Bedrock is chipped from outcrops, or sampled after trenching or shallow drilling has exposed the rock. The sampling density depends on the type of deposit sought and ranges from the occasional rock sample taken to calibrate results from other geochemical surveys to systematic grids which follow up soil geochemical anomalies and serve to define drill targets. Rock geochemical surveys (Govett 1983) provide direct evidence about the geochemical characteristics of the rocks that cause the anomaly, and help in the geologic interpretation of stream sediment and soil surveys.

A variety of other sampling media or geochemical methods may be used locally, for example analysis of manganese- and iron-oxide coatings in stream sediments for an enhanced detection of absorbed zinc, copper, and lead, or determination of oxygen and carbon isotopes which form halos around a variety of ore deposits. Effective geochemical exploration involves combinations of the methods described above, each one of which has to be examined and adjusted to local conditions to assure an exhaustive and successful evaluation of a region's mineral potential.

2.3.5.7 Analytical Methods

The analytical methods applied in geochemical exploration depend on the requirements of particular stages of exploration. We may contrast, for example, an orientation stream sediment survey in which many elements are analyzed semiquantitatively, rapidly, and at low cost with a quantitative evaluation of a mineral prospect, in which one or a few elements must be analyzed quantitatively and accurately. The choice of the proper analytical method further depends on the availability of instrumentation. Only rudimentary field kits or field instruments are available in remotely located field laboratories while sophisticated instrumentation is available in company-run research or commercial laboratories. The sample preparation and analytical facilities must be able to handle a variety of materials – rock, soil, stream sediment, water, plant ashes and ore, and a short turnaround time between sampling and receipt of the analytical results is essential to assure continuity and coherence of the ongoing exploration program.

Only the most important analytical methods are described below; for more comprehensive coverage, the reader is referred to Fischer (1981), and Reeves and Brooks (1978).

Colorimetry is one of the classical methods of analysis. It has largely been replaced by instrumental methods in general exploration, but is still used for quantitative determination of some elements. Element concentrations are measured by observing concentration-specific colors resulting from chemical reactions of sample solutions with standard reactants. Field kits are available for qualitative and semiquantitative colorimetric determination of some elements, for example copper, lead, zinc, and molybdenum, by use of element-specific reagents.

Emission Spectroscopy is the oldest and a widely used instrumental method for multi-element analysis. A small, pulverized sample loaded into carbon electrodes and vaporized in an electric arc emits an electromagnetic spectrum in which the wavelengths identify the elements and the spectral line intensities reflect element concentrations. Emission spectroscopy gives rapid, inexpensive, semiquantitative determinations of over 30 elements simultaneously. Because it is semiquantitative, the method is useful mainly for orientation surveys and for stream and soil geochemical reconnaissance.

Atomic Absorption Spectrometry (AAS) is the most widely used analytical method in exploration geochemistry because it provides rapid, reliable, and low-cost semiquantitative or quantitative analyses. Samples are ground and dissolved, and the sample solution is vaporized into a gas flame. Vaporization transforms ions contained in solution into atoms which absorb element-specific wavelengths of electromagnetic radiation that passes from a source tube through the vaporized sample into a detector. The degree of absorption gives a measure of element concentration. Several elements can be analyzed simultaneously and procedures for combinations of element analyses can be tailored to the needs of particular exploration programs. Simple atomic absorption spectrometers for use in large field camps, and highly automated spectrometers for quantitative work in commercial laboratories, are available.

X-Ray Fluorescence (XRF) measures fluorescence induced in pulverized samples by X-ray radiation. The method is accurate at high concentrations and finds routine application in the major-element analysis of rocks, but it is not widely used in exploration because of its low sensitivity for minor and trace elements. Portable instruments in which the X-rays are emitted by radio-isotopes have been used for specialized exploration purposes.

Inductively Coupled Plasma Spectrometry (ICP) is a recently developed method for multi-element analysis. It combines the capability of analyzing many elements simultaneously with good sensitivity and accuracy, and has found widespread application in the analysis of geologic materials. ICP instruments are still expensive, but their ability to provide rapid, accurate determinations of trace element concentrations at low cost promises increasing use in exploration geochemistry.

These analytical methods cover the full range of capabilities needed for successful exploration and evaluation of ore deposits, from semiquantitative analysis of target and pathfinder elements during regional reconnaissance and follow-up exploration, to quantitative, single-element assays of ore materials. Most commercial laboratories deliver not only the analytical data but also their statistical evaluation if desired. Results may be transmitted by direct computer link, assuring a fast turnaround time and maximum usefulness of the data to guide the active exploration program.

2.3.5.8 Interpretation of Geochemical Surveys

The data set from a geochemical exploration program consists of sample locations, geologic notations made during the sampling campaign, and values of element concentrations in samples. This information is to be interpreted and presented in such a way that ore-related geochemical anomalies are reliably recognized. Interpretations may use simple frequency diagrams or more complex, computer-supported statistical calculations and plotting routines.

The first task in the interpretation of geochemical data is the discrimination of background element compositions from anomalous compositions which may reflect the presence of a mineral deposit. In a traditional approach to the choice of an appropriate threshold value between background and anomaly, the mean and standard deviations of element concentrations of the sample population were calculated, and all concentrations more than two standard deviations above the mean were arbitrarily taken as anomalous. Because this method gives no indication as to the true distribution of sample populations, it is not very satisfactory and no longer in use. Histograms, in which the number of samples in different concentration ranges are plotted, give a better indication of the distribution of sample populations (Fig. 2.13). These frequency versus concentration plots often approximate a bell-shaped distribution curve with a normal distribution for most of the abundant elements, and a lognormal distribution for most trace elements. The concentration ranges in background and anomalous sample populations are reflected by deviations from the bell-shaped curve or by the presence of separate populations at the high-concentration end of the histogram. For a more sensitive discrimination of different sample populations, element concentrations are plotted on cumulative frequency diagrams drawn on normal or lognormal probability paper (Sinclair 1976). Normally or lognormally distributed populations plot as straight line segments, and transitions between populations plot as curved lines or inflection points between straight segments on the corresponding probability diagrams (Fig. 2.13). In the presence of two populations, for example one population of samples derived from country rock and another derived from a massive sulfide deposit, the concentration at the inflection point is taken as the threshold concentration separating rock-related background from ore-related anomaly. Once background and threshold values have been established and concentration ranges to reflect the intensity of anomalies have been chosen, the data are plotted by symbols or colors on topographic and geologic maps in preparation for geochemical and geologic interpretation.

The simple statistical approaches and interpretations described above are complemented by complex statistical evaluations which can add substantially to the geologic interpretation of geochemical data, as described in comprehensive overviews by Davis (1986) and McCammon (1974, 1980). Statistical methods rely heavily on computer processing of data, and in the STATPAK, the US Geological Survey has assembled a set of the statistical programs which are most often used in geologic applications to facilitate the interpretation of large arrays of geochemical data.

Univariate statistics relate to each element separately, for example in the calculation of the mean and standard deviation of element concentrations, or in the determination of regional trends in their distribution. *Multivariate statistics* relate several elements to each other and facilitate the geologic interpretation of multi-element data.

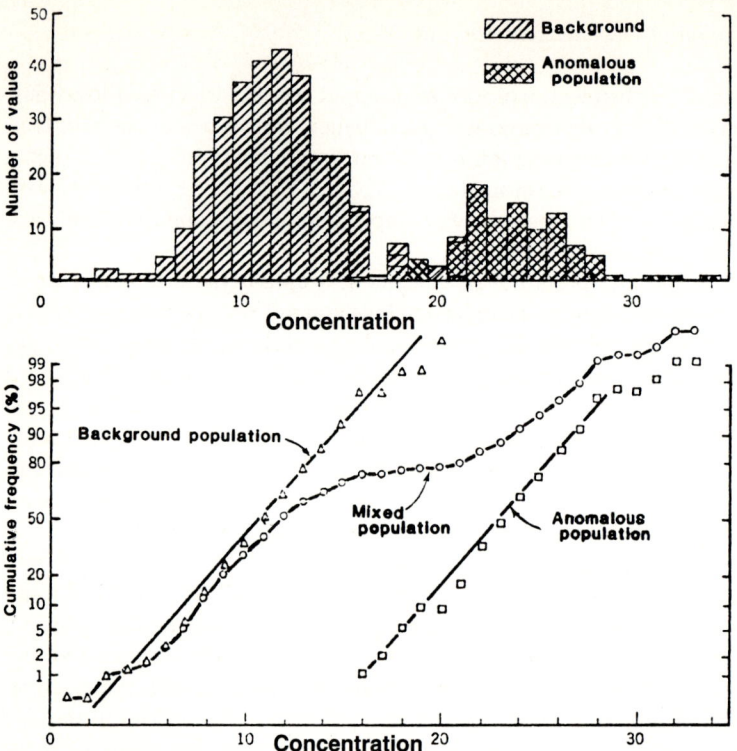

Fig. 2.13. Histogram and cumulative-frequency plots to illustrate the differentiation of background and anomalous sample populations

Since mineral deposits are characterized by closely correlated assemblages of minerals and metals, a statistical calculation of the degree of inter-element correlations in geochemical samples enables conclusions about their source material – rock or ore. The degree of similarity in the trends displayed by two elements is calculated by *R-mode analysis* on the basis of their mean concentrations and standard deviations, and is displayed as correlation coefficients in a correlation matrix. Zinc, lead, and copper, for example, are likely to show a good positive correlation in soil samples taken over a massive sulfide deposit because they make up a large proportion of the sulfides which cause the anomaly; they are not likely to be correlated significantly in soils over country rock, however, because they are held independently from each other in separate silicate minerals which react differently to weathering.

Further statistical manipulation based on correlation coefficients aims at reducing the number of relevant variables. *Cluster analysis* provides an objective comparison of inter-element correlation coefficients by grouping elements which are highly correlated with each other, and by indicating which elements exhibit the greatest within-group correlation. The results of cluster analysis are presented in dendrograms which highlight element associations. These can be interpreted in terms of geologic processes, as for example the clustering of limonite-facies elements in a typical, iron-bearing sediment, which are identified by statistical and geologic coherence in Fig. 2.14. The correlation coefficients between elements and element clusters are given on the y-axis. High positive correlations (close to +1) suggest similar behavior of two elements; values close to 0 suggest unrelated behavior; and negative values close to -1 suggest antithetical behavior. Through further statistical manipulation, *R-mode factor analysis* mathematically identifies *factors* which give rise to the observed correlations. These factors are mathematical concepts, but they may correspond to well-known geologic relationships; for example, one "sulfide factor" in soil samples may explain the positive correlations between copper, lead, and zinc, which are derived from sulfide minerals in a massive-sulfide deposit. The number of factors is likely to be much smaller than the number of variables, and the factors can be plotted and interpreted more easily on geochemical maps than the full data set because more geochemical information can be summarized at each sampling point.

While R-mode analysis evaluates the relationships between elements in sample populations, *Q-mode analysis* statistically describes the compositional similarities between samples on the basis of all measured element concentrations. Q-mode analysis measures the similarity of samples quantitatively by defining each sample as a vector in M-dimensional space, where M is the number of elements analyzed. Factors, expressed as groups of elements which account for the variation between samples, are extracted mathematically, and the proportion to which each factor contributes to the variation is calculated. These factors allow the differentiation, for example, of rock- and sulfide-induced anomalies in stream sediment exploration. Once the Q-mode factors for different rock and ore types have been determined for an area, single samples

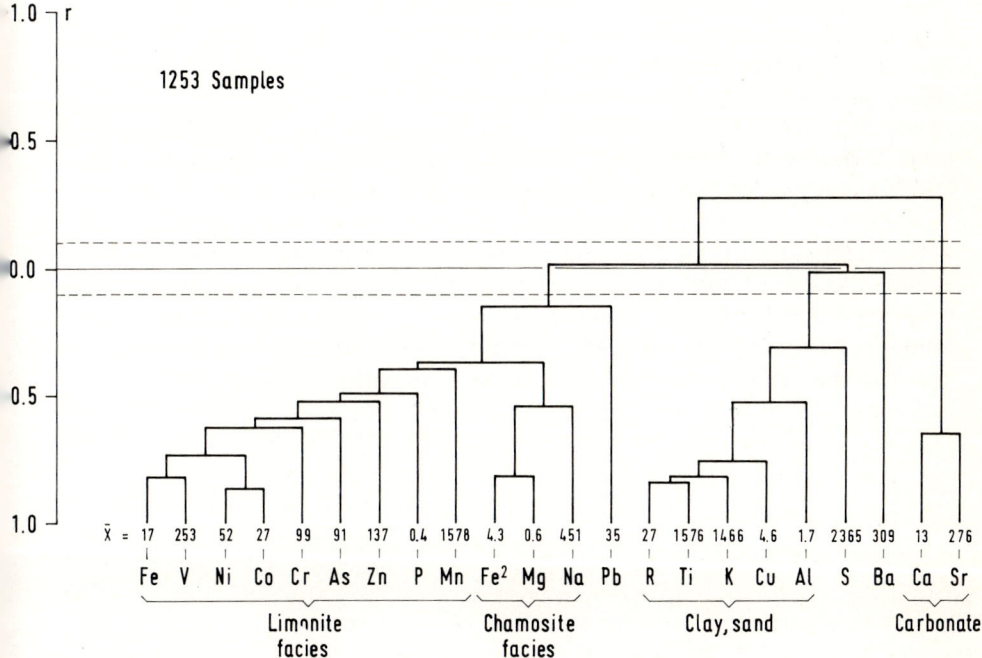

Fig. 2.14. Dendrogram illustrating groupings of geologically related elements in the Minette iron ores, Luxembourg, Lorraine. (After Siehl et al. 1978)

can be compared to these types by *discriminant analysis.* Davis (1986) gives an example in which one group of stream sediment samples from an area of active mining, and a second group from a geologically similar area without known mines, are analyzed by Q-mode factor analysis. The discriminant function provides not only the possibility of assigning samples of unknown association to one of these two groups, but also of measuring the degree to which each of the variables contributes to the classification. The better understanding of the relevance of individual variables, made possible by R-mode and Q-mode analysis, translates into time and cost savings in exploration programs by identifying those variables that should be measured and taken into consideration, and those that may safely be ignored because they do not add significantly to the discovery of anomalies.

A careful coordination and interpretation of geologic, geophysical, and geochemical exploration fulfills the main objective of exploration: to assess the mineral potential of a region, and to identify mineral targets for detailed evaluation.

Notes on the Literature

Peters' book Exploration and Mining Geology (1986) gives a comprehensive account of all aspects of exploration and mining geology. The lucid presentation of topics reflects the lifelong experience of a geologist who is deeply involved in the world of exploration and mining. Exploration strategies and methods are summarized in a series of papers which comprise the second part of the Economic Geology 75th Anniversary Volume edited by Skinner (1981), and a mathematical treatment of the design of optimal exploration strategies is given by DeGeoffroy and Wignall (1985). The applications of remote sensing to the exploration of mineral deposits and hydrocarbons are clearly illustrated in the joint NASA/Geosat Test Case Project edited by Paley (1985). Griffiths and King (1980) summarize geophysical methods in easily readable form, Telford et al. (1976) give a much more comprehensive treatment, and Hood (1977) has assembled and edited an impressive array of papers on the principles and applications of geophysical and geochemical exploration methods, including a series of case studies. Geochemical exploration methods are covered in textbooks by Rose, Hawkes and Webb (1979) and Levinson (ed.) (1974, 1980). Background on the statistical methods of data interpretation are given in the textbooks on geochemical exploration, and discussed with more mathematical rigor by Davis (1986). A brief and concise economic analysis of mineral exploration is presented by Eggert (1987).

Chapter 3 Quantitative Assessment of Mineral Potential

Previous sections have dealt with regional reconnaissance exploration and the identification of mineral targets by geologic, geophysical, and geochemical follow-up surveys. We now turn to the next step in the search of an ore deposit: the search for the source of an anomaly, and, if it is an ore body, the determination of its ore grade and tonnage by systematic sampling. This type of assessment may also take place, without prior regional exploration, in areas of known mining or exploration history which become attractive because of rising commodity prices, cost reductions in mining and processing, improvements in recovery by more effective processing technology, or the recognition of new geologic models which warrant renewed exploration. This quantitative stage of exploration represents a transition from the investigation of secondary and primary dispersion halos to ore; from indirect exploration to direct target definition; from relatively low cost to high cost in absolute terms and per unit area; and from the geologic domain of exploration and grade-tonnage assessment to the engineering and economic domains of mining, mineral processing, financial evaluation, and marketing studies. Because of the large expenses involved in these evaluations without assurance that a project will be profitable, this is the highest-risk stage of geologic exploration and the stage at which vigorous management support is most needed. Exclusive exploration and exploitation rights are secured to protect the investment, and access and land use is negotiated with surface owners. The environmental impact of exploration and future exploitation must be assessed, and the equipment and labor for development and mining must be secured.

Depending on the type of mineral sought and on the depth of its occurence, evaluation begins by surface work or by drilling; in either case, drilling is ultimately needed for a three-dimensional evaluation of the target. Examination and revision of the exploration model and exploration strategy during surface sampling and drilling assure that a potential deposit is not missed. Excellent examples of the power of this approach are given, as discussed in the section on geologic models, by successful deep drilling of the Henderson molybdenum deposit in Colorado (Wallace 1975, 1978), or the Kalamazoo deposit in Arizona (Lowell 1968).

3.1 Accessing the Ore

If a mineral deposit or the fringes of its primary dispersion halo crop out at the surface, much information can be gained by rapid and inexpensive methods of removing soil and weathered overburden to examine primary material. If there is no outcrop or if weathering is too deep to allow adequate surface access, drilling provides the sought-after information.

3.1.1 Trenching and Pitting

Soil and overburden can often be removed by trenching to expose, for example, concentrations of gold or tin at the base of unconsolidated gravels, or metal sulfides at the base of oxidized surface outcrops. Depending on the geometry of the mineralized zone, trenches may form a network at 50–200 m spacings for reconnaissance and 10–50 m intervals for evaluation of equidimensional zones, or they may trend across the longitudinal axis of elongated zones with wide spacing along the axis and close spacing across it. The spacing depends on: (a) the detail sought in the program; (b) the regularity of distribution of the minerals or metals sought; and (c) the unit value of the commodity, which in turn determines the minimum size of minable mineral concentrations. A lens of only 100000 t (100 m x 100 m x 10 m) of high-grade gold ore, for example, may be a very profitable, if short-lived, deposit, while millions of tons are needed to make deposits of copper, lead, and zinc attractive. Trenching is best done by backhoe, which can dig several hundred meters of trench about 1 m wide and up to 5 m deep per day and can rapidly restore the surface to its former state after sampling and mapping are completed. In remote areas and in the absence of mechanized equipment, trenches are routinely made with pick and shovel. For geochemical interpretation and quantitative assessment, trenches are sampled in vertical or horizontal sample channels. Depending on the size of sample needed, these channels may be 10 to 20 cm wide, 5 to 10 cm deep, and as long as dictated by the regularity of mineral distribution and the minimum dimensions needed for mining.

Trenches provide not only the opportunity for systematic sampling, but also an excellent, continuous geologic exposure for geologic mapping of rocks, structures, ores, and ore-related features such as alteration zones and quartz veins. This information is exceedingly important for the prediction of ore characteristics and ore distribution at depth.

Where trenching is not feasible because overburden is too thick or too unstable, vertical exploration pits, which can be supported more easily, may be dug. These pits, square or round, with a diameter of about 80–120 cm, may reach up to 40 m in depth and are generally designed to cut through the overburden to bedrock. Pits expose fresh surfaces for sampling and geologic observation, but they do not provide the continuous exposure afforded by trenching.

3.1.2 Drilling

Drilling penetrates fresh rock through overburden, and drill grids give a three-dimensional picture of the geology and ore content of mineral prospects. Drilling may be used (a) before a precise target is defined, for geologic information; (b) when the target has been defined, for the discovery of a potential ore deposit; (c) for the delineation and preliminary geologic evaluation of such a deposit; (d) for a systematic ore reserve estimation; and (e) for control of ore grade and mining strategy during exploitation. Different drilling methods are available to evaluate, for example, mineral accumulations in unconsolidated rocks at shallow depths of less than 100 m; metallic and nonmetallic minerals in hard rock to depths of more than 2000 m; and for petroleum and natural gas to depths of over 7000 m. Drill mud, drill cuttings, or drill core provide a record of the rock types encountered, and geochemical analyses and down-hole geophysical measurements help to define the distribution of the ore and its quality and quantity. In the drilling for petroleum and natural gas, the drill hole often serves for both exploration and production. A variety of types of drills are available for different purposes:

- *Augers* are hand-held or truck-mounted spiral drills used for shallow sampling of soft materials. Their most common application is in soil and placer sampling.
- *Churn Drills* have a heavy drilling tool which is suspended on a rope or cable and fractures the rock by impact. Samples are retrieved from the drill hole by bailer or sand pump. Churn drilling is the oldest and simplest drill method available and is useful particularly for shallow ores, unconsolidated overburden, or in areas where more advanced types of drills are not available. The Bangka drill, designed especially for the exploration of placer accumulations of tin, has an outer drill casing which is continuously rotated and lowered while the drill bit fractures the rock. The simultaneous progress of casing and drill tool assures than there is no contamination of the sample from upper layers of unconsolidated material.
- *Percussion Drills* are powered by air pressure and are extensively used in surface and underground mining to prepare blast holes. These drills can be used for hard-rock exploration down to about 100 m depth.
- *Rotary Drills* cut the rock by rotation of large diameter (124–900 mm) drill bits studded with steel, tungsten carbide, or diamond cutting tools. The broken rock material is brought to the surface by circulating drill water or mud, which also serves to cool the drill bit. Rotary drills, predominantly used in the exploration for petroleum and natural gas, progress rapidly and routinely reach depths of 5000–7000 m. Geologic information is gained from drill cuttings and from down-hole geophysical surveys.
- *Core Drills* or *Diamond Drills* cut rocks with a ring-shaped drill bit which is attached to the head of a hollow core barrel. The diamond-studded drill bit cuts out a cylindrical core of rock which is accommodated by a nonrotating inner core tube. A wedge-shaped core catcher closes tightly around the lower part of the core, pulling it off its base and transporting it securely to the surface when the drill barrel is lifted. The inner core tube is from 3 to 6 m long and, in modern drills, can be brought to the surface through the string of drill pipe by a wireline, without the need to lift the whole drill string. The drill bit and drill rods are cooled and lubricated by water or drill mud pumped through the rods. Diamond drill rigs are available for a wide range of purposes, including small units that can be carried by backpack and reach 20 m depth; large units, most of which can be carried by helicopter and reach depths of several 1000 m; and units which are designed to drill in underground workings. Diamond drills, whose diameters typically range from 30 to 146 mm, easily penetrate hard, crystalline rock and perforate at any angle. Deep drill holes are generally started at a large diameter and tapered with depth; to protect the walls from collapse, casing is installed in the upper levels. Diamond drills are the most essential tool in the final exploration and evaluation of mineral projects. Study of the drill core yields a three-dimensional geologic picture of ore and host rock, and splits of the drill core provide samples for chemical analysis, mineral recovery tests, and rock stability tests which are essential for mine planning.

The spacing of drill holes depends on the stage and the aim of the drilling program. In the early stages of exploration, orientation drill holes may be spaced far apart and drilled mainly to obtain information about the rock types to judge their favorability for ore formation. In the later stages of target delineation, drill holes are spaced at intervals of 100's to 10's of meters, depending on the size and extent of the mineral deposit and the regularity of the ore distribution.

Drill cuttings are taken at regular intervals for inspection or analysis, particularly if core recovery is poor. The core, which is retrieved at the intervals dictated by the capacity of the core barrel, is logged geologically to assemble a record of rock types, structures, ores, and hydrothermal alteration patterns encountered in the subsurface. After logging, the core is split and one half is sent for chemical analysis while the other is kept for future geologic or chemical reference. The need to arrange adequate and long-range storage of drill core cannot be overemphasized. The core is extremely valuable for a re-evaluation of the prospect and of the geologic models used in its exploration; all too often, it is no longer available when needed to guide the search for new reserves in mines which are close to exhaustion, or for a re-interpretation of past exploration efforts.

3.1.3 Drill-Hole Geophysics

Because drilling itself is very expensive, it is imperative that as much information as possible be derived from every drill hole. In many cases, geophysical logging can eliminate the need for coring and thereby speed up exploration and decrease its costs. Rotary drilling, generally of soft, sedimentary rocks in the search for hydrocarbons, is fast and less expensive per meter drilled than diamond core drilling, but yields less geologic information about rock characteristics. Core drilling, more often done in crystalline rock for the discovery of mineral deposits, is slower and more expensive per meter, but the core gives direct geologic information. Geophysical drill hole logging is therefore much more essential in rotary drilling and, consequently, in the search for hydrocarbons. However, the methods of geophysical logging are increasingly applied to the exploration for metallic mineral resources as well, and some of the geophysical methods discussed in section 2.3.3 for airborne and ground surveys are also applied in drill hole logging to identify types of rocks and ores. Rock magnetism, electric conductivity, radiation, density, seismic response, and rock composition are measured.

- Magnetic logs help to characterize rock types but, more importantly for mineral exploration, help to detect magnetic disturbances due to ore in the immediate vicinity of drill holes.
- Electromagnetic (EM), resistivity, and induced polarization (IP) logs respond to electric conductors and are very effective in the direct detection of metallic minerals. They indicate the presence of massive sulfides or disseminated sulfides in metallic ore deposits, and of clay layers or brines contained in porous sandstones in the exploration of hydrocarbons.
- Gamma-ray spectrometry indicates the presence of uranium-, thorium-, and potassium-bearing minerals and is especially applicable to the subsurface exploration of uranium deposits.

– Neutron activation can be used to measure concentrations of copper, lead, zinc, gold, and silver in the walls of the drill hole (Schneider 1981). While still experimental, the ability of the method to measure metal concentrations underground, without the need for core recovery, is very attractive.

A wide array of geophysical logging methods and instruments is used to measure rock composition, density, porosity, water content, and oil content in the exploration of hydrocarbons. In the exploration of mineral deposits, resistivity, IP, and electromagnetic methods are most relevant. A three-dimensional interpretation of a target by geophysical means begins with measurements within drill holes, followed by measurements between drill holes, and between drill holes and the surface. Because most geophysical anomalies can be interpreted in a variety of ways, geologic control is needed to assure that the geophysical models are consistent with geologic observation and evidence.

Drilling projects conclude with a final report consisting of (a) a technical log in which drilling progress, core recovery, and technical matters are recorded; (b) a geologic log supplemented by geologic maps and cross sections; (c) geochemical logs of rock and ore assays, and (d) geophysical logs. The results of a successful drilling program are a three-dimensional picture of the geology, geometry, ore grade, and tonnage of a mineral prospect.

3.1.4 Underground Exploration

The evaluation of a mineral prospect often requires underground mine workings during the final stages of exploration, to reach the ore for direct inspection and systematic sampling. The large-sized samples obtained by tunnels, inclined adits, or vertical shafts yield much essential geologic, chemical, and engineering information. Geologic inspection and chemical assays reliably reflect mean ore grades and the distribution of the ore; engineering tests address questions of mineral processing and the recoverability of minerals and metals, often in a pilot plant; and geomechanical studies aid in the choice of the best mininig procedures. Expenditures of 5 million to over 10 million at this stage of exploration and evaluation are not unusual.

3.2 Sampling and Assaying of the Ore

Systematic sampling campaigns mark the beginning of quantitative geologic evaluations and feasibility studies. While semiquantitative analyses were satisfactory for exploration geochemistry, accurate, quantitative chemical assays of ore are required for final evaluation. The suite of elements analyzed is specific for each type of deposit and includes, besides the major elements or metals, the concentrations of byproducts which can be recovered, for example cobalt in some copper deposits, and the concentration of "poison" elements for which a penalty is deducted by the smelter, for example arsenic or antimony in silver-lead-zinc ores.

Metallurgical investigations determine the recoverability of the metals in a mineral deposit, for example by ore microscope studies of mineral compositions and textures to indicate the grain size to which ore has to be ground for mineral concentration, and which concentrating procedures are most effective. Thorough processing tests are generally needed to find the methods that are best suited to particular types of ore. Geomechanical studies, done on the whole-rock samples or on drill core, measure rock strength and are essential to the planning of mining methods and the estimation of mining costs.

3.2.1 Sampling

The sampling program for a grade/tonnage evaluation of a mineral deposit must be designed so that the small samples which are taken and analyzed are representative of the whole deposit. Random and systematic errors involved in the collection, preparation, analysis, and evaluation of samples must be recognized and accounted for.

Samples may be taken from easily accessible surfaces in trenches, shafts, or underground workings, or they may come from drill cuttings and drill core. For preliminary evaluations, random grab samples, point samples, or chip samples are taken from accessible rock exposures. For more systematic work, shallow channels, about 10 cm wide and 2 to 3 cm deep, are made by use of hammer and moil. For placer deposits, channels must be wider and deeper to obtain a sample that is large enough for the recovery of heavy mineral concentrates. The spacing and length of individual channel samples depends on the inhomogeneities in the distribution of the ore, the amount of material needed for analysis, or the minimum mining width of prospective mine workings; different parts of a vein, for example, may be sampled individually, or the full width may be combined into one sample (Fig. 3.1). If possible, channels are made at right angles to the trend of the ore, but if oblique or longitudinal channels are more convenient, results must be re-calculated to represent the true width of the ore-bearing structure.

Drill samples are taken from percussion and diamond drill holes during exploration and mining. The drill dust and cuttings from pneumatic drills, recovered in special dust bags or in cyclones, and from drills with water circulation, collected in settling tanks, are analyzed for mineral and metal concentrations. In core drilling, core recovery should be 80% or more for reliable evaluation, but even at this level of recovery, it is essential to determine whether losses are random or whether specific types of ore or gangue are lost preferentially, yielding a systematically biased result. A combination of sludge and core analysis helps to overcome such systematic errors in case of poor core recovery.

Sampling, sample preparation, analysis, and interpretation in the final stages of exploration and mining are planned and carried out by a staff of geologists, chemists, statisticians, and engineers, who contribute their expertise to the interpretation of the sampling data. The importance of thorough, joint planning and interpretation is obvious because they form the basis for an economic and technical evaluation of the mineral prospect, and because of the large financial commitment that the development of a potential ore deposit requires.

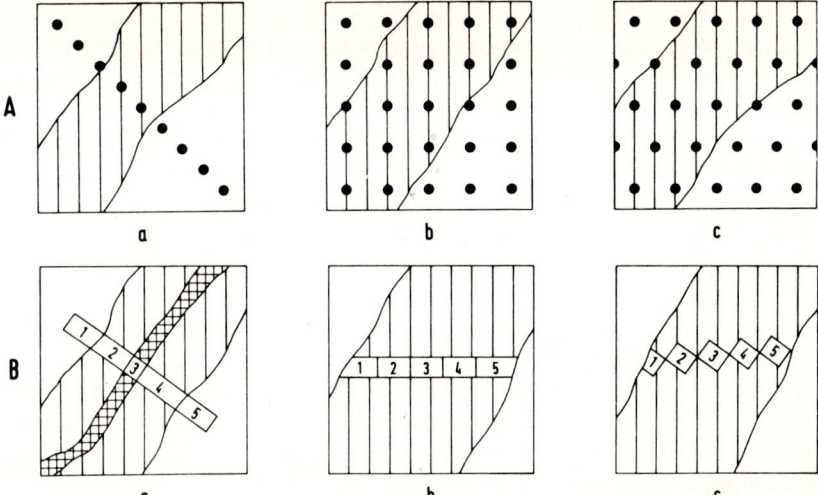

Fig. 3.1 A, B. Sampling patterns employed in the evaluation of vein ore. **A** Point samples (*a* along channel; *b/c* over full area.) **B** Channel samples (*a* according to mineral type; *b* horizontal width; *c* true width)

3.2.2 Sample Preparation

The preparation of samples depends on their size, physical properties, and on the analytical method to be used. All sample preparation involves three steps: (a) crushing and grinding of the sample, followed by screening and re-grinding as needed, to assure a fine and uniform grain size; (b) homogenization of the sample; and (c) splitting to an amount that can be assayed conveniently.

Depending on the initial size of fragments, the sample must be crushed in jaw crushers and then ground and pulverized to the final analytical size in rotary mills or disk mills. Splitting is done in a variety of ways: bulk samples are heaped to a cone and divided into two (or four) subsamples by use of a sampling board or a sampling cross. Subsamples are heaped and divided again until a small enough sample size is achieved to continue the splitting process in commercially available splitters. Homogenization of the subsamples is necessary at every step to insure that particles of different grain size or density have not segregated.

3.2.3 Chemical Assays

Of the many analytical methods available, only a few find application in ore grade estimation because highest accuracy, precision and efficiency of procedure are needed. The main principles involved in the different methods are described under exploration geochemistry. Because of their accuracy, some of the classical wet-chemical methods, which are no longer used in exploration geochemistry, remain mainstays of quantitative analytical work, for example gravimetric, volumetric, and colorimetric methods. These methods rely on reactions between solutions containing the unknown elements with reagents of known compositions. The chemical reactions are interpreted quantitatively on the basis of the weight of precipitate that forms, the volume of solution needed to achieve chemical equilibrium, or colors produced in the solutions by the reactions. Among the instrumental methods, atomic absorption spectrometry (AAS) and inductively coupled plasma spectrometry (ICP) are most widely used. A variety of special methods is available for specific purposes, for example fire assay for accurate analysis of precious metals and radiometric methods for the analysis of uranium.

The statistical evaluation of analytical data and the calculation of ore reserves based on these data are generally done with the help of complex statistical procedures and computer programs. The geologic setting of a mineral deposit must be thoroughly understood and incorporated in the evaluation to account for any geologic constraints in the procedures of evaluation. A thorough understanding of the geologic setting of a mineral deposit at this stage will help to reach a geologically reasonable evaluation.

3.3 Ore Reserve Estimation

The ore reserve estimation is based on grades and tonnages of the mineral prospect determined during final geologic sampling and evaluation. The estimate takes into account the geologic controls on mineral distribution and the influence of mining and metallurgical factors which bear on the recoverabilities and successful processing of the ore. The magnitude of any technical or metallurgical correction factors and the reliability of estimations applicable to the reserve figures enter into the assessment. Economic considerations, for example commodity prices and price trends, and financing, mining, and processing costs, will dictate what ore grades are needed to make a mining operation profitable.

The methods of ore reserve estimation depend on a variety of factors: (a) the type of commodity; (b) the type of mineral deposit; and (c) the geometry, distribution, and homogeneity of the ore. These factors determine the size of samples and the sampling density needed for an estimation of the desired reliability. The degree of certainty and probable deviations from calculated averages are assessed for all steps in this quantitative evaluation, from sampling and assaying to the calculation of grade and tonnage of the ores. In other words, the evaluation seeks the average ore grade, the spread of grades, and the confidence level of these values. Three fundamentally different methods are used to estimate ore reserves: (a) geometric methods; (b) statistical methods; and (c) geostatistical methods. The following discussion refers mainly to metallic ore deposits, but the procedures are applicable to the full range of non-metallic mineral resources and coal as well.

3.3.1 Geometric Methods of Ore Reserve Estimation

The main parameters of a geometric ore reserve estimation are: (a) the volume of ore, which may be calculated by simple or more complex geometric models, depending on the regularity of ore distribution and stage of the evaluation; (b) the average density of the ore; and (c) the losses during mining, given as an estimated correction factor. These data provide an estimate of the ore reserves (Q, in metric tons t):

$$Q = V \times d \times l_o,$$

where V is the volume in m^3; d is the density of the ore in t/m^3; and l_o is an estimated correction factor to account for losses of ore during mining ($0 < l_o < 1$). The metal content (M, in metric tons t) is:

$$M = Q \times z \times l_m,$$

where Q are the ore reserves, z is the ore grade in %, expressed as a decimal (e.g., 20% = 0.20), and l_m is an estimated correction factor to account for the losses of metal recovery during processing ($0 < l_m < 1$).

In geometric reserve estimations, each sample or drill intersection of the ore body is assigned an area and volume of influence by construction of simple geometric models, for example rectangles, triangles, polygons, or profiles, and their three-dimensional equivalents (Fig. 3.2). These volumes and assay values are used to calculate ore grades and tonnages of the blocks represented by a sample or drill section. The assumption in geometric estimations is that the area of influence of each sample is a function of its distance to adjacent samples, an assumption which is not realistic in the case of nonhomogeneous ore deposits. To estimate the needed sampling density and to minimize any errors, the grade and tonnage of an experimental ore block can be calculated on the basis of a progressively denser grid of sampling points. The sampling density above which there is no significant change in the grade and tonnage estimate may be taken as the minimum density needed.

Because of the uncertainties and subjectivities involved in assigning areas of influence in geometric ore reserve estimation, these methods, while still very useful for preliminary assessments, are now largely superseded by statistical and geostatistical methods of evaluation.

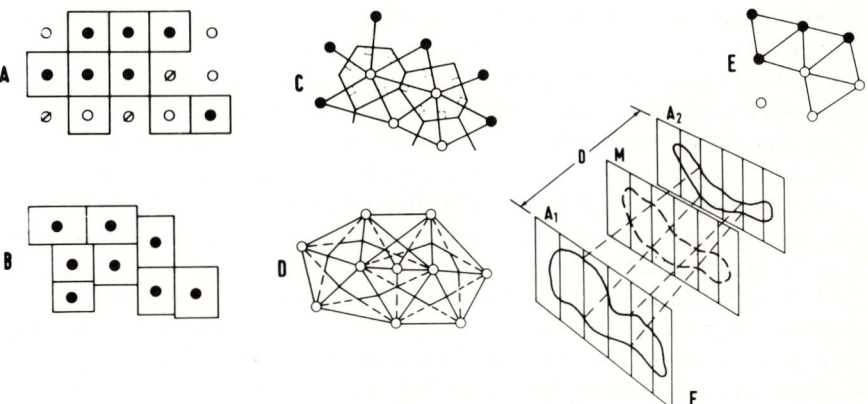

Fig. 3.2 A–F. Geometric approaches to ore reserve estimation. **A** Square blocks for regular sampling or drilling patterns. **B** Rectangular blocks for irregular sampling patterns. **C** Polygonal blocks constructed by joining the mid-points between samples. **D** Polygonal blocks defined by angular bisection. **E** Triangular blocks produced by joining three adjacent sampling blocks. **F** Profiles through a hypothetical ore body. (After Patterson 1959)

3.3.2 Statistical Methods of Ore Reserve Estimation

The most sensitive problem in ore reserve estimation is the determination of average grades, range of grade values, and confidence levels of the estimates which are derived from the sampling and assaying program. Statistical ore reserve estimation allows optimization of the sampling program by calculation of the minimum size and number of samples required to assess the grade and tonnage of a mineral deposit with the precision desired for a reliable economic evaluation. While *classical statistical* calculations involving frequency distributions and standard deviations of sample assays yield satisfactory estimates of the grade and tonnage of a prospect for pre-feasibility studies, more advanced methods of *geostatistical* evaluation play a much more important role during the final stages of quantitative feasibility studies.

The original data set for an ore reserve estimation are sample locations and sample assay values. The sample size of a representative sample is a function of the distribution and grain size of mineral constituents. From a rock which has a grain size of 5 cm, for example, about 250 kg must be taken to obtain a representative sample, while 1 kg is sufficient from a rock which has a grain size of 0.3 cm (Gy 1982). The distribution of assay values may be normal, recognized graphically by a symmetrical frequency curve when plotted on an arithmetic base; log-normal, recognized by a symmetrical curve when plotted on a logarithmic base, or irregular. The average grade, statistical variation in grade, and confidence levels of the calculated average grade can be determined by the statistical methods described in section 2.3.5.8 for the interpretation of geochemical results.

The next paragraphs present the main aspects and capabilities of statistical and geostatistical ore reserve estimation for the case of normal distributions. If the distributions are log-normal, the approaches are mathematically more complex, as discussed in David (1977). In the case of mixed or irregular distributions, the ore reserve calculations become more complex, but a knowledge of the geologic controls of different sample populations may enable recognition and separation of the different populations for purposes of reserve estimation. A full discussion of the statistical calculations involved in ore reserve estimations is beyond the scope of this summary. The emphasis here is on the application of the statistical methods; for a more comprehensive treatment of the mathematical background and calculations, the reader is referred to the specialized literature (David 1977; Davis 1986; Harris 1984; Verly et al. 1984).

The purpose of statistical ore reserve estimations is twofold: (a) to determine the number of samples needed to achieve a desired precision, and (b) to assess the ore grade and ore quantity. The mean and standard deviation of sample assay values from an ore block are the basis for a calculation of the block's average ore grade and standard deviation, and the number of additional samples needed to achieve a desired level of precision for the grade estimate may also be computed. Pre-condition for these calculations is that the samples be random variables.

These applications of classical statistics do not answer questions as to the influence of the sample size on the estimate of the mean grade of the ore, the effect of a nonrandom distribution of the samples, and the spatial distribution of ore grades. These questions are addressed by *geostatistical methods* of ore reserve calculation.

3.3.3 Geostatistical Methods of Ore Reserve Estimation

Geostatistical ore reserve calculations take into account the spatial relationships of samples to each other and are essential tools in the calculation of representative and optimal sampling densities and of ore reserves. In geostatistical calculations, the degree of correlation between adjacent samples is determined statistically as a means to assign an area of influence to each. Noncorrelation between two samples does not give any indication as to the relationship between them. If, for example, there is a sudden termination of ore grades between two sampling points due to a fault, no meaningful area of influence can be assigned these points on a statistical basis. If there is a correlation, as shown, for example, in a gradual change in composition between adjacent samples from a porphyry copper deposit, an area of influence can be assigned. High positive correlation coefficients of metals between adjacent samples, i.e., close to $+1$, reflect a high degree of correlation, indicating a sufficient sampling density; coefficients close to 0 indicate noncorrelation and insufficient sampling density for meaningful statistical evaluation.

A more precise tool to describe the spatial relationships of variables to each other and to measure their degree of randomness is given by *variograms*, which take into account ore grades and their distribution in space. An example of a variogram from a lead-zinc deposit is given in Fig. 3.3. Distance is plotted on the x-axis, and the variogram function gamma(h), which describes the similarity between ore grades of adjacent sample points, is plotted on the y-axis. There is a positive correlation between ore grades up to a distance between samples of approximately 4.6 m, the *range*. At this point the variogram function reaches the *sill*, which marks non-correlation between metal values in samples which are outside the distance given by the range. The sampling density should be within the range, and a sampling interval of about 2/3 of the range is generally adopted; closer sampling yields a higher precision but increases effort and cost, while sampling at a larger interval exceeds the statistically determined maximum area of influence for each sample. Figure 3.3 includes both the experimental variogram derived from exploration and a mathematically derived variogram calculated according to the spherical model or "Matheron model", which applies well to a wide variety of geological situations.

The *dispersion variance* of the metal grade of a mining block is a measure of the reliability of the estimated mean ore grade and can be calculated on the basis of the variance of the samples, in combination with the variogram. The *extension variance* determines the error which results from estimating the ore grade of a mining block on the basis of one central sample, or on the basis of several samples distributed in the block. In ore reserve estimation by Kriging (named after Krige), each block is evaluated not only on the basis of the samples in the block itself, but also of samples from adjacent blocks which are considered and weighted by use of the variogram function, in such a way that errors of estimation are minimized. Kriging optimizes the ore grade estimation for ore blocks and has a series of advantages as compared to other evaluations: (a) it allows the interpolation between sampling points and the representation of data on a contour map; (b) it condsiders the spatial distribution of sampling points (regular spacing, irregular spacing) in the ore reserve estimation; and (c) it enables the estimation of error at every point in the deposit or for every mining block.

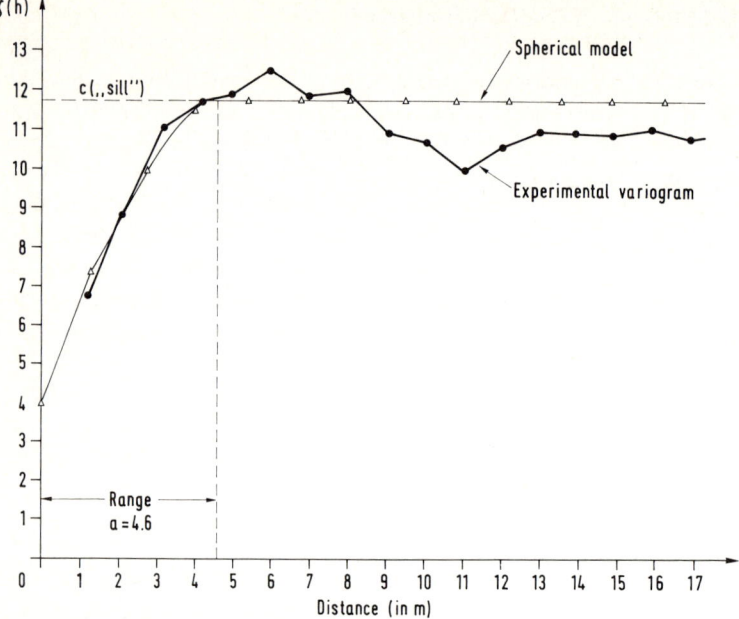

Fig. 3.3. Experimental and modeled variogram for a lead-zinc deposit in the Harz, Germany

The type of information that goes into a reserve estimation, and the output, are summarized in Table 3.1. Statistical packages for geostatistical evaluations are commercially available, and applicable programs are described, for example, by David (1977) and Journel and Huijbregts (1981). Data preparation, calculation of the experimental variograms, development of model functions, interpretation of the ore reserve calculation, and the interpretation of the error of estimation require a thorough integration of geologic and geostatistical knowledge to maximize the reliability of the ore reserve estimate and to avoid misinterpretations.

3.3.4 Classification of Reserves and Resources

Because of the high cost involved in exploration and evaluation of mining projects, only the minimum ore reserves needed to justify mine development are determined initially, while other parts of the deposit or district are left for later evaluation. The degree of certainty of the grade and tonnage values of ore discovered, which is a direct function of the efforts made in the estimation, is the basis for a classification of ore reserves and, more generally, of resources for present or future use.

Geologic assurance, technical feasibility, and economic viability are the main criteria for a classification of ore reserves and resources. Classification schemes were developed in the 1970s by the US Geological Survey, the US Bureau of Mines (McKelvey 1973, 1986; Fig. 3.4), the Canadian Department of Energy, Mines and Resources (Zwartendyk 1981), and European organizations (Fettweis 1981). The

Quantitative Assessment of Mineral Potential

Table 3.1 Procedure of statistical ore reserve calculation.

Geologic input
(1) Exploration data (geologic, geophysical, and geochemical)
(2) Sample data (location, coordinates, level)
(3) Mineralogic and assay data (ore mineral compositions and metal grades)
(4) Geometry of the ore body (dimensions, ore distribution; ore blocks)
(5) Units and correction factors (physical and chemical units; recovery coefficients)

Ore Reserve Calculation (geometric, statistical, and geostatistical methods)
(1) Frequency distributions of assay data
(2) Areas of influence of assay data
(3) Statistical grade-tonnage values for ore blocks and the deposit (according to metals contained, grade levels, and reserve categories)
(4) Range of values and confidence levels

Technical and economic input
(1) Extraction coefficients
(2) Cost factors and costs
(3) Revenue

Geologic, technical and economic ore reserve estimation
(1) Minable grade, cut-off grade
(2) Tons of ore (according to ore blocks and reserve categories)
(3) Recoverable metal contents

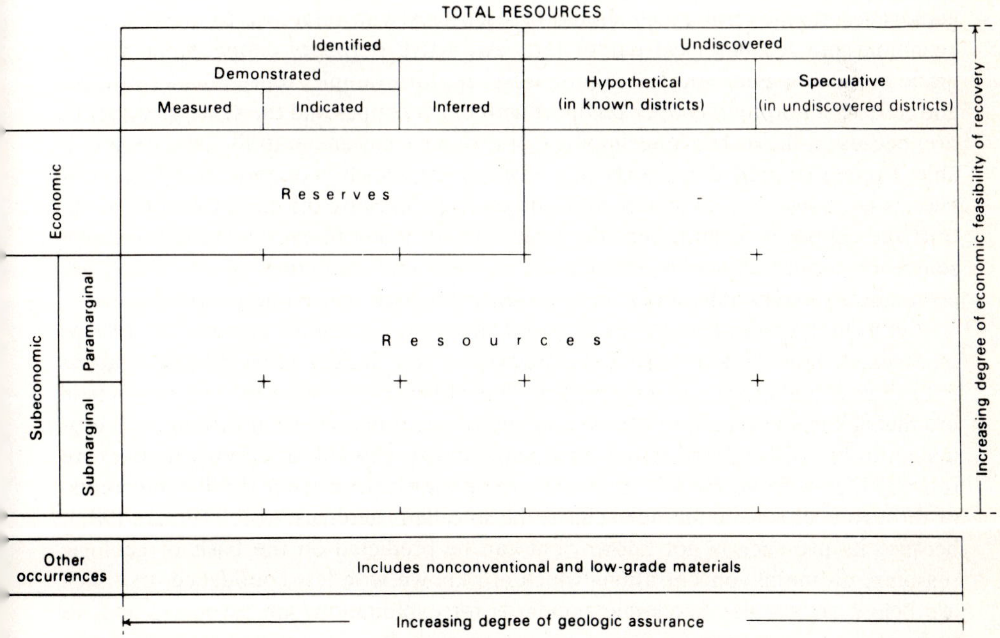

		TOTAL RESOURCES			
		Identified		Undiscovered	
		Demonstrated		Hypothetical	Speculative
		Measured	Indicated	Inferred	(in known districts)	(in undiscovered districts)
Economic		Reserves				
Subeconomic	Paramarginal	+	+	Resources +	+	
	Submarginal					
Other occurrences		Includes nonconventional and low-grade materials				

← Increasing degree of geologic assurance →

↑ Increasing degree of economic feasibility of recovery

Fig. 3.4. Classification of mineral resources and reserves. (McKelvey 1986)

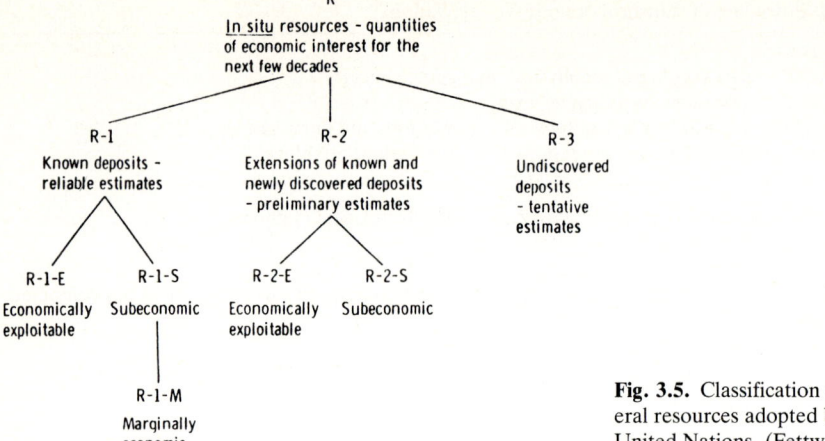

Fig. 3.5. Classification of mineral resources adopted by the United Nations. (Fettweis 1981)

United Nations developed and recommended a classification scheme for international adoption in 1978/79 (Fig. 3.5), but little international consensus was achieved and a large and confusing variety of classifications and terms is used by national and international government agencies, and by different groups of professionals, for example mining geologists, mining engineers, and mineral economists involved in resource assessments.

The comprehensive term that defines currently or potentially exploitable mineral commodities is *resources*. This term was defined by Harris and Skinner (1982) as a geologic-economic term which should be used only with reference to a given set of economic and technological parameters. An active porphyry copper mine with a grade of 0.8% copper and 500 million tons ore, for example, represents a resource, and so does a porphyry copper prospect with 0.3% copper and the same quantitiy of ore, because a slight shift in technologic or economic parameters will make it exploitable. Copper contained in granite at an average composition of about 0.001%, however, is not a resource no matter what the tonnage, because the energy needed to extract the copper is so large that the granite is not likely to ever represent a usable source of copper. *Reserves* are defined as measured quantities of ore which are economically exploitable under present-day technologic and economic conditions.

Let us look first at reserves as seen and named by the mining community. *Proven* or *measured reserves* define ore that is blocked out on all sides and estimated to about 10% of its actual grade and tonnage (Fig. 3.6). *Probable* or *indicated reserves* include ore that is known on two or three sides from adjacent ore blocks or drilling, and estimated to about 20–30% of actual grade and tonnage. *Possible* or *inferred reserves* are reserves known from one side of the mining operation, from some drilling intercepts, or through geological inference. This is the so-called "geologist's ore" (Peters 1987), because its presence is not known, but can be predicted on the basis of geologic reasoning. Mineral concentrations which are known with less confidence, or which are below present-day economic standards for exploitation, are resources and, as such, exploration targets for future evaluation. The discovery or "making" of a mineral deposit by skillful exploration and geologic reasoning traces the sequence from

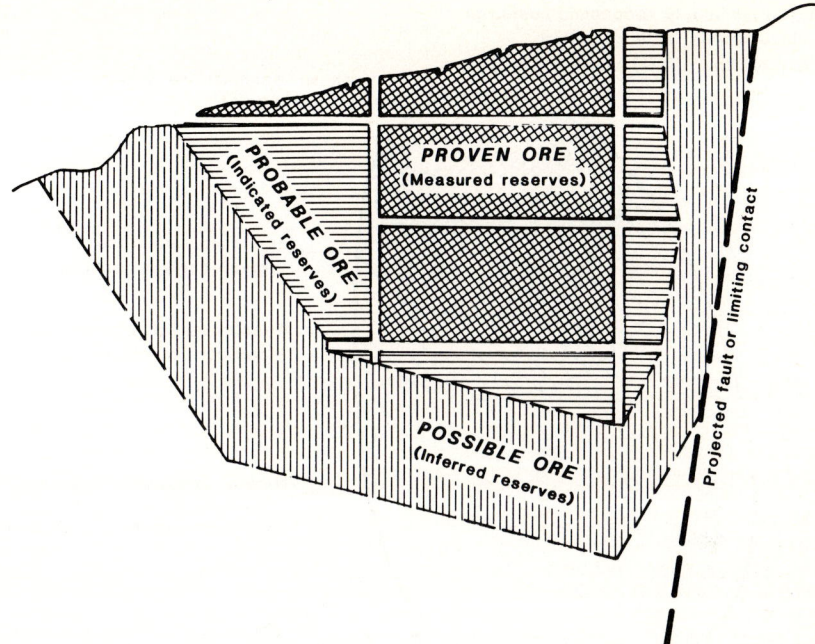

Fig. 3.6. Categories of ore reserves in an active mining operation. (After Peters 1987)

undiscovered speculative resources (see Fig. 3.4) to measured economic reserves; from an initial non-economic geophysical or geochemical anomaly to the economic core of a deposit; or from "geologist's ore" to "mining engineer's" and, finally, "mineral economist's" ore.

The simple and very practical classification of reserves given above is not adapted to the needs of government agencies which look not only at the present economic potential of particular exploration targets, but to the long-term availability of resources in order to develop suitable strategies and policies of exploitation. The McKelvey box (Fig. 3.4), which includes categories for resources of possible future interest, may serve that purpose. Increasing degree of geologic assurance of a resource is marked on the x-axis from right to left; increasing degree of economic (and technologic) feasibility is marked on the y-axis, from bottom to top. The mining categories measured, indicated, and inferred reserves are in the upper left-hand corner of this classification matrix. Roughly speaking, the classification separates reserves, defined as identified and economically exploitable ore, from resources, defined as undiscovered or noneconomic mineral accumulations. Nondiscovered resources are classified as *hypothetical* if there is a geologic reason to suspect their presence or *speculative* if not. Subeconomic (but geologically known) mineral accumulations are subdivided into *paramarginal* and *submarginal,* depending on the likelihood of their exploitability in the foreseeable future.

The United Nations classification (Fig. 3.5), while very similar to the McKelvey box, measures geological assurance as a certainty of estimate and the economic feasi-

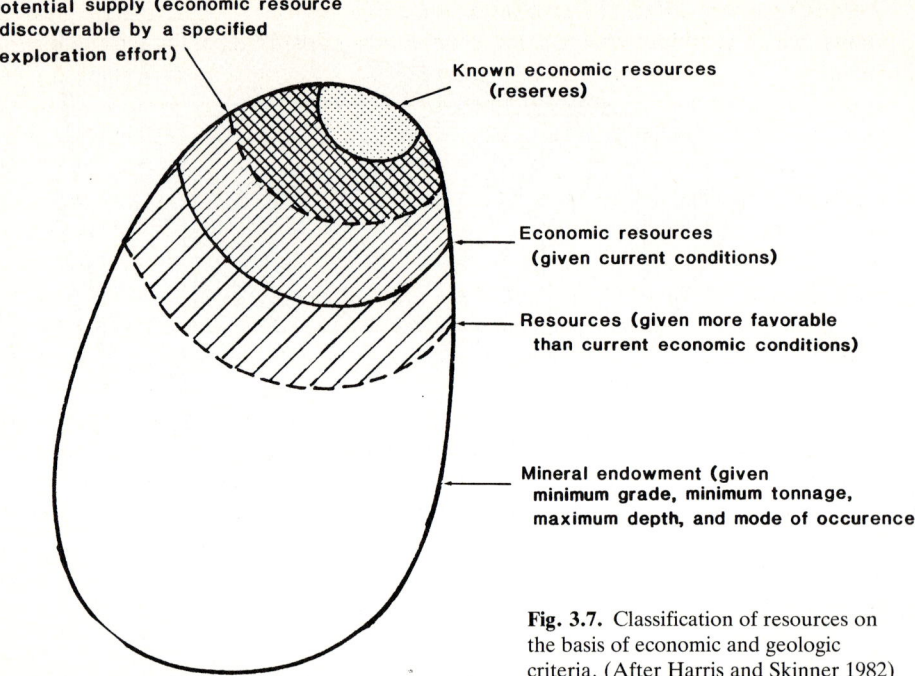

Fig. 3.7. Classification of resources on the basis of economic and geologic criteria. (After Harris and Skinner 1982)

bility as exploitability. Letter-number combinations rather than reserve-resource expressions identify different types of resources. The resources are classified by certainty of geologic occurrence as:

Category R-1, which denotes reserves estimated with a statistical error of less than 50%. The chemical and physical characteristics of the reserves are sufficiently known to serve as a basis for the planning of exploitation and mineral processing. The R-1 category, therefore, covers the categories described above under measured and indicated reserves.

Category R-2, which includes resources which have been estimated with a statistical error of more than 50%. These resources are considered worthy of more intense exploration. The R-2 category may correspond approximately to inferred reserves.

Category R-3, which includes resources which have been estimated on the basis of geologic analogy, geophysical and geochemical indicators, and statistical extrapolations. The R-3 category, therefore, denotes hypothetical resources.

According to exploitability, resources are subdivided into:

Category E ("Economic"), which includes the R-1 and R-2 resources considered exploitable at the time of the assessment under local technical, economic, and social conditions. The full UN classification for these resources is R-1-E or R-2-E.

Category M ("Marginally Economic"), which defines those R-1 resources which may become exploitable in the near future as judged by projected changes in technology or commodity prices (R-1-M).

Category S ("Subeconomic"), which includes all R-1 and R-2 reserves which are not yet economically exploitable, but which may be of mining interest in the more distant future.

A refreshingly different classification for use in regional or global resource assessments, which combines geologic and economic considerations in its terminology, is recommended by Harris and Skinner (1982) (Fig. 3.7). Here, resources are divided into *known economic resources,* defined as discovered and economically exploitable resources, and *economic resources,* which include economically exploitable resources which have not yet been discovered. Intermediate in quantity between the known economic resources and the economic resources is the *potential supply,* or the supply of a mineral resource of stipulated economic parameters which is discoverable by a specified exploration effort. A more speculative resource category is *mineral endowment,* the estimated quantity of a commodity at a given minimum grade and tonnage, maximum depth, and specified mode of occurrence. The boundaries between the different categories shift with new discoveries, changes in price of mineral commodities, and technological progress.

A purely hypothetical resource category, which is occasionally used in global resource assessments, is the *resource base,* defined as the quantity of a given commodity in the earth's crust, whether exploitable or not. Since the category "resource base" includes nonexploitable resources, for example copper, lead, or zinc contained in the silicate minerals of granite, this category is meaningless for practical purposes and should be avoided in any realistic discussion of the availability or long-range supply of mineral resources.

Notes on the Literature

For a general discussion of the geologic aspects of quantitative assessment of mineral potential, the reader is referred to Peters (1987). Mathematical and statistical approaches to reserve and resource estimation are covered in excellent and comprehensive volumes by David (1977), Journel and Huijbregts (1981), Harris (1984), and Verly et al. (eds.) (1984). Reserve and resource classifications are presented by McKelvey (1973, 1986), and long-range assessments of the availability of mineral resources are discussed by Skinner (1979), Skinner and Harris (1982), and Harris (1984).

Chapter 4 Mining and Mineral Processing

Successful exploration leads to the development of a mine for extraction of the ore and of a processing facility to transform the ore into a marketable product. The evolution toward this stage, and some of the costs involved in developing a mine, are discussed in Chapter 2 and illustrated in Figures 2.1, 2.2, and 2.3.

4.1 Mining

The mining methods and the scale of mining are determined by (a) geologic factors relating to the geometry and distribution of the ore; (b) engineering factors involving rock stability, technology, and equipment; and (c) economic considerations, including the magnitude of the investment, availability of financing on reasonable terms, payback period, and mine life, as discussed more fully in Chapters 5 to 7. In this chapter, the geologic factors which determine the choice of the most adequate mining methods will be addressed, and the mining methods themselves will be described briefly.

As discussed in Chapter 1, minerals and metals occur in distinct types of ore deposits which are characterized by their geologic setting, geometry, and range of grades. Each type of deposit or individual mine requires specific mining methods depending on the setting of the ore. Copper, for example, may occur in relatively small massive-sulfide deposits which grade from 1% to 4% copper, or in large porphyry copper deposits with ore grades of 0.5–1.0% copper. Gold may be hosted in narrow veins containing 10–20 g (0.3–0.6 oz) gold per ton, or as disseminations in rock containing as little as 1 g (0.03 oz) gold per ton of ore. High-grade deposits, or deposits of a commodity of high unit value like gold, can be mined by expensive, selective mining methods while low-grade deposits, or deposits of commodities of a low unit value, must be large enough to justify large-scale, more cost-effective bulk mining methods. The cost of mining per ton of ore ranges from very low, for example for mining of unconsolidated rock on the surface or for leaching of very low-grade gold and copper ores, to very high, for example for selective mining of precious metals at depth as illustrated by many of the rich silver-lead-zinc mines of Mexico and Peru. Intermediate costs apply to open pit and underground mining in hard rock by bulk mining methods.

All mining methods involve (a) removal of the ore, and host rock if needed; (b) separation of the ore from the host rock; (c) crushing of the ore to a size that is convenient for handling; and (d) transportation to a processing plant. Inexpensive removal of ore by earth moving equipment is possible for deposits hosted in soft, unconsolidated rock, for example placer deposits of gold and tin, or near-surface deposits of tar

Mining and Mineral Processing

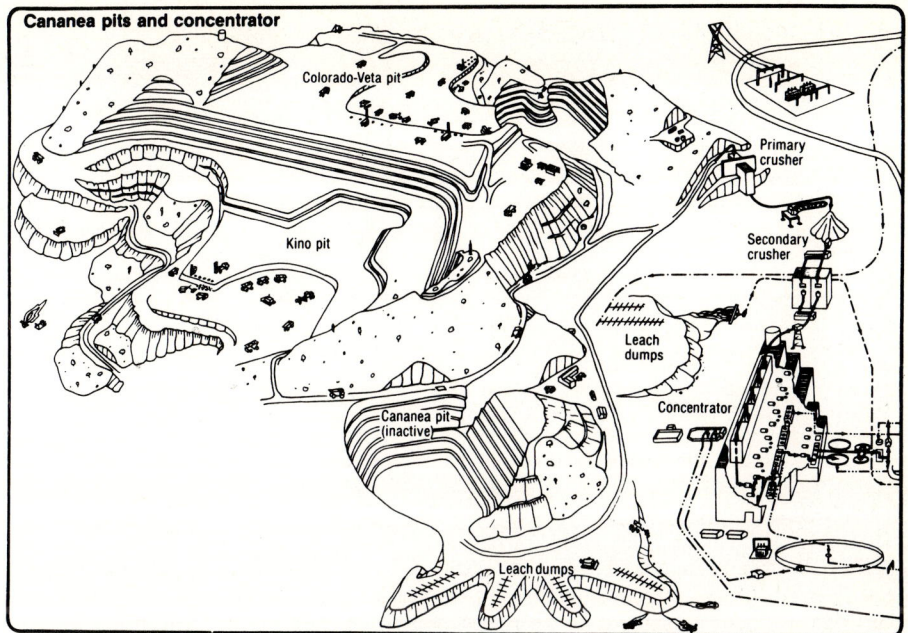

Fig. 4.1. Layout of an open pit copper mine, Cananea, Mexico. (White 1980)

sands and brown coal. Mining of ore hosted by hard rock requires the added expense of drilling and blasting. As a general rule, surface mining is less expensive per ton of ore than underground mining. However, the amount of waste rock that must be moved in surface operations increases as mining proceeds to greater depth, to the point at which underground mining becomes the less costly alternative.

4.1.1 Surface Mining

The ore-to-waste ratio is the main factor in the choice between surface and underground mining, and in any later decision to change to underground mining methods to recover deeper portions of an ore body, as shown for example by the combination of surface and underground mining adopted for the Mt. Isa deposit in Australia. The selction of the most suitable surface mining method depends on the physical and chemical characteristics of the ore.

- Placer deposits of gold, platinum, cassiterite (tin), rutile/ilmenite (titanium), monazite (thorium), and diamonds hosted in unconsolidated gravels or beach sands can be exploited by use of large earth-moving equipment, including dredges and bucket-wheel excavators. This equipment is used both to remove overburden and to mine the ore, which is then concentrated by gravity methods at the minesite. Large operations to extract heavy minerals from beach sands along the coasts of Brazil and Australia are examples. For environmental reclamation, the waste material can be replaced to its former location as mining proceeds.

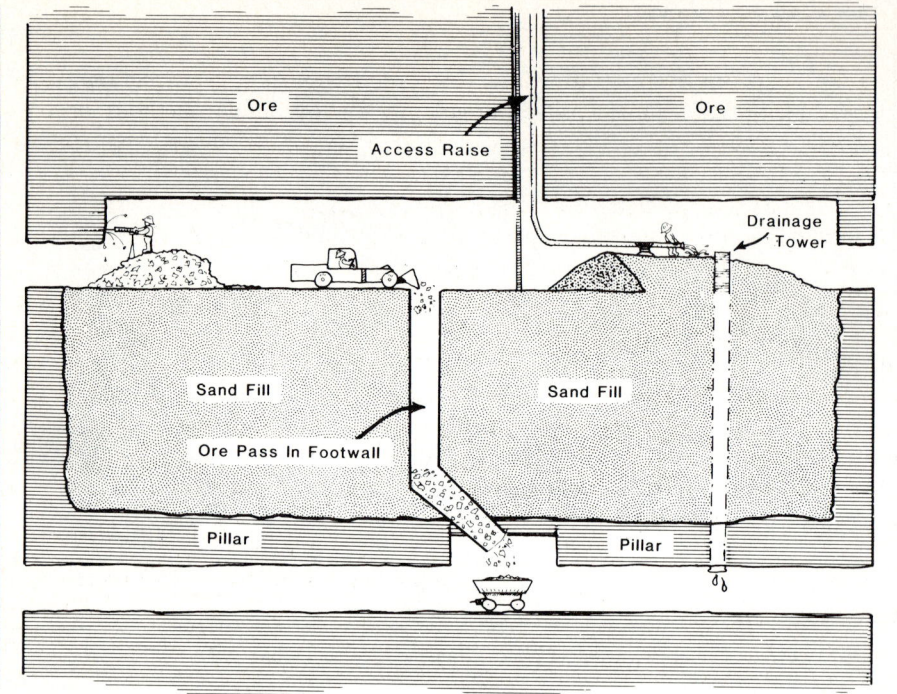

Fig. 4.2. Underground stoping of a tabular ore body

- Near-surface coal seams located under cover of unconsolidated or poorly consolidated overburden, for example in the western US or the Federal Republic of Germany, or tar sands, for example in Athabaska, Canada, are extracted by strip mining. The methods and equipment are very similar to those described above, except that heavier machinery must be used to handle the tougher material and larger tonnages involved.
- Near-surface, high-tonnage metal deposits in hard rock are mined by open pit methods. The grade and tonnage may be high, for example in iron, manganese, or aluminum deposits; intermediate, for example in massive-sulfide deposits of copper, lead and zinc; or low, for example in porphyry Cu-Mo or disseminated-gold deposits. The development of an open pit operation generally requires stripping of overburden. Once exposed, the ore is extracted by drilling and blasting of the rock, transported to a central crushing facility, and the ore minerals are concentrated in a processing plant prior to smelting. The ore is mined along benches (Figs. 1.3; 4.1), and transported, generally by diesel equipment, over ramps which connect the benches.

4.1.2 Underground Mining

Underground mining is used to extract ores of relatively high grade and unit value, for example ores of lead, zinc, silver, or gold, by selective mining methods, or to extract disseminated, low-grade ores, for example of copper and molybdenum, by bulk mining methods. The cost of underground mining has decreased significantly over the last decade because of the development of larger and more versatile diesel equipment for underground use, and of an attendant increase in the scale of operations made possible by progress in mining technology.

Selective mining or *stoping* is used to exploit tabular ore bodies, for instance layers of massive sulfides of lead-zinc-copper, or veins of gold and silver. The ore is removed horizontally by longwall or room-and-pillar methods, or upward in inclined ore bodies by one of a variety of stoping methods (Fig. 4.2). As the ore is removed, the open stope is supported by pillars of ore which are left in place, or with timber or waste rock. In modern mines, stopes are often filled with tailings from the processing plant which are transported to the underground workings as a slurry. To extract all the ore without leaving ore pillars, mines are often divided into a system of access tunnels on one hand, and the actual mining areas or "stopes", on the other. Stopes can be mined out completely because there is no need to support and preserve drifts and cross cuts for access after the ore has been removed.

Large bodies of disseminated ore, for example of disseminated porphyry copper and molybdenum deposits, are mined by underground bulk mining methods in which large blocks of ore are prepared by the construction of draw points and haulage tunnels on the level just below the blocks (Fig. 4.3). In block caving, the ore at the base

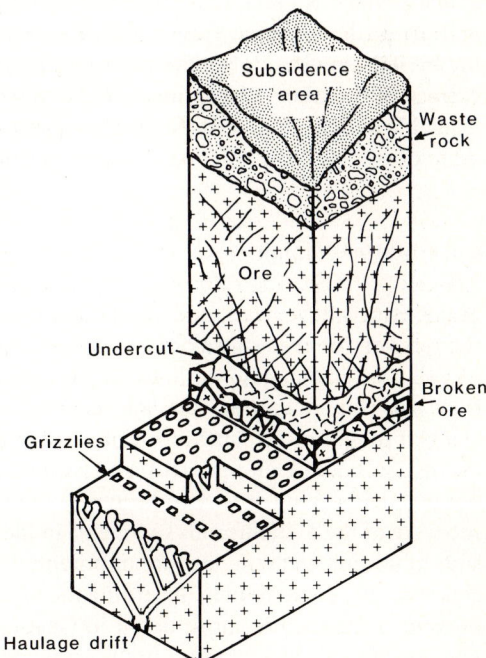

Fig. 4.3. Block caving of a large, low-grade porphyry copper deposit

of the block is drilled, blasted, and removed through the system of draw points which connect with the haulage levels. The remainder of the ore collapses into the large cavity thus formed, to be drawn out at the base. In panel caving, one side of the ore block is removed and the ore is blasted sideways, again to be recovered at the base of the block. No attempts are made to fill the mine openings. The overlying, barren rock collapses, leading to the formation of large collapse craters on the surface.

As these methods illustrate, gravity is used to the extent possible to facilitate the removal and transport of ore underground. Depending on the size and depth of the ore deposit, access from the surface is provided by horizontal adits, inclined ramps, or vertical shafts. Levels analogous to the benches used in open pit operations are developed to reach and transport the ore to a central, underground crusher for comminution. The crushed ore is then brought to the surface, vertically by a shaft or horizontally by an access tunnel, depending on the topography and the relative cost of the two options.

4.1.3 Solution Mining

Solution mining is suited to the exploitation of easily dissolved materials, for example sodium and potassium-bearing evaporites or sulfur, and has also been successfully applied to the recovery of uranium ores hosted in porous sandstones. In a wider sense, coal gasification and oil shale retorting by underground combustion may be placed in this category. A variant of solution mining which is rapidly gaining in importance is dump leaching of copper and heap leaching of gold and silver. For the leaching of low-grade ores of copper, acid is percolated through ore dumps to dissolve the copper, which is precipitated electrically on cathodes. In heap leach mining of gold and silver, the low-grade ore is mined, crushed, and placed on leaching pads. Leaching by sodium cyanide solution chemically dissolves the gold and silver. The gold is extracted by carbon adsorption and electrowinning, and the silver by precipitation as insoluble sulfide (Fig. 4.4). Small, low-grade gold deposits containing as little as one gram of gold per ton (0.03 oz/t) can be exploited profitably.

4.1.4 Marine Mining

Major efforts have been made in the last two decades to develop mining methods for the recovery of near-shore heavy-mineral deposits of tin, diamonds or monazite, and of deep-sea manganese nodules which cover large areas of the Pacific deep ocean floor and contain appreciable amounts of copper, nickel, and cobalt. These nodules can be brought to the surface for processing by dredges and suction tubes. Similarly, massive sulfides of copper, lead, and zinc which form in spreading centers in the deep ocean have become attractive targets, and tests have already been conducted for the recovery of the metalliferous sediments in the Atlantis Deep of the Red Sea. The pursuits of marine metal mining are of potential rather than immediate economic significance. Lack of agreement concerning the legal aspects of ownership and rights to production, addressed in more detail in Chapter 8, and the currently depressed metal markets have slowed progress in the implementation of marine mining. This does not

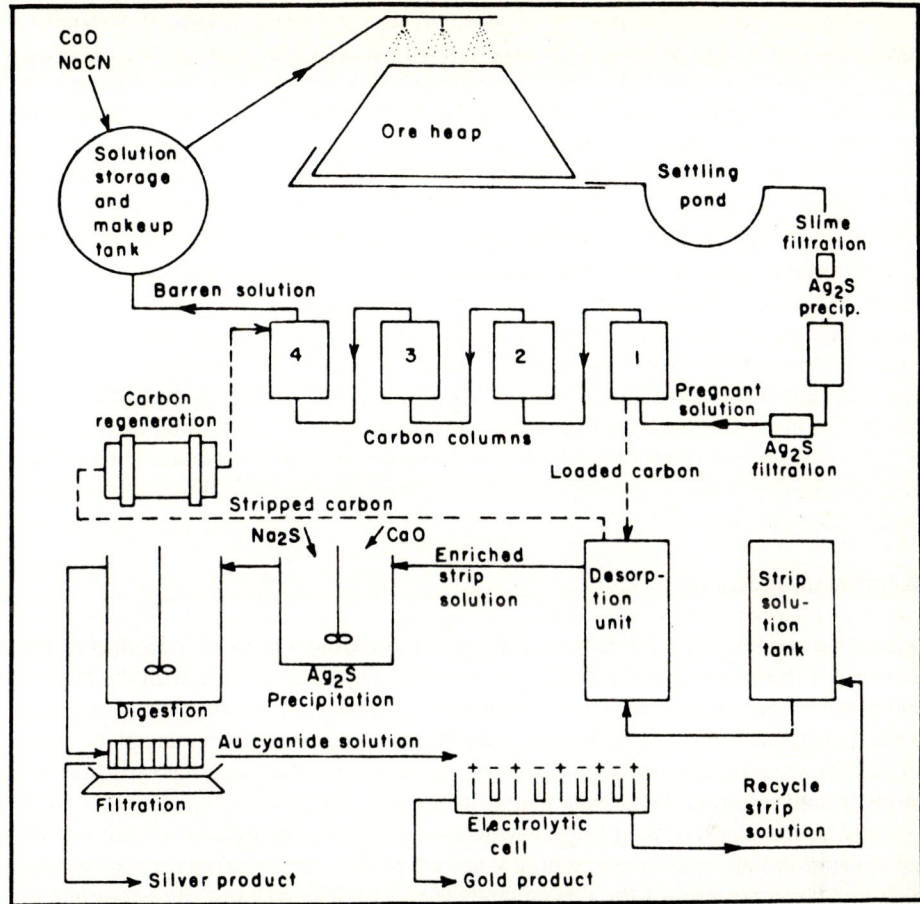

Fig. 4.4. Flow sheet of a gold-silver heap leach mining operation. (Eveleth 1979)

apply to the exploitation of oil and gas on the oceanic shelves, which is of great significance in the supply of fossil fuels.

4.1.5 Strategies and Trends in Mining

The mining scale, payback period for the investment, and mine life are planned to yield a maximum metal recovery and a maximum profit for a company or country. Every mining venture represents a compromise between these two goals and under certain conditions, maximum recovery will be sacrificed in favor of a larger, short-term profit, for example to accelerate the initial payback of the investment, or to avoid long-term economic and political risks. However, mining of the best ore initially, to achieve higher profits, results in decreased overall metal recovery because any lower-grade material, which can be processed only if it is mixed with high-grade ore, will have to remain in the ground.

The progressive exhaustion of the most easily discovered near-surface deposits and of the high-grade deposits worked in the past causes a shift from surface to underground mining and from the mining of high-grade or to low-grade ore. This shift is mirrored by changes in mining methods, from selective to bulk mining, and from labor-intensive to highly mechanized mining. Mechanization has been aided by the development of trackless diesel equipment for ore transport, and of efficient drilling, loading, and hauling equipment which makes increases in production possible.

Mining geologists and exploration geologists use the same approaches and tools, but the mining geologist's main task is to secure the short-range and long-range reserves which are needed to sustain the mining operation, in the mine and district-wide. Mining geologists are also responsible for grade and tonnage control during mining, achieved by systematic sampling of the ore before it is extracted, to assure that the mill receive a blend of ores of the optimum ore grade from different ore blocks, while at the same time assuring maximum recovery. The direction and sequence of mining is decided on the basis of detailed geologic work and engineering considerations.

4.2 Mineral Processing

The aim of mineral processing is to upgrade the ore from a mine to a product of the purity and characteristics required by the market. This is to be done with the highest possible recovery and at the lowest possible cost. Large blocks composed of ore minerals, gangue minerals, and host rock from the mine reach the processing plant. Mineral processing transforms this material by selectively concentrating the valuable constituents and rejecting the admixed impurities.

Most mineral processing begins with a thorough disaggregation of the ore by crushing, grinding, and screening until it has the proper size required for best separation and concentration of the ore minerals (Fig 4.5). The coarsest ore is crushed in jaw crushers, ground to centimeter size in rotary mills, and pulverized in large rod or ball mills which consist of rotating cylinders half-filled with steel rods or balls. Screens between the grinding mills recycle oversized material, to assure a constant feed of properly sized material at each step of the process. The optimum grain size for separation of the ore minerals is determined by microscope studies of ore textures and by metallurgical tests.

Some ores are used without the need for concentration, or need only agglomeration or pelletization before use, for example ores of iron, manganese, and phosphates. Most minerals and metals, however, have to be concentrated and smelted to extract the metals. Concentration of minerals takes advantage of differences in their physical and chemical properties. Density separation and magnetic separation are the main methods which rely purely on physical properties. Density separation is employed, for example, in the concentration of gold and the heavy minerals cassiterite (tin), rutile (titanium), ilmenite (titanium,), monazite (thorium), and zircon (zirconium), traditionally in gold pans, sluices or shaking tables, but now mostly in cones, spiral concentrators, jigs, and cyclones. All of these devices use the high density of the metals and minerals to separate them from less dense gangue or rock material. Magnetite and ilmenite from beach sands are further concentrated by magnetic

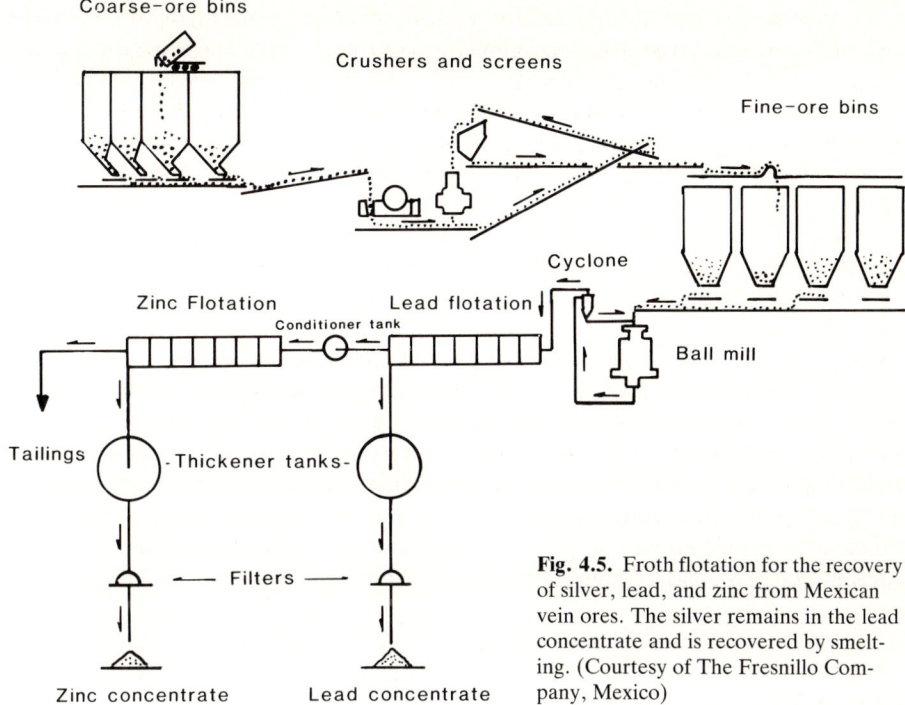

Fig. 4.5. Froth flotation for the recovery of silver, lead, and zinc from Mexican vein ores. The silver remains in the lead concentrate and is recovered by smelting. (Courtesy of The Fresnillo Company, Mexico)

separation. A reverse process is used in heavy media separation, for example of iron and manganese ores, in which dense ore is separated from less dense barren rock in a heavy liquid, generally a water suspension of a finely ground, dense mineral. The density of the liquid is adjusted so that gangue and rock fragments float and the denser ores sink and collect on the bottom of the flotation chamber.

Of much more economic importance than these physical methods of purification are froth flotation and solvent extraction (Fig. 4.5) to recover base-metal sulfides and precious metals. The principles employed are simple but the details of procedure are complex. The ore is pulverized, mixed with water to form a pulp, and ore minerals are chemically conditioned to attach to air bubbles which are circulated through the pulp in flotation tanks. The air bubbles and attached ore minerals are removed from the upper surface of the tanks. The separation of different mineral types by this process is made possible by differences in their surface chemistry. Hydrophobic particles will attach to the bubbles and rise to the surface while hydrophilic particles do not and sink to the bottom of the tanks. The flotation cells are designed in such a way that the ore pulp is kept in constant motion, and that there is a constant stream of bubbles flowing upward. The separation process is controlled by a number of reagents: *frothers* increase the surface tension of the bubbles so that they are stable and can carry the mineral fragments upwards; *collectors* are polar molecules which attach to the mineral particles and render them hydrophobic so that they are carried upward with the gas bubbles; *modifiers* enhance collector selectivity for different minerals; *activators* cause flotation of certain ore minerals while *depressants* discourage flotation; *dispersants* clean mineral surfaces and *pH regulators* adjust the pulp to the best pH range for separation of specific minerals. A proper choice and sequence of application of reagents in conditioning tanks allows selective concentration, in sequence, of ore minerals of copper, lead, zinc, or other sulfides. During froth flotation, original metal concentrations of 0.5–5% of the base metals are increased to 25–30%. The final extraction of the metals is achieved by smelting of the concentrates or by chemical processing.

Notes on the Literature

Mining approaches are covered in a textbook by Thomas (1979). Collections of papers about surface and underground mining have been assembled for the American Institute of Mining Engineers by Cummins and Given (1973), Crawford and Hustrulid (1979), and Hustrulid (1982). A three-part Library of Operating Handbooks has been published by the Engineering and Mining Journal covering surface mining (Hoppe 1978), underground mining (Sisselman 1978), and mineral processing (Thomas 1977). Summaries of our knowledge of marine resources are given by Cronin (1980) and McKelvey (1986), and a detailed account of solution mining is found in Schlitt (1982).

Part II Mineral Economics

Part I of this book reviewed the technical concepts important for understanding the geology of ore deposits, the methods of exploration and deposit evaluation, and the activities of mining and mineral processing. The book turns now – in Part II – to the economic and related concepts important for understanding mineral development (Chap. 5), the evaluation of exploration and mining projects (Chap. 6), and mineral markets and models (Chap. 7).

Chapter 5   The Economic, Institutional, and Legal Framework for Mineral Development

This chapter discusses important economic, institutional, and legal aspects of mineral development, broadly defined to include exploration as well as actual deposit development. It begins by considering mineral development as an economic activity, and then discusses the roles that various types of people and organizations play in mineral development. The chapter concludes by reviewing important aspects of mining legislation.

5.1 Mineral Development as an Economic Activity

In one sense, mineral exploration and development (in this chapter referred to together as mineral development) can be thought of as the first two stages in a four-stage sequential process of mineral supply. During the first stage, exploration, a variety of techniques described in Part I – including geologic, geochemical, and geophysical surveying, and drilling – are used to identify mineral deposits and then to evaluate the economic potential of these deposits, as well as to evaluate the potential of previously known prospects and deposits. Following identification and initial evaluation, a deposit that continues to be economically attractive is prepared for mining in the second stage, deposit development. Activities at this stage include estimating ore reserves, designing and constructing the mine and any processing facilities, and arranging for infrastructure (for example, roads, town sites, water and power supplies) and financing. During the third stage, mining, ore is extracted by surface or underground methods. In the fourth stage, mineral processing, the ore is transformed from material containing a relatively small amount of metal to nearly pure metal. At this point, the metal is ready to be used by fabricators and then final consumers of metal-containing products. A similar sequence of activities applies to the supply of other resources, such as oil and gas, coal, and nonmetallic minerals.

In a broader sense, however, mineral development is part of a dynamic world of mineral supply, in which mining companies and society respond to increases in demand or the depletion of known mineral deposits in a variety of ways. Exploration for previously unknown deposits is one response, appropriate when the expected costs of exploration, development, and mining for a new deposit are lower than the expected costs of other possible responses. A second response is to develop known but previously undeveloped deposits. In other words, an inventory of undeveloped deposits exists at any point in time as a product of past exploration. At the time of discovery, these deposits had higher costs per unit of metal for development and mining than other ways of expanding supply (such as expanding existing operations to extract lower-quality resources). But at a later date, some of these deposits are developed and mined because (1) demand for a metal increases, as well as its price, calling into production deposits whose unit production costs were previously higher than the existing price, or (2) total costs per unit of metal for these deposits fall below the total costs of other means of satisfying demand, as higher-quality reserves at existing operations are depleted. A third response is simply to move to lower-quality resources at existing operations. This leads to higher costs of production and mineral prices, signalling over the longer term increasing scarcity, and encouraging the substitution of relatively abundant materials for the increasingly scarce material (see Tilton 1983).

In practice, these three responses define a continuum of responses to the depletion of known deposits. In other words, the distinctions among discovering previously unknown deposits, developing previously known but undeveloped deposits, and expanding to lower-quality resources at existing operations are often blurred. The discovery of a previously unknown deposit, for example, may be based on detailed examination of known mineralization, which in the past had not been explored sufficiently to identify a deposit. Expanding reserves at existing operations often involves delineating mineralization in areas adjacent to current reserves, resembling in many respects the development of a known deposit.

These responses to the depletion of known deposits assume that technology – the techniques and processes used to obtain outputs from inputs of raw materials, labor, and capital – remains fixed. But over time technologies change and significantly influence all of these responses. Technological changes in exploration make it less costly to discover mineral deposits, compared to discovery costs in the absence of change. Such changes stimulate exploration at the expense of the other potential responses to depletion, other factors remaining the same. Airborne geophysical techniques, for example, were first applied on a large scale following World War II and were instrumental in a large number of base metal discoveries in Canada in the 1950s (Cranstone 1988). In the 1980s, gold explorationists benefit from improved geochemical techniques that reliably measure trace elements associated with gold mineralization in the parts per billion (ppb) range.

Technological changes in extraction and processing reduce the total costs of producing refined metal, and thus encourage increased output from existing mines able to use these new technologies. Moreover, these changes encourage exploration and development of those geologic types of deposits that can use the new methods or processes. In the early part of the twentieth century, for example, new large-scale mining techniques and the froth-flotation method of concentrating copper ore encouraged the development of large, low-grade copper deposits (called porphyries) at a time when copper demand was rising quickly with the spread of electrification in the United States and when high-grade vein deposits were close to depletion (see Joralemon 1973). More recently, in the 1970s and 1980s, improvements in heap leaching and carbon-in-pulp recovery of gold have significantly reduced the costs of gold extraction and have made exploitation possible at the low-grade deposits typical of recent discoveries.

Technological changes also result in new products and new and better materials, such as higher-quality metal alloys, plastics, ceramics, and composite materials. These changes, in turn, alter the demand for metals and nonmetallic materials. In essence, materials compete with one another to provide a set of physical and chemical properties at the lowest total cost. One material will replace another in a particular application if it provides the same properties at a lower cost than the other, or if it provides more desirable properties at the same cost. Over the last 100 years, for example, steel and other materials have substantially replaced wood in construction. Over the last 40 years, aluminum has replaced tinplate in a number of container and packaging applications (Demler 1983). Over the last 20 years, aluminum and plastics have replaced carbon steel, cast iron, and zinc die castings in the automobile (Eggert 1986; Compton and Gjostein 1986).

As shown above, a variety of options are available to companies and society to respond to increased demand or the depletion of known mineral deposits: explore for and discover previously unknown deposits, develop known but undeveloped deposits, expand facilities at existing operations to extract lower-quality resources, reduce the total cost of producing refined metal through technological advances in mining and processing, and develop alternative materials so that relatively scarce materials can be replaced with more abundant ones. These options are not mutually exclusive. At any point in time, all are being pursued.

The important lesson for mineral-producing countries and companies is that mineral development is much more than the first part of a sequential process of mineral supply. Rather, it is one element of a dynamic world of mineral supply and demand. As such, it is both influenced by and responds to changes in mineral prices, demand, and technologies of exploration, extraction, processing, and material use. Moreover, as will be discussed later, political and legal factors are also important.

5.2 Participants in Mineral Development

Various types of people and organizations undertake mineral exploration and development. This section identifies these people and organizations and then evaluates their roles in mineral development.

People. Prospectors with a keen eye for surface mineralization were undoubtedly responsible for discovering most of the mineral deposits mined through the centuries. But during the twentieth century, the prospector's role has declined, especially since World War II. Exploration has moved away, gradually at first, from visual identification of surface mineralization to systematic, scientific discovery of deposits with little or no surface exposure. The prospector with little formal training in the geologic sciences has given way to teams of specialists with training in geology, geochemistry, and geophysics (for more detail on exploration methods, see Chap. 2).

This evolution in the scientific sophistication of exploration and the associated decline in importance of the lone prospector are reflected by Harris and Skinner's (1982) suggestion that exploration moves over time through three stages: (1) visual detection of surface mineralization, or prospecting, (2) detection of near-surface deposits with geophysical and geochemical methods, and (3) detection of concealed deposits at depth relying largely on geologic inference that in essence predicts where ore should occur in areas with little or no surface mineralization or geochemical/geophysical anomalies (see Sect. 2.3.3). Prior to World War II, most discoveries belonged to the first category. But since then discoveries have fallen increasingly into the second and third categories, at least in the United States and Canada (Tables 5.1 and 5.2).

Table 5.1 Principal discovery methods for Canadian metal deposits, 1920–75

Deposits discovery date	Conventional prospecting no. percent		Geophysical anomaly no. percent		Geochemical anomaly no. percent		Geologic inference no. percent		Total number of deposit discoveries
Pre–1920	26	93	0	0	0	0	2	7	28
1920–1929	12	80	0	0	0	0	3	20	15
1930–1939	13	87	0	0	0	0	2	13	15
1940–1950	13	76	0	0	0	0	4	24	17
1951–1955	16	46	5	14	0	0	14	40	35
1956–1960	6	25	14	58	0	0	4	17	24
1961–1965	4	27	5	33	2	13	4	27	15
1966–1970	2	10	13	65	1	5	4	20	20
1971–1975	1	4	15	58	3	11	5	19	26*

Source: Office of Technology Assessment, Congress of the United States, *Management of Fuel and Nonfuel Minerals in Federal Land: Current Issues and Status* (Washington, D.C., Government Printing Office, 1979), based on D.R. Derry, "Exploration Expenditures, Discovery Rate and Methods", CIM Bulletin vol. 63, no. 362 (1970); and D.R. Derry and J.K.B. Booth, "Mineral Discoveries and Exploration Expenditure-A Revised Review 1965–1976", paper prepared for 1977 CIM Symposium.
* No principal method listed for two discoveries in 1971–75.

This is not to say that prospectors and identification of surface mineralization no longer play a role in deposit discovery. On the contrary, conventional prospecting continues to be important in those parts of the world where relatively little prospecting has occurred. But it is much less important in areas such as the western United States and Europe, which have been more intensively explored. Moreover, not all minerals and deposit types are at the same stage of exploration sophistication. Most iron ore and bauxite discoveries since World War II, for example, fall into the first two categories. It is unlikely that iron ore and bauxite exploration will move into the third stage of geologic inference any time soon, because of the abundance of reserves and resources at existing mines and known deposits. Most base metal discoveries since the 1960s, however, belong in the second and third categories; this is particularly true for volcanogenic massive sulfide and porphyry copper and molybdenum deposits.

Table 5.2 Principal discovery methods for US metal deposits, 1951–70

Deposits discovery date	Conventional prospecting no. percent		Geophysical anomaly no. percent		Geochemical anomaly no. percent		Geologic inference no. percent		Total number of deposit discoveries
1951–1955	1	8	2	17	0	0	9	75	12
1956–1960	2	13	2	13	1	7	10	67	15
1961–1965	0	0	2	13	0	0	13	87	15
1966–1970	0	0	2	11	2	10	15	79	19

Source: Office of Technology Assessment, Congress of the United States, *Management of Fuel and Nonfuel Minerals in Federal Land: Current Issues and Status* (Washington, D.C., Government Printing Office, 1979), based on Paul Bailly, "Changing Rates of Success in Metallic Exploration", paper presented at the Geological Association of Canada-Mining Association of Canada-Society of Exploration Geologists-Canadian Geophysical Union Annual Meeting, Vancouver, British Columbia, April 25, 1977.

Moving from exploration to deposit development, people with an even wider range of specialties are involved. Engineers estimate reserves and design and then oversee construction of the mine, mill, and any required infrastructure. Financial analysts arrange for financing. Marketers line up potential customers. Thus people with a variety of scientific and other specialties are involved in exploration and development.

Organizations. A variety of private, government, and international organizations are active in mineral development. The private sector includes large multinational mining and oil companies, and small and medium-sized domestic companies (including small companies that raise exploration funds in regional stock markets and typically generate geologic prospects to be sold to larger and more established companies). The most important objective of these organizations is to maximize profits, although other motives exist as well, including increasing sales, maintaining market share, and leading a quiet, stable life. Government organizations include (1) geological surveys, which in many countries provide much of the basic scientific information on which reconnaissance and more detailed types of exploration are based, and (2) state-owned mining companies. Compared with private organizations, state-owned mining companies often have a wider range of less clearly specified objectives, including providing employment, earning foreign exchange, training skilled workers, and promoting regional economic development, as well as maximizing profits. At times, one goal can be achieved only at the expense of another. Finally, several international organizations, including the United Nations and the World Bank, provide financial and technical assistance for mineral development in the developing countries (see Chap. 8). The relative importance of these private, government, and international participants in mineral development varies greatly from country to country.

In the United States, Canada, Australia, and many other industrialized countries, government's role is limited largely to providing basic geologic information through geological surveys. The private sector then is responsible for conducting actual exploration and development, sometimes with the help of government subsidies or other forms of financial assistance. However, in a few countries – including France, Sweden, and Canada – state-owned mining companies are active in mineral development. The private sector in industrialized countries includes all of the types of participants identified above. The relative importance of these participants varies over time and across countries, as well as with respect to particular minerals. The role of the prospector, for example, certainly has declined, as noted earlier. Small and medium-sized domestic mining companies (the so-called junior companies) apparently have been more important in Canada than in other countries, at least in part because of less stringent requirements for raising monies on the Canadian stock exchanges. More generally, smaller exploration and development companies have become more important in the 1980s relative to larger companies in those geographic areas favorable for gold mineralization; many types of gold deposits can be found, developed, and operated with much smaller capital investments than most other metal deposit types.

In the developing countries, it is difficult to generalize. Private companies (both domestic and foreign), government-sponsored geological surveys, and state-owned mining companies are all active to varying extents in mineral development. Compared to the industrialized countries, state-owned companies in general play a larger role in the developing countries, and their role has increased considerably in the last several decades. In 1950, governments owned virtually none of the equity in metallic mineral production in the developing world, whereas by 1980 these governments owned approximately half of the equity in these industries (Radetzki 1985). In the 1980s, however, the role of the private sector in mineral development has increased in a number of developing countries, including Peru, Thailand, and Indonesia; but state-owned companies continue to be much more important than in the 1950s and 1960s. International organizations are more important participants in mineral development in developing countries than elsewhere. The United Nations (UN) has two programs of exploration assistance: the UN Development Program and the UN Revolving Fund for Natural Resources Exploration. The World Bank provides both technical and economic assistance to developing countries. The subject of international organizations and developing countries is covered in Chapter 8.

In the socialist economies of the Soviet Union, eastern Europe, and China, government organizations control and conduct all mineral exploration and development. One exception is China, where some multinational companies have been invited in the 1980s to participate in the development of coal and offshore oil and gas. Nevertheless, this is the exception, rather than the rule, in the socialist countries.

5.3 Mining Law

The economic attractiveness of an exploration or development project depends on legal, as well as purely economic, factors. Thus this section reviews mining law around the world. These laws typically include provisions for: (1) ownership or control of mineral resources, (2) conditions under which mineral exploration, deposit development, and mining can occur, and (3) mineral taxation. The intent here is not to provide comprehensive descriptions of the mining laws of particular countries, but rather to briefly discuss the historical roots of mining law and then to compare and contrast several important types of mining codes.

Rules and customs governing mining date back to at least several centuries B.C. and the Egyptian civilization. The first modern mining laws were enacted in the 19th century, coincident with the opening of a number of mines providing raw materials for the Industrial Revolution. Two of these laws – French mining law under the Napoleanic Law of April 21, 1810, and the General Prussian Mining Law of June 24, 1865 – reflect the tradition of regalian law (sometimes called civil law). Under regalian law, originating with the Romans, the government owns or at least controls subsurface mineral resources, regardless of who owns the surface above the minerals. The state authorizes mineral exploration, development, and mining through some type of concession system and typically has some degree of discretionary authority over whether or not to permit exploration or mining. In addition, the government usually collects a tax on mine production. The French and German laws reflected this legal tradition, as did Spanish mining ordinances that importantly influenced mineral development in Spanish America. British mining law of the nineteenth century, in contrast, reflected the second important source or tradition of mining law: common law. The tradition of common law provides for private ownership of land and the right to extract minerals from this land; in other words, it does not separate surface rights from subsurface mineral rights.

Most current mining laws around the world reflect one or a combination of the regalian and common law traditions (these traditions, obviously, do not apply to the centrally planned economies of the Soviet Union and eastern Europe, where the state owns all mineral rights and controls all phases of mineral development). The British, French, and Spanish mining laws have had a considerable impact on the mining laws of many other countries, not only in the colonial regions. Even today, the mining laws of such important mining countries as Canada, Australia, and South Africa resemble in many respects the British Mining Act of 1880.

The United States today provides the best example of the common law tradition. It is the most important mining country in which the rights to subsurface minerals are generally held by the owner of the surface estate. It is important to distinguish between privately held lands and public – or government owned – lands. On private lands – accounting for some 60% of US lands – the owner of rights to the surface in most instances owns the subsurface mineral rights and sets the conditions for exploration and development. On public lands owned by the federal government – about 32% of the United States – the situation is more complicated. (The remaining 6 to 7% of US land is owned by state and local governments, which have their own mineral-disposal systems.) Access to most metallic minerals is governed largely by the remarkably durable General Mining Law of 1872, which provides for free and open access for mineral exploration. By staking claims and performing annual assessment work, explorationists obtain exclusive exploration rights to an area for an indefinite period of time. Claimholders can acquire full title (a patent) to both surface and subsurface resources if they show that a valuable deposit exists, but a patent is not required for mining. Patented claims, as private property, are no longer federal lands. The federal government collects no royalty or other form of production tax on minerals extracted from federal lands (other than the normal income tax).

A leasing system exists, in contrast, for most energy-resource exploration on US federal lands. This reflection of the regalian legal tradition dates back to the Mineral Leasing Act of 1920. The government grants both competitive and noncompetitive leases to private companies to explore for and extract oil, natural gas, coal, and other resources and then collects royalties on any production. Title to the land remains with the federal government.

Thus, in the United States, the common-law tradition in mining law was modified in 1920, and the Mineral Leasing Act established at that time reflects the regalian tradition. In England and most of the Commonwealth nations, the commonlaw tradition in mining law has also been modified. In England, ownership of coal and uranium resources was transferred to the government in 1938 and 1946, respectively. In Canada, explorationists have free and open access to stake claims in metallic mineral exploration, as in the United States. But rather than having the right to purchase full mineral title, they instead have the right to a mineral lease for most metals; mineral titles remain with the provincial governments in the provinces and with the federal government in the Northwest Territories and the Yukon Territory.

The current mining laws of most countries have been more strongly influenced by the regalian, rather than the commonlaw, tradition. The government is viewed as the rightful owner or controller of mineral wealth. Over the centuries, this view has persisted, but for a variety of reasons. In primitive societies, the use of metals was limited primarily to the manufacture of weapons. Rulers were concerned with defending their power, and thus they found it in their self-interest to control metal deposits. In Roman times, most mines were in conquered territories and thus became the property of the republic (and later the empire) as spoils of conquest. Similarly, during the Spanish conquests of the 16th, 17th, and 18th centuries, mineral wealth was viewed as part of the King's estate, captured for the glory of the state. Today in many countries the rationale for government control of mineral resources has a similar but slightly different twist: mineral resources are viewed as part of the patrimony of the people as a whole, rather than just the ruler, and as a natural right, rather than a reward of conquest.

This view prevails in many developing countries, some of which have gained their independence from colonial control only in the last several decades. In a number of countries the regalian tradition of ownership or control of mineral resources, as well as discretionary authority over mineral-development activities, has been extended to include negotiations between host governments and foreign investors. These negotiations typically cover a wide range of issues, including the terms under which exploration and mining concessions will be granted, work commitments and schedules, taxation, and local employment and training (see Mikesell 1984). As discussed by Brown (1986), these new mining codes are characterized by the consensual nature of agreements, in contrast to the open-access nature of many older mining codes (such as the US General Mining Law of 1872).

Many of the new mining codes are products of a period of confrontation between governments and multinational corporations during the 1960s and 1970s (see Chap. 10). Many countries had recently gained their political independence from colonial powers, and others, even though politically independent, sought their economic independence from North American and European companies, particularly in South America. They were all eager to exercise greater control over the exploitation of natural resources that in many instances were crucial for a country's economic well-being. Most multinational corporations, on the other hand, vigorously opposed allowing host governments to have a larger role in mineral-development decisions. As a result, a number of countries nationalized the properties of foreign companies and also revised their mining codes. In the 1980s, the mood of confrontation has given way to one of greater cooperation between governments and companies, in part because companies have learned to live with the new demands of host governments and in part because governments realize that to continue to attract foreign investment requires less confrontation and more cooperation and compromise.

Notes on the Literature

The discussion of mineral development as an economic activity draws on the work of Adelman, Houghton, Kaufman, and Zimmerman (1983), *Energy Resources in an Uncertain Future.* Readers interested in exploration as an economic activity are referred to Eggert (1987a), *Metallic Mineral Exploration: An Economic Analysis,* and Tilton, Eggert, and Landsberg (eds.) (1988), *World Mineral Exploration: Trends and Economic Issues,* and the references cited therein.

Those interested in state mineral enterprises in developing countries should consult Radetzki (1985), *State Mineral Enterprises: An Investigation into Their Impact on International Mineral Markets.* Radetzki examines the motivations for establishing state mineral enterprises and compares these enterprises with their private-sector counterparts. He also presents case studies of Venezuelan iron, Indonesian tin, and Zambian copper.

The Economic, Institutional, and Legal Framework for Mineral Development 93

The section on mining law is based on a number of sources. For the historical development of regalian and common law traditions, see in particular Ely (1964), "Mineral Titles and Concessions", in Robie (ed.) *Economics of the Mineral Industries* (second edition), and Kaufman (1976), "Mineral disposal systems", in Vogely (ed.) *Economics of the Mineral Industries* (third edition). Eggert (1987) compares US and Canadian policies governing exploration on public lands in "Exploration and Access to Public Lands", in Cordes (ed.) *Public Policy and Competitiveness of U. S. and Canadian Metals Production*. Leshy (1987) provides a detailed history of the US General Mining Law of 1872.

Chapter 6   Economic Evaluation of Mineral Deposits

Economic geology and mineral economics encompass a variety of activities. One important aspect of these disciplines is the identification and evaluation of mineral deposits. Part of this work is largely *technical* in nature. Explorationists, for example, use the tools of geology, geochemistry, geophysics, and other fields to identify mineral deposits. They then study the deposits to determine if they exhibit the physical and chemical characteristics typical of similar deposits that have been developed into mines; and, in the case of a new type of deposit, to draw inferences on the basis of new scientific knowledge. Other parts of this work, however, are largely *economic* in nature. Economists, engineers, and others work together to evaluate a mineral deposit's economic potential by comparing the expected revenues from mine production with the associated expected costs of further exploration, development, and production. It is their task to determine if the expected revenues from developing a deposit will be sufficiently in excess of costs to provide investors with an adequate return on their investment. Still other parts of the work may be *socio-economic* in nature, if a government is involved; the concern here is how the development of a deposit will affect the overall economic and social development of a region or country.

The relative importance of each type of evaluation – technical, economic, socio-economic – at any point in time depends on the stage of development. The early stages of reconnaissance exploration, target selection, and target drilling, for example, draw most heavily on the geologic sciences, while the later stage of feasibility analysis relies more on the engineering sciences and economics. The socio-economic evaluation is carried out only when actual development of a mineral deposit is being considered. Moreover, rather than being independent of one another, these types of evaluation are inter-related. They are often carried out in parallel. The results of the technical evaluation serve as important inputs to the economic evaluation, and together the technical and economic evaluations serve as a starting point for the socio-economic evaluation. Finally, these evaluations are never done in a once-and-for-all manner, but rather are constantly revised in the light of new information.

This chapter discusses a variety of methods for evaluating mineral deposits. The emphasis is on economic evaluation, but defined broadly enough to recognize the important technical aspects of mineral development. Section 6.1 identifies the range of goals pursued by both private and government organizations, and the strategies used to achieve these goals. Section 6.2 reviews the basics of investment analysis. With these two sections as background, Section 6.3 focuses specifically on the evaluation of exploration projects and mineral deposits. Finally, Section 6.4 reviews the financing of mineral exploration and development.

6.1 Goals and Strategies

All mineral exploration and development projects are investments of one sort or another: expenditures are made today in delineating a deposit, designing the mine and mill, or constructing the production facilities, but the associated revenues or benefits come in the future. This is true regardless of whether private or government organizations undertake a project. The most important objective of decision makers should be to obtain the highest level of net benefits for their company or country. How these net benefits are defined may, however, differ between the private and public sectors.

Private organizations make decisions based largely on financial criteria. Economic theory usually assumes that these organizations strive to maximize profits, although as noted in the last chapter the goal of profit maximization may be tempered by several other objectives, including increasing sales revenue or maintaining market share. To achieve its primary or long-run objective of maximizing profits, a company must set several intermediate and more manageable goals, because there are many ways of striving to maximize profits. These intermediate goals take an almost infinite number of forms, but might include achieving target production levels, reducing labor costs per unit of output, increasing the recovery rate at an existing operation, broadening the company's expertise in a particular area of extraction or processing technology, or discovering or purchasing a mineral deposit of a certain type, size, or grade.

Government organizations active in mineral development often have nonfinancial – as well as financial – goals. In addition to striving to maximize profits, a state-owned mining company may have economic-development goals, such as earning foreign exchange, reducing unemployment, or serving as a catalyst for overall economic development for a particular region of a country. Such a company may also have mineral-supply goals, such as increasing security of supply or reducing the impact of mineral price fluctuations on domestic markets. As a result of the simultaneous pursuit of a variety of goals, it is not surprising that the set of objectives of state-owned mining companies or other government organizations is sometimes blurred. Moreover, one objective may often be achieved only at the expense of another. Thus it is important for government organizations, and for private companies as well, to periodically review a project's set of goals to determine if they are compatible with one another and if the set of goals is logical, consistent, and realistic.

Those involved in mineral development use a variety of strategies to achieve their objectives. The choice of strategy is shaped by a variety of factors, including the historical experience of a company with particular minerals or geographic regions, stockholder interests, the opinion of public officials, or the economic policy of a country. Factors such as these establish a philosophical framework in which an organization identifies a range of alternative strategies for achieving its goals. One family of strategies concerns an organization's degree of specialization. An organization might, for example, choose to concentrate its efforts on particular minerals – such as copper or aluminum – or particular geographic regions – such as the western United States or Chile – because it has experience or special expertise with these minerals or regions. Or it might choose to concentrate its efforts on a particular stage of extraction or processing, such as smelting and refining rather than mining and concentration. Alternatively, rather than specializing in a few minerals or countries, an organization might choose to diversify its activities among a larger number of minerals or countries. The general rationale for such a diversification strategy is to reduce fluctuations in earnings due to (1) mineral price instability and (2) unforeseen government actions or other events in a particular country.

Another family of strategies relates to the size of the organization. An organization might follow a growth strategy of expanding output at existing operations and bringing new facilities into production if it perceives opportunities to profitably do so. It might follow a steady-state strategy if it perceives a relatively stable future market. Finally, an organization might follow a retrenchment strategy when it is performing poorly financially or during recessions. Such a strategy is short-term, and the goal is to weather the storm and then reorganize and return to financial health using one or a combination of methods: improving the efficiency of existing operations, selling off parts of the organization, or liquidating or terminating the existence of part of the organization.

Other types of strategies exist as well. Two organizations may merge into a single entity (a merger strategy) or one organization may acquire another organization (an acquisition strategy) for a number of reasons, including making better use of existing facilities, entering a new line of business, or efficiently reducing the total production capacity of both organizations. Similarly, a strategy of participating in joint ventures – separate corporate entities owned by two or more parent organizations – may be followed for a number of reasons. An organization, for example, may desire to increase the number of projects and countries in which it is involved (that is, to diversify its activities) and chooses to do so through joint ventures. Or a private firm investing in a foreign country may be required by government regulation to have partners from the host country. Joint ventures may take place between private companies or between a combination of private and government organizations.

Thus private and government organizations make decisions about a particular exploration or development project in light of these broader organizational goals and the strategies used to achieve them.

6.2 Methods of Investment Analysis

The general procedure for evaluating investment opportunities is to compare the benefits of any particular opportunity with the associated costs, and then to invest in those projects that are worth more than they cost. But what factors should be included when estimating benefits and costs? And how should exploration and development costs incurred in the early years of the typical mineral project be compared with the benefits of mineral production that are received in later years (Fig. 6.1) and are, furthermore, uncertain? The answers to these questions vary considerably depending on the context. What follows are the principles of investment analysis for private and public organizations, providing a foundation for the more specialized, subsequent discussions of the economic evaluation of exploration projects and mineral deposits.

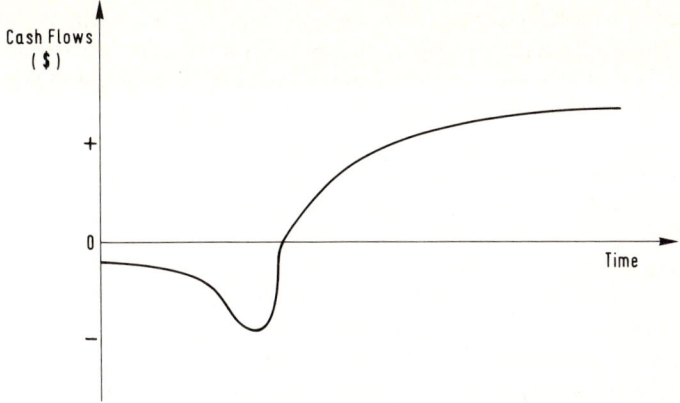

Fig. 6.1. Schematic diagram of cash flows for a typical mineral project

6.2.1 Cash Flows and the Time Value of Money

For both private and public organizations active in mineral development, investment analysis starts with estimating the amounts of money to be spent on exploration, development, operations, taxes, and other items and activities; and the amounts to be received as revenues from mineral production. These amounts of money are referred to as cash flows.

Organizations in the private sector, as noted earlier, generally strive to maximize profits, the difference between revenues and costs. To do so, organizations invest in those activities that have expected revenues in excess of costs. But it would be incorrect to directly compare the costs incurred early in the life of a project with revenues received only later. Cash flows must be adjusted for the time value of money. Money has a time value because a dollar received today is worth more than a dollar received tomorrow: at the very least, today's dollar can earn the going rate of interest in a savings account, and more generally it can grow in value through other types of investment. In other words, financial capital (money) is a productive resource. Investors demand to be compensated for accepting payment tomorrow rather than today. Therefore, the value of future cash flows must be discounted before being compared with current costs and revenues.

The rate of discount can be viewed in two ways. In the first case, the discount rate is seen as the cost of capital associated with raising funds from outside the company. It is the percentage rate of return that the firm must generate to compensate those outside investors, usually lenders or stockholders, who supply funds to the company rather than investing their money in another company or activity. In the second case, the discount rate is the opportunity cost of using internal funds, such as retained earnings, for the project under consideration rather than for another project or investment. The company has many potential uses for its internal funds. For example, rather than invest in the development of a particular gold deposit, the company could use its money to develop another gold deposit, to develop some other type of mineral deposit, to increase dividends to stockholders; to purchase stock in another company, or to buy US Treasury bills. The discount rate is the opportunity cost of capital, or rate of return forgone by not investing in these other activities.

The discounted, or net present, value of a stream of revenues and costs is given by the following formula:

$$\text{NPV} = (R_O - C_O) + \frac{R_1 - C_1}{1 + r} + \frac{R_2 - C_2}{(1 + r)^2} + \ldots \frac{R_n - C_n}{(1 + r)^n}.$$

The net present value in year 0, NPV, is the sum of revenues (R) minus costs (C) in years 0 through n, adjusted back to the present using a discount rate, r, that adjusts each annual cash flow for the time value of money. The farther cash flows get into the future, the more their values are discounted. When evaluating a potential investment, a positive NPV indicates that expected revenues exceed expected costs after accounting for the time value of money; the investment is worth more than it costs and thus should be undertaken. A negative NPV indicates that costs exceed revenues after accounting for money's time value; a firm's money would be better spent elsewhere. According to the method of NPV, organizations maximize profits by investing in those projects with net present values greater than zero, or in the case of two mutually exclusive projects, choosing the project with the largest positive net present value.

The following example illustrates the arithmetic of the discounting procedure. The data (Table 6.1) represent a hypothetical mineral project with exploration expenditures in years 0 (the present) and 1, development expenditures in year 2, and production revenues and operating costs in years 3 through 6. The discount rate is assumed to be 10%. The net present value of these cash flows is:

$$\begin{aligned} \text{NPV} &= (-5) + (-10/1.1) + [(-20)/(1.1)^2] + [(5)/(1.1)^3] + [(10)/(1.1)^4] \\ &\quad + [(20)/(1.1)^5] + [(20)/(1.1)^6] \\ &= -5 - 9.09 - 16.53 + 3.76 + 6.83 + 12.42 + 11.29 \\ &= 3.68 \text{ million US-\$}. \end{aligned}$$

Table 6.1 Cash flow data for a hypothetical mineral project

Year	Revenues (million US-$)	Expenditures (million US-$)	Net cash flow* (million US-$)	Remarks
0	0	5	− 5	Exploration
1	0	10	−10	Detailed exploration
2	0	20	−20	Development
3	10	5	+ 5	Initial production
4	20	10	+10	Production increases to 50% of capacity
5	40	20	+20	Production at full capacity
6	40	20	+20	Production

* Net cash flow = revenues − expenditures.

Thus, this hypothetical investment should be undertaken because the net present value is greater than zero. Actual NPV calculations for a mining project would be much more complicated. The stream of revenues and costs would cover a longer period of time. It also would contain a more detailed breakdown of cash flows. Specific cash outflows might include the costs of drilling, design and construction of the mine and mill, wages and salaries, maintenance, income taxes, production royalties, and interest on debt. Cash inflows might include revenues from the sale of main products and byproducts, as well as any salvage value of the equipment.

Only cash revenues and costs are included in the calculation of NPV. In other words, only those revenues actually received or costs actually incurred are included in the cash flow for a particular time period. Examples of noncash costs to be excluded are depreciation and depletion allowances. Depreciation is an accounting rule allowing firms to spread the costs of equipment over the life of the equipment, even though the cost of the equipment was a cash outflow to the firm at the time it was purchased (assuming that the company paid cash). Although depreciation is not a cash cost and thus is excluded from cash-flow analysis, it does indirectly affect the analysis through its effect on tax payments. Depreciation is deducted from gross income before income taxes are calculated, and thus the existence of depreciation reduces tax payments and cash outflows. Depletion is also a tax deduction that reduces a firm's taxable income and tax payments.

Note that no previous expenditures are included in the cash-flow example. Previous expenditures, often called sunk costs, are not affected by today's investment decision regardless of whether a project is accepted or rejected. Sunk costs thus are completely irrelevant to the investment decision concerning future revenues and costs. Investment decisions made today should be made from today's perspective, considering the net impact of future expenditures and receipts. (It is sometimes said that investments should be made an a "money-forward" basis, while managers should be promoted on a "scorecard" basis. That is, investments should be based on expectations about the future, while promotions should be based on past performance).

Determining the discount rate to be used when calculating the NPV of a possible investment is arguably the most difficult aspect of cash-flow analysis. If a company is raising funds from external sources, the discount rate is appropriately viewed as a cost of capital, as noted previously. External capital is a combination of debt (money borrowed) and equity (money raised by selling shares of stock). The cost of debt frequently is well defined, as the after-tax rate of interest on a bank loan or the rate of interest to be paid on a bond. The cost of equity, however, is less clearly defined. It is the rate of return expected by stock purchasers to compensate them for investing in one company rather than another. This required rate of compensation depends critically on expectations about the future, which are very difficult to quantify. If a company will use internal funds (cash on hand), then the appropriate discount rate is the opportunity cost of not using these funds for other purposes. This opportunity cost, therefore, is the rate of return the company could earn in the next best use of its money. Calculating the opportunity cost of capital requires comparing expected returns of all possible uses of funds. In practice, the discount rate to be used in the economic evaluation of mineral projects usually is determined by the office of the corporate treasurer and then handed down to the departments responsible for actual evaluation of projects.

Table 6.2 Ranking investment projects using net present value and present value ratios

Investment opportunity	Net present value* (US-$)	Present value of revenues (US-$)	Present value of costs (US-$)	Present value ratio**
1	100	150	50	3.00
2	70	100	30	3.33
3	50	70	20	3.50

* Net present value = (present value of revenues) − (present value of costs).
** Present value ratio = (present value of revenues)/(present value of costs).

Discounting is completely different than adjusting cash flows for the effects of inflation (increases in the general price level that reduce purchasing power). However, inflation can significantly influence the economic attractiveness of an investment opportunity, and thus must be considered explicitly in the investment analysis in one of two ways. The first and quicker method of accounting for inflation is to state all cash flows and the discount rate in terms of constant inflation-adjusted dollars (sometimes called real dollars). This method implicitly assumes that inflation has the same impact on all future revenues and costs. The second method utilizes current (nominal) dollars, rather than constant dollars. Calculations by this method are more time-consuming than the first method because all cash flows and the discount rate must reflect assumptions about future inflation rates. But this method is better able to account for the fact that inflation does not necessarily affect all types of cash flows equally. For example, labor costs may be expected to increase more rapidly than raw-material costs in some instances.

As noted earlier, the method of net present value (NPV) tells firms to invest in those projects with NPVs greater than zero (or in the case of mutually exclusive projects, the project with the highest NPV). But what if a firm has insufficient funds to undertake all projects with positive NPVs (that are not mutually exclusive)? The obvious solution, ranking projects according to NPV and then choosing those projects with the highest NPVs until the available funds are used up, will not necessarily be in the best interests of the firm, particularly when the alternative projects have significantly different investment requirements. Table 6.2 illustrates such an example. A firm is faced with three investment opportunities, but only has 50 US-$ (in present-value terms) to spend. If the firm ranked the projects according to size of NPV, it would choose project #1 because it is the project with the largest NPV; the firm would be unable to undertake projects #2 and #3 because project #1 requires all of the available funds. However, the firm would be better off investing its 50 US-$ in projects #2 and #3, which are smaller in terms of both size of investment and NPV and yet when combined have an NPV of 120 US-$, which is greater than project #1's NPV of 100 US-$. One way to choose projects #2 and #3 is to rank these projects according to present-value ratios, the present value of revenues divided by the present value of costs. Present-value ratios control for size of investment. Projects #2 and #3 both have larger ratios than project #1.

Calculation of net present value is not the only method used by firms to evaluate investment opportunities. Perhaps the two most widely used alternatives are internal rate of return and payback period. The internal rate of return (IRR) of an investment opportunity, sometimes referred to as the discounted cash flow rate of return, is simply the discount rate or rate of return at which NPV equals zero. Calculating IRR thus requires solving the following equation:

$$NPV = 0 = (R_O - C_O) + \frac{R_1 - C_1}{(1 + IRR)} + \frac{R_2 - C_2}{(1 + IRR)^2} + \ldots \frac{R_n - C_n}{(1 + IRR)^n}.$$

The variables R and C are revenues and costs, respectively, over the life of the project in years 0 through n. The decision rule is that a firm should undertake those projects with IRRs greater than its minimum-acceptable rate of return (equivalent to the discount rate, discussed above). Using the data from the previous NPV example (Table 6.1) yields an IRR of 13.59% (the arithmetic is tedious, but with a programmable calculator or computer the calculation is simple). The IRR of 13.59% is greater than the firm's discount rate of 10% and thus the project should be undertaken. Note that when a particular stream of cash flows yields a positive (negative) NPV, the IRR will be greater than (less than) the discount rate used in the NPV calculation.

Calculating the IRR in this example yielded the same decision recommendation as calculating NPV. Although calculations of IRR and NPV usually give the same accept or reject recommendation, there are some important exceptions that make applying the method of IRR trickier than relying on NPV calculations. If, for example, a project will have significant costs late in life, such as those for mine reclamation, it may have more than one IRR, or perhaps none at all. If different cash flows have different costs of capital, such as when short-term interest rates differ from long-term rates, the method of IRR evaluation has no simple way to handle the situation (while calculation of NPV merely requires discounting different cash flows with different discount rates).

Payback period is the number of years required for the net cash flows from mineral production to pay back the initial investment costs. Companies that evaluate mineral projects by payback period accept investments with payback periods less than an arbitrarily defined length of time. For instance, the payback period for the NPV example (Table 6.1) is 5 years because the net cash outflows of 35 million US-\$ in years 0 through 2 are paid back with the cumulative net cash inflows of 35 million US-\$ in years 3 through 5; a firm would undertake this project if its cutoff payback period were 5 years of more. Payback period is simple to calculate and thus a useful preliminary indicator of a project's economic attractiveness. But the evaluation completely ignores cash flows after the payback period (both revenues and expenditures) and the time value of money during and after the payback period. Moreover, the method provides no guidance in the choice of acceptable payback period. The choice is essentially arbitrary; one company may choose 3 years, for example, while another chooses 6 years under the exact same set of circumstances. Thus, payback period should not be used as the primary determinant of an investment decision.

6.2.2 Benefit-Cost Analysis

The method of net present value allows private companies to evaluate the economic attractiveness of delineating and developing a mineral deposit. A similar method, referred to as benefit-cost analysis, allows government organizations active in mineral development to evaluate investment alternatives and their overall impact on society. Benefit-cost analysis is similar to net present value in that discounted benefits are compared with discounted costs. But the evaluation is different in at least two respects. First, both benefits and costs are defined more broadly to include not only financial revenues and costs directly associated with the mining venture but also indirect effects on society as a whole. Indirect benefits might include stimulating broader economic development in a region, earning foreign exchange, providing employment, or training a more technically skilled workforce (see Sect. 10.6). Indirect costs might include harming environmental quality. Quantifying these indirect effects of mineral development on society is much more difficult than quantifying the direct financial revenues and costs of a private company. The difficulty arises because there are no market values assignable to these indirect effects, whereas market forces determine the direct revenues and costs of the private firm. Second, it is often argued that governments should have a lower discount rate than private firms. One basis for this argument is that governments should be more concerned about the future than the private sector. Greater concern for the future implies a lower discount rate because the lower the discount rate the greater the present value of benefits and costs far in the future. Another justification for a lower government discount rate is that governments can borrow money at lower interest rates than private firms because investors consider loaning money to a government as less risky.

This view of decision making by government organizations using benefit-cost analysis is simplified. Some state-owned mining companies, for example, operate essentially like private companies, and therefore the net present value model is relevant. Other government organizations clearly pursue nonfinancial goals, and in these cases, the benefit-cost framework for analysis is appropriate. In reality, there is a wide range of intermediate approaches to the assessment of mining ventures.

6.2.3 Risk Analysis

The methods of net present value and benefit-cost allow private and public organizations to evaluate investment alternatives by comparing benefits with costs after adjusting for the time value of money. These methods of evaluation assume that future benefits and costs are known with certainty at the time of investment. Clearly, this is a questionable assumption for many types of investments, especially mineral exploration and development. During initial exploration, for example, many outcomes are possible, ranging from no indication of commercial mineralization to geologic evidence that eventually leads to a producing mine. During the development of a deposit, initial ore reserve estimates may have to be revised, thus altering estimates of future production and revenues. During production, mineral prices may be higher or lower than predicted at the time of investment, leading to higher or lower revenues than anticipated.

Economic Evaluation of Mineral Deposits

Table 6.3 Risks important in mineral development

Category and type	Major impact on:
Market risks	
Price	Prices received for mineral output
Demand	Mine production
Foreign exchange	Revenues, costs
Technical risks	
Reserve	Mine production, costs
Completion	Mine production, costs
Production	Mine production, costs
Political risks	
Currency convertability	Revenues
Environmental	Costs
Tax	Costs
Other regulatory	Costs
Nationalization	Mine production

Sources: Tinsley (1985) Analysis of risk sharing, in: Tinsley et al. (eds) (1985) pp 419–426.
Jütte-Rauhut (1986) Project financing in the mining industry. In: Gocht, Jütte-Rauhut (eds) (1986) pp 266–274.

More specifically, risk is best thought of as a measure of the degree of variability of possible future revenues and costs. A low-risk investment has a lower variability of potential financial returns than a high-risk investment. For example, investing a certain amount of money in US government securities is a low-risk investment because these securities pay a fixed rate of interest known at the time of investment; thus, the variability of possible financial returns is low. Investing the same amount of money in the development of a mineral deposit involves a higher degree of risk, because of the wider range of possible financial outcomes for this investment. Future revenues and costs associated with mineral development are not known with certainty because the factors that determine these revenues and costs are impossible to know with certainty at the time of investment.

These factors can be grouped into three categories of mineral-development risk according to the cause of the risk (Table 6.3). The first category is market risk, sometimes called economic or business risk. This type of risk is determined by the economic system as a whole. Several types of risk are included here. Price risk is the variability of possible future mineral prices. Mineral prices, together with production levels, determine revenues from mining. Thus to the extent that actual future prices differ from the prices expected at the time of the cash-flow analysis, actual revenues and – in turn – profits will differ from those expected. Another type of market risk is demand risk, the variability of future demand for minerals. To the extent that actual and expected mineral demands differ, actual mine production and revenues are affected. Foreign-exchange risk is the variability of possible foreign-exchange rates in the future. Rates of foreign exchange importantly influence the revenues of firms operating outside their home country, and the revenues of firms selling products that are priced in terms of foreign currencies (for example, Australian gold producers whose revenues are based on the price of gold in US-$), as well as the costs of firms importing equipment from outside the country of operation. Mineral producers have some control over these market risks in those instances where they can lock in prices, sales, or exchange rates in futures markets. Futures markets are not a perfect answer to market risks. They do not exist for all commodities, and they tend not to be useful for periods more than about 1 year into the future. Nevertheless, futures markets enhance the ability of mineral producers to control market risks (see Kaufman 1984).

The second category or family of risk is technical in nature and encompasses several specific types of risk. Reserve risk, determined both by nature (the distribution of minerals in the earth's crust) and the quality of ore-reserve estimates, reflects the possibility that actual reserves will differ from initial estimates. If actual tonnage is larger (smaller) than anticipated, actual production – and, in turn, revenues – will be larger (smaller) than expected. If actual ore grades differ from expected grades, rates of production and costs of extraction and processing are affected. Completion risk reflects the possibility that a mineral-development project will not make it into production as anticipated because of, for example, cost overruns, construction delays, or engineering or design flaws. Once a deposit is brought into production, there is production risk, which reflects the possibility that production will not proceed as expected, either because of problems with equipment or extraction processes, or because of poor management. Technical risks are at least partly under the control of the organizations active in mineral development.

The third category of risk is political and is determined by the actions of governments. Political risks reflect the possibility that unforeseen government actions will affect the profitability of an investment. Potential actions include nationalization and changes in regulations concerning, for example, the environment, taxation, currency convertibility, or import duties. The extent to which organizations active in mineral development are able to influence government actions varies significantly from case to case.

Any method of evaluating exploration projects and mineral deposits, therefore, also must consider the risk involved in the investment. Investors demand to be compensated for investing in a risky project rather than in a safe alternative. In other words, a "risky" dollar is worth less than a "safe" one. Therefore, the value of "risky" dollars must be adjusted before being compared to the value of "safe" dollars. Within the framework of the methods of net present value and benefit-cost, the adjustment can be made in one of three ways. The first is to raise the discount rate. The rationale is that income from successful risky projects must be sufficiently greater than the normal (or riskless) rate of return to make up for losses incurred from unsuccessful risky projects. Rather than discounting risky future cash flows by a rate reflecting the forgone opportunity to invest in safe government securities, companies should discount risky cash flows by a higher rate that reflects the extent to which the future cash flows are less certain than the interest to be received from government securities. The appropriate discount rate thus has two determinants: the riskless rate of return, and a risk premium. Each project will have its own degree of risk, and thus each project will have its own discount rate. (If different cash flows within a particular project have different degrees of risk, then these cash flows should be discounted with different discount rates). The greater the risk, the higher the discount rate should be. Raising the discount rate reduces the NPV of a set of cash flows that are negative at the beginning and positive at the end of a project (typical of mineral investments; see Fig. 6.1). This reduction of NPV, as risk is factored into an evaluation, is illustrated in Table 6.4,

Table 6.4 Adjusting for risk in the method of net present value

Adjust the discount rate

Discount rate	NPV
10% (risk-free rate)	$3.68 million
13%	$0.56 million
16%	−$2.03 million
20%	−$4.74 million

Adjust the cash flows

	Revenues		Expenditures		Net cash flow	
Year	Unadjusted	Certainty equivalent	Unadjusted	Certainty equivalent	Unadjusted	Certainty equivalent
0	0	0	5	5	−5	−5
1	0	0	10	10	−10	−10
2	0	0	20	20	−20	−20
3	10	9	5	5	+5	+4
4	20	18	10	10	+10	+8
5	40	35	20	20	+20	+15
6	40	34	20	20	+20	+14

NPV (discount rate = 10%)
Unadjusted $3.68 million
Certainty equivalent −$4.94 million

Sensitivity Analysis	NPV (discount rate = 10%)
Upside case (increase revenues by 10%; reduce costs by 10%)	$17.02 million
Base case	$ 3.68 million
Downside case (reduce revenues by 10%; increase costs by 10%)	−$ 8.93 million

Source: calculations by author, based on data in Table 6.1.

based on the cash flows from the previous NPV example (Table 6.1). NPV is positive at a discount rate of 10%, as calculated earlier. At a discount rate of 13%, which might reflect a reserve risk of 3%, NPV is smaller yet still positive. At the higher discount rates of 16% and 20%, reflecting progressively greater degrees of risk, NPVs are negative. The decision rule is unchanged from the earlier discussion of the NPV method of evaluation: invest in those projects with positive NPVs. In this case, the investment decision would depend on perceptions of risk. If investors believe that a risk-adjusted discount rate of 10% or 13% is appropriate, then they should undertake the project. If, however, they perceive a higher degree of risk such that a discount rate of 16% or 20% is correct, then the investors should reject the investment opportunity.

The second way of accounting for risk within the framework of net present value and benefit-cost analysis is to adjust the future cash flows, rather than the discount rate. The value of a risky future cash flow is reduced to its certainty equivalent: the smallest certain cash flow the investor would be willing to exchange for the risky cash flow. The riskier or more uncertain the future cash flow, the smaller its certainty equivalent. A simple example of using certainty equivalents to account for risk is shown in Table 6.4, again based on the cash flows from the previous NPV calculation (Table 6.1). The revenues expected in years 3 through 6 are not known with certainty; the certainty-equivalent values of these revenues are the smallest certain cash flows the investor would accept in exchange for the risky revenues. Expenditures are assumed to be known with certainty, and thus their certainty-equivalent values are equal to the original estimates. The next step is to discount the certainty equivalents of the net cash flows by the risk-free discount rate of 10%. The resulting, risk-adjusted net present value is -4.94 million US-$. In this case, investors should reject the investment opportunity.

The third and perhaps most widely used method is sensitivity analysis. A base case is established using the most likely values for each variable in the NPV or benefit-cost analysis. Then a range of other possible NPVs or benefit-cost ratios is calculated using other possible values of each variable. The range of possible outcomes often includes best-case and worst-case scenarios, reflecting the best and worst combinations of other possible values of each variable influencing NPV or the benefit-cost ratio. Sensitivity analysis also may involve testing the extent to which individual variables influence the economic attractiveness of an investment; a series of calculations is made using various values of the variable under consideration while holding all other variables constant. Table 6.4 illustrates a simple example of sensitivity analysis, based on the NPV calculation in Table 6.1. The base case is simply the NPV calculated previously using the most likely future cash flows. The upside case (best-case scenario) assumes that revenues are 10% greater than expected and that costs are 10% lower than expected: not surprisingly, the NPV is much greater under these assumptions. The downside case (worst-case scenario) assumes that revenues are 10% lower than expected and that costs are 10% higher than expected; the resulting NPV is lower than the base case and in fact is negative. These cases or scenarios define for a decision maker the range of possible outcomes of the investment. This example provides no clearcut decision: although two of the three calculations have positive net present values, the third is negative. The decision maker will have to decide whether the risk of losing 8.93 million US-$ is worth the possibility of earning 17.03 million US-$. If all three scenarios had either positive or negative NPVs, the investment decision would be obvious.

The three methods of risk analysis described above – adjusting the discount rate, adjusting the cash flows, and conducting sensitivity analysis – are indirect in the sense that they do not directly consider the variabilities inherent in the variables that determine NPV or the benefit-cost ratio. In other words, these methods do not directly account for the possibility that variables such as the date of mine start-up, production levels, and mineral prices will be different from the most-likely values assumed at the time of the initial investment decision. Two methods, however, more directly account for risk by forcing the decision maker to think about the underlying determinants of NPV or the benefit-cost ratio in terms of probabilities. Although these direct methods are not yet commonly used in the metal mining industries, they are widely used in the petroleum industry and undoubtedly will be applied more extensively in metal mining in the coming years.

The first direct method combines the concepts of expected value and decision-tree analysis. Expected value equals the probability of an outcome's occurrence multiplied by the value of the outcome should it occur (the conditional value). Suppose, for example, that for a particular exploration project a geologist earns either 100 US-$ if he discovers an ore deposit or –15 US-$ (his expenses) if he discovers nothing. Further suppose that the probability of making a discovery is 0.2 out of 1 and that the probability of finding nothing therefore is 0.8 (the probabilities must sum to 1.0). The expected value of this exploration project is (0.2) (100 US-$) + (0.8) (–15 US-$) = 8 US-$. The expected-value concept tells decision makers to undertake those investments with expected values greater than zero. Another way of looking at this example is to draw a decision tree (Fig. 6.2). The choice – to explore or not to explore – is denoted by the square box (A), sometimes referred to as the decision node of the decision tree. The possible outcomes of a decision to explore are shown around the circle (B), called the chance node. The decision is made by comparing the expected values of the decision branches.

Most investments, however, are not as simple as the previous example suggests. One investment decision typically is but the first of a series of investment decisions, and these subsequent decisions and their possible outcomes should be included in the initial decision. For example, the first decision may require choosing to drill a geophysical anomaly, to conduct additional geophysical tests, or to abandon the project. This initial choice leads to subsequent choices. If, for example, the geologist chooses to drill the geophysical anomaly, then once initial drilling is completed the geologist has to decide what to do next. Subsequent choices might include drilling close-spaced holes if initial drilling intersected promising mineralization, drilling wider-spaced holes in other parts of the anomaly if initial drilling did not intersect sufficiently promising mineralization to warrant closer-spaced drilling, or walking away from the project if initial drilling intersected nothing of interest. To evaluate these choices, a decision tree is constructed with several decision and chance nodes reflecting the sequence and possible outcomes of the series of investment decisions.

The method of expected value and decision-tree analysis, therefore, accounts for risk by using the probability of an outcome's occurrence directly in the calculation of

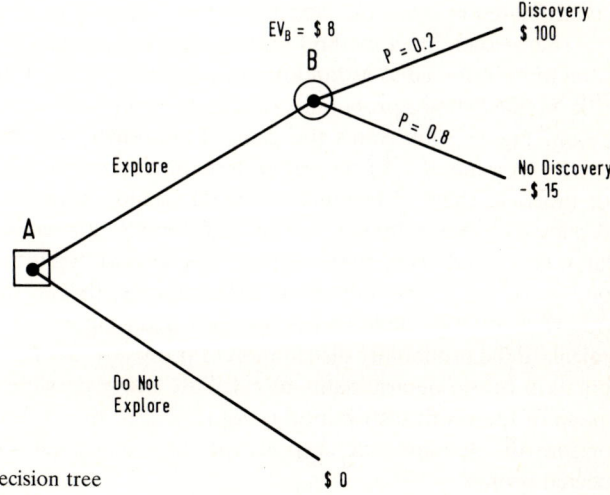

Fig. 6.2. Simple example of a decision tree

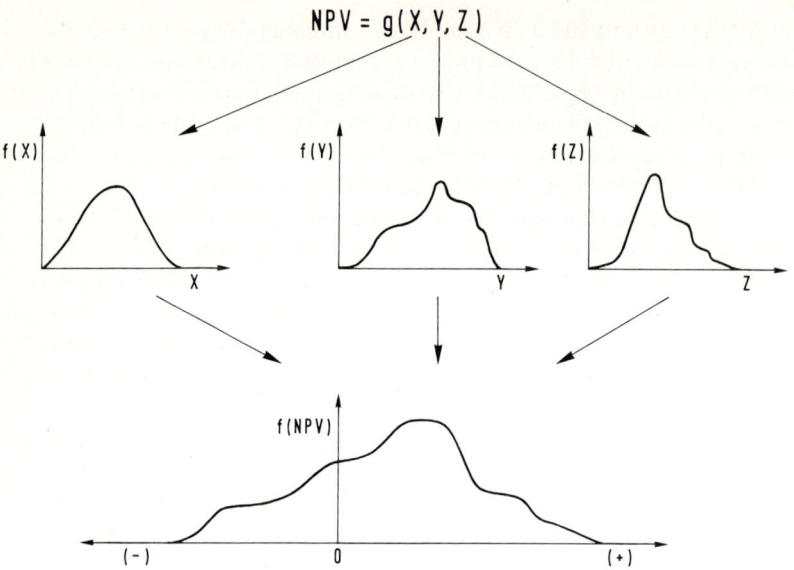

Fig. 6.3. Schematic diagram of a simulation model for an investment decision. (After Newendorp 1975)

a project's worth. The method forces decision makers to think about the range of possible outcomes of an investment decision, and encourages them to deal with risks in a logical and consistent manner. The biggest difficulty in applying this method is the difficulty in estimating probabilities with any degree of reliability. Nevertheless, estimating probabilities of occurrence is crucial to any method of evaluation that directly deals with risk and thus this problem is not unique to expected value and decision-tree analysis. Expected values and decision trees are not inconsistent with NPV or benefit-cost analyses described earlier: rather, conditional values are most appropriately stated in terms that account for the time value of money.

The second direct method for handling risk is simulation modeling. This method, sometimes referred to as random simulation or Monte Carlo simulation, accounts for risk by calculating a probability distribution of possible profits from an investment decision. Fig. 6.3 illustrates the general approach. The first step is to define profits (NPV) as a function of several underlying determinants (X, Y, and Z in the figure). In practice, these determinants would include a variety of factors influencing revenues (reserve estimates and price forecasts, for example) and costs (capital and labor costs, and taxes, for example). The second step is to define probability distributions for each of the underlying determinants (that is, the range and frequency distribution of all possible values for each determinant). The third and final step is to calculate the probability distribution of profits by, in effect, using every possible combination of the determinants to calculate every possible profit level. The statistical mean of the profit distribution is equivalent to the expected value of the investment option; the decision rule is to accept those investments with positive means or expected profits.

Simulation modeling, therefore, accounts for risk in a continuous, rather than discrete, manner. It considers all possible values of the underlying determinants of profitability, rather than just several specific values as in the methods described earlier. Another important advantage of the method is that subjective assessments about possible future values for each determinant of profitability are made one at a time at the beginning of the analysis, reducing the likelihood of the biases likely when risks are evaluated all at once in some vague or general sense at the end of an analysis. Finally, simulation is not limited to modeling profitability and in fact can be used to describe any distribution, including tonnage and grade of ore reserves, mine production, or sulfur dioxide emissions from a smelter.

6.3 Evaluating Exploration Projects and Mineral Deposits

Exploration and evaluation are information-gathering processes. Information gathered at each stage is used to determine whether to proceed to the next, more detailed stage of analysis, to conduct additional tests at the same stage, or to abandon the project. As described in Part I, explorationists gather information in the program design stage to help them select particular minerals, geologic deposit types, and geographic areas for exploration; during reconnaissance exploration to select small areas for more detailed examination; and during detailed exploration and target evaluation to identify and delineate mineral deposits and to assess a deposit's commercial viability. In other words, exploration is a sifting process, the goal of which at each stage is to narrow the focus of the search and to eliminate from further consideration areas with little ore potential, while conducting further tests on areas or sites that continue to be promising.

The degree of quantitative rigor with which the methods of net present value, benefit-cost, and risk analysis can be applied varies significantly as a project moves from initial exploration to detailed drilling to development. In the early phases of exploration, when information to assess future benefits and costs is extremely scant, these methods at best serve as qualitative or conceptual frameworks within which to consider investment options. But the closer a project is to actual mining, the more rigorous and quantitative the analysis can and must be. At later stages of exploration and development, after a mineral deposit has been identified, the evaluation techniques described in the previous section can be used more formally and quantitatively to determine the economic worth of a project. Thus, it is appropriate to discuss economic project evaluation within the context of the various stages of a multistage exploration process.

6.3.1 Program Design

During the first stage of the generalized sequence of activities – Program Design – explorationists choose what minerals to explore for and where to explore. To make these decisions, they conduct a number of activities. They identify or review the goals and strategies of the organization, search the existing geologic literature for favorable areas, and perhaps conduct initial field evaluations. A private organization also would review government policies and other literature to determine the favorability of policies and the availability of land in the geologically favorable areas.

Evaluation at this stage is based largely on how particular minerals or geographic areas fit into an organization's goals and strategies. Some organizations, for example, limit their exploration to a small number of countries or minerals in which they have experience, whereas others are willing to explore for a large number of minerals all over the world, regardless of their lack of experience with all of the minerals or countries. Exploration by state-owned companies usually is limited to the home country.

Risk analysis is largely informal at this stage, although some private companies use formal analyses of political risk to define the universe of countries in which they will explore.

6.3.2 Reconnaissance Exploration

During the second phase – often referred to as Reconnaissance – explorationists perform regional geologic, geochemical, and geophysical surveys over large areas of land. The goal is to select smaller targets for subsequent more detailed examination (see Sect. 2.2.2).

At this stage, the potential revenues and costs associated with a project are so uncertain and far in the future that they cannot be estimated with any degree of certainty. Thus, the methods of net present value and benefit-cost cannot be used formally to evaluate a project's economic potential. Instead, an area's geologic characteristics tend to be compared to idealized sets of geologic criteria, often referred to as geologic models, which serve as proxies for the economic values necessary for more formal investment analysis (see Sect. 2.3.3). These sets of evaluative criteria vary considerably from one geologic type of deposit to another because each deposit type has a unique set of characteristics which influence potential revenues and costs. If what is known about an area is similar geologically to the characteristics of areas with deposits that are currently profitable, then the project most likely will advance to the next, more detailed phase of evaluation.

In effect, geologic models predict where an ore deposit should occur. Rarely, however, do new discoveries exactly mimic the generalized models. Rather, they typically are different in one respect or another. The challenge for explorationists is to use these geologic models for guidance in reconnaissance and target selection, while avoiding becoming myopic and thus missing important opportunities for discovery. Increasingly since the 1950s, explorationists have relied on these sorts of models as exploration's focus has shifted toward concealed deposits. Previously, exploration was less scientific and systematic. Most discoveries were based on visual identification of surface mineralization.

When exploring in remote areas with potentially high development costs for infrastructure or transportation, evaluation standards are most exacting. In other words, in a remote region it takes more geologic promise to advance to the next stage of exploration than in a region with lower development costs for infrastructure or transportation. The remote deposit will have to overcome its disadvantage in terms of higher infrastructure or transportation costs by having lower production costs.

Risk analysis at this stage tends to be highly informal, although it can be argued that explorationists would benefit from more formal application of the risk-analysis methods described previously.

6.3.3 Detailed Exploration and Initial Target Evaluation

After identifying a promising target, exploration moves to its third stage, Detailed Exploration and Initial Target Evaluation. Explorationists typically begin by examining the surface of a target with a combination of geologic, geochemical, and geophysical techniques, and then based on the results of these activities they plan and conduct a program of drilling that will yield information on the subsurface geology. Geologic information collected during drilling permits explorationists to construct three-dimensional geologic models of the target area, to identify and begin to delineate mineral deposits, and to begin to evaluate a deposit's economic potential. This phase typically concludes with a prefeasibility study summarizing what is known about the deposit and how it might be developed, including geologic and metallurgical characteristics of the deposit, initial estimates of ore reserves, preliminary plans for the mine and any processing facilities, infrastructure requirements, and initial estimates of revenues and costs.

The geologic criteria used during reconnaissance continue to guide project evaluation early in this stage of exploration. But as explorationists identify and begin to delineate a mineral deposit, engineering and economic criteria become progressively more important. The geologic characteristics of a deposit form the basis for the engineering and economic evaluations. Geologic data importantly influence engineering plans for the mine and mill and, in turn, estimates of development and extraction costs. Geologic estimates of ore tonnage, grade, and recoverable reserves, along with mineral-price forecasts, determine potential revenues from mine production.

During the course of these geologic, engineering, and economic analyses, several reference numbers and calculations often are used to guide decision making and become elements of the prefeasibility study.

6.3.3.1 Minimum Acceptable Reserves – Tonnage and Grade

Each target-evaluation project has a minimum tonnage and quality of reserves that will permit commercial development and extraction of metal from the mineral deposit. This reference number serves as a standard against which to compare actual ore-reserve estimates as they evolve over the course of target evaluation. The minimum tonnage and quality of reserves are partly dependent on technical factors because mines and mills must have minimum scale of operation to be efficient. Just as important, however, are more purely economic factors, because what is defined as reserves must be exploitable under current economic conditions, which include not only expected revenues and costs of development, extraction, and processing, but also any costs for infrastructure and transportation. Total costs will vary significantly from project to project, even for those projects with the same level and quality of mineral resource. Thus, minimum acceptable reserves will vary from project to pro-

Table 6.5 Representative minimum-acceptable reserves for selected types of ore deposits (metric tons of contained metal)

Metal	Ore deposit type	Metric tons
Iron	Taconite ore (sometimes called iron formation, jaspilite, itabirite, or banded ironstone)	1,000,000
	Oolitic ore (Clinton type)	500,000
Manganese	Marine-sedimentary ore	30,000
	Hydrothermal ore	5,000
Chromium	Magmatic ore	50,000
	Laterite ore	5,000
Copper	Porphyry deposits	50,000
	Sedimentary ore (Katanga type)	30,000
Lead/zinc	Hydrothermal ore	25,000
	Volcanogenic massive sulfide deposits	50,000
Nickel	Sulfide ore	10,000
	Laterite ore	20,000
Tin	Magmatic ore (primary)	5,000
	Placer or alluvial ore	3,000
Antimony	Hydrothermal ore	1,000

Source: Gocht (1983).

ject. A remote site, far from existing power sources, population centers and consumers, will require larger and higher-quality reserves than a site with lower infrastructure and transportation costs. Despite the project-specific nature of minimum acceptable reserves, average or representative examples of the minimum level of reserves indicate that the minimum size varies significantly from one commodity or deposit type to another (Table 6.5).

One reason why minimum sizes vary for different deposit types of a particular metal is that the quality of the metal deposit also defines what is commercially viable. In other words, reserves are measured not only in terms of quantity (tons of ore) but also in terms of quality (ore grade). The higher the grade or metal content of a deposit (usually expressed as a percentage of total deposit tonnage), the smaller the deposit needs to be to be commercially viable. Thus, minimum acceptable reserves for any particular mineral deposit actually will be alternative combinations of ore tonnage and grade that will permit commercial operations, as shown in Fig. 6.4. In this schematic example, a deposit with 42% chromite ore requires 5 million tons of ore to be profitable, whereas one with 48% ore requires only 2 million tons.

Comparing minimum acceptable reserves with actual estimates of reserves in a deposit creates at least one practical problem because various categories of reserves exist, distinguished largely by the degree of geologic certainty with which they are known (see Sect. 3.3). One solution to this problem is to calculate a weighted sum of the various reserve categories, with those reserves known with a greater degree of certainty receiving higher weights than those known with lesser certainty. For example, according to a system of the German Association of Metallurgical and Mining Engineers, proved reserves are multiplied by a factor of 0.9 (corresponding to a geologic certainty of 90%), probable reserves by a factor of 0.8 (80% certainty), and possible reserves by a factor of 0.6 (60% certainty). Then the adjusted reserve esti-

Economic Evaluation of Mineral Deposits

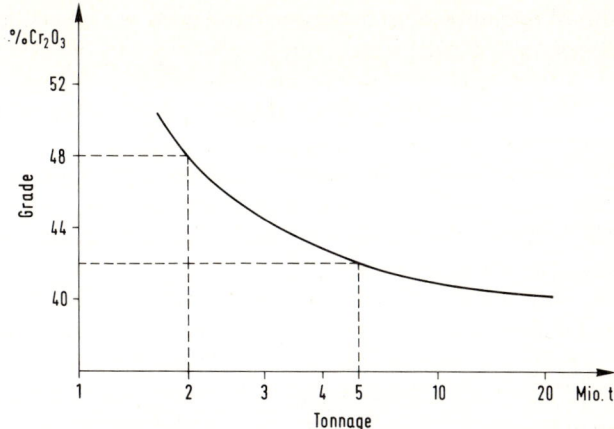

Fig. 6.4. Schematic indifference curve between tonnage and grade of reserves

mates are added together to obtain an estimate of total reserves to be compared to the minimum acceptable level of reserves.

6.3.3.2 Cutoff Grade

The previous section defined ore reserves in terms of tonnage and grade. Grade, in that instance, referred to the *average* grade or metal content of a deposit, expressed as a weight percentage. Another important concept is cutoff grade, the minimum or lowest grade of ore that can be included in the reserve estimates.

Cutoff grade is a geologic/technical measure that embodies the important economic aspects of mineral production from a deposit. In other words, it is defined not only by a deposit's geologic characteristics and the technological limits of extraction and processing, but also by costs and mineral prices. As mineral prices and the various costs of extraction, processing, transportation, and other factors vary over time, cutoff grade evolves to reflect these changes. If, for example, mineral prices rise and all costs stay the same, then the cutoff grade will fall, because extraction of metal from lower-quality rock now will be profitable. If, on the other hand, energy costs rise and all other costs, as well as prices, stay the same, then the cutoff grade will rise because extraction from lower-quality rock no longer will be profitable.

Another important implication of this definition is that cutoff grades vary significantly from deposit to deposit, even in those that are very similar geologically because of differences among deposits in a wide variety of factors including: mining and processing techniques, energy costs, labor costs, distance to market (and transportation costs), and taxes.

6.3.3.3 Optimal Scale of Operation

Every mining operation has an economically efficient scale of operation, expressed, for example, in tons of ore mined per day or year. The optimal scale is, of course, the size of operation with the lowest total costs per unit of metal produced. In practice this optimal size depends on the level and quality of available reserves and the available technologies of extraction and processing.

The first step in calculating the optimal scale of operation consists of defining the range of possible sizes. The larger the scale of operation, the shorter the mine life will be. Alternative mine lives usually range from approximately 8 to 30 years, depending on the type of mineral deposit and the mining and processing methods to be used. The next step is to calculate the total costs (capital and operating) for each alternative size of operation. The optimal size from a purely economic standpoint is simply the one with the lowest costs per ton of material. The graphical relationship between total unit costs and annual production rate is typically U-shaped (Fig. 6.5).

Total unit costs usually fall initially as size increases, because of economies of scale. Less than optimal scales of operation, with correspondingly longer mine lives, are uneconomic because larger equipment for mining and processing with lower unit operating costs cannot be used. At some point, however, costs increase because of diseconomies of scale. Greater than optimal scales of operation, with correspondingly shorter mine lives, also are uneconomic because (1) capital costs per additional unit of output are much higher and (2) costs for water and energy can climb disproportionately steeply.

Calculating alternative scenarios for total costs of course requires precise estimates of the types of mining and processing facilities necessary. Whenever economic geologists or mineral economists have no data on comparable projects on which to base preliminary estimates, they must work closely with mining engineers and others to come up with the appropriate project design.

The cost structures for mining and processing facilities vary substantially from mine to mine and country to country, as shown by data on copper production costs (Table 6.6).

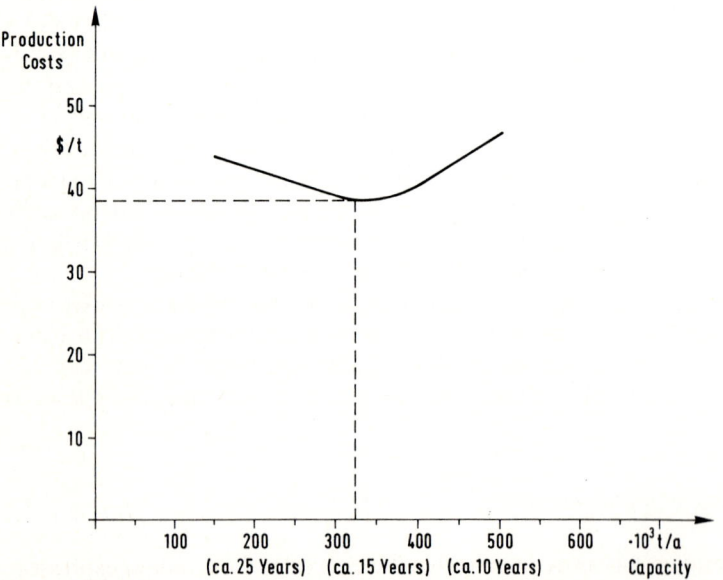

Fig. 6.5. Schematic relationship between total production costs and annual production rate

Economic Evaluation of Mineral Deposits

Table 6.6 Estimated production costs for producing copper mines, 1983 (per pound of refined copper)

Type of operation and country	Number of mines	Mine cost	Mill cost	Smelter refinery cost[a]	Total operating cost[b]	Taxes	Byproduct credits	Net cost[c]	Total cost[d] FOB refinery
Surface:									
Canada	10	$0.42	$0.44	$0.27	$1.13	$0.03	$0.29	$0.87	$1.04
Chile	3	0.28	0.15	0.26	0.69	0.10	0.19	0.60	0.76
Peru	4	0.23	0.28	0.31	0.82	0.12	0.06	0.88	1.12
Philippines	6	0.33	0.36	0.27	0.96	0.09	0.14	0.91	1.04
United States	19	0.30	0.28	0.38	0.96	0.05	0.16	0.86	1.02
Zaire	3	0.19	0.11	0.30	0.60	0.22	0.12	0.70	0.79
Other Countries	12	0.24	0.30	0.27	0.81	0.14	0.34	0.61	0.81
Underground:									
Australia	4	0.22	0.12	0.39	0.73	0.08	0.09	0.73	0.89
Canada	10	0.32	0.14	0.83	1.29	0.07	0.92	0.44	0.61
Chile	4	0.36	0.19	0.20	0.75	0.04	0.06	0.72	0.79
Philippines	5	0.41	0.33	0.28	1.02	0.10	0.40	0.72	0.85
United States	5	0.54	0.31	0.35	1.20	0.07	0.10	1.18	1.26
Zaire	3	0.42	0.16	0.38	0.97	0.24	0.15	1.05	1.14
Zambia	6	0.41	0.20	0.27	0.87	0.17	0.03	1.01	1.09
Other Countries	20	0.57	0.24	0.58	1.38	0.12	0.88	0.63	0.92

Source: Rosenkranz et al. (1983).
[a] Includes smelting and refining charges, transportation costs to the smelter and refinery, and post-mill charges for noncopper commodities.
[b] Total of mine, mill, and smelter-refinery costs.
[c] Total operating costs plus taxes minus by product credits.
[d] Net cost plus recovery of capital at a 15% rate of return.

6.3.3.4 Cash-Flow Analysis

Preliminary cash-flow analysis is an important element of the prefeasibility study, which is the final product of initial target evaluation, the third of four exploration stages. The general method was described earlier in Section 6.2: Future expenditures and receipts must be estimated and adjusted to account for the time value of money, and perhaps the riskiness of the project. The result is an indication or measure of the economic attractiveness of the project. More specifically, cash-flow analysis requires (a) translating the geologic characteristics and engineering aspects of the project into costs of development and extraction; (b) converting preliminary estimates of reserves into potential revenues from mining, making assumptions about recovery rates and future mineral prices; (c) estimating the timing of these costs and revenues; and (d) adjusting the cash flows for time and, if desired, for risk. The input data are admittedly preliminary, but a formal investment analysis is required at this point because the next steps of exploration and feasibility analysis are much more expensive than any of the activities up to this point. A more reliable evaluation of the deposit requires data of much better quality, which are obtained by increasingly expensive exploration methods. Perfect information will not be available until after a deposit has been depleted. Investment decisions, of necessity, are based on incomplete information.

The following example illustrates the important aspects of cash-flow analysis during initial target evaluation. It is based on a hypothetical, moderate-sized tin mine in southeast Asia. Table 6.7 contains the important input data and estimates of annual cash flows. Table 6.8 lists the cash flows by year after adjusting for the assumed infla-

Table 6.7 Basic data for a prefeasibility cash-flow analysis of a hypothetical tin mine in southeast Asia

Time schedule	
Preliminary work	3 years
Development and mine preparation, construction of processing facilities	2 years
Production period	8 years
Costs (nominal US-$, millions)	
Exploration	0.5 (year 1)
Infrastructure	0.5 (year 1)
Initial mine investment	1.0 (year 1), and 2.0 (year 2)
Initial investment in processing facilities	2.0 (year 1), and 1.0 (year 2)
Mine maintenance investments (principally in vehicles)	1.0 (year 5)
Floating capital	0.2 (year 1)
Operating costs	2.0 per year
Revenues (nominal US-$, millions)	
Revenues from concentrate sales	3.5 per year beginning in year 2
Liquidation value of plant and equipment at end of mine life	1.0 (year 10)
Discount Rate	8% per year

Calculation:
Net present value = $(-1.00 \times 1) - (1.61 \times 0.9259) - (1.36 \times 0.8573) + (1.29 \times 0.7938) + (1.23 \times 0.7350) + (0.39 \times 0.6806) + (1.12 \times 0.6302) + (1.07 \times 0.5835) + (1.02 \times 0.5403) + (0.97 \times 0.5002) + (1.65 \times 0.4632) = \$1.66$ million.
Note: see Table 6.8 for the timing of cash flows and Table 6.9 for the discount factors.

Table 6.8 Discounted cash-flow data for hypothetical tin mine in southeast Asia (constant US-$, millions)

Year	0	1	2	3	4	5	6	7	8	9	10
Costs											
Exploration	0.5										
Infrastructure	0.5										
Mine investments		0.95	1.81			0.78					
Investment in processing facilities		1.90	0.91								
Floating capital		0.19									−0.12
Operating costs		1.90	1.81	1.73	1.65	1.57	1.49	1.42	1.35	1.29	1.23
Total costs	1.00	4.94	4.53	1.73	1.65	2.35	1.49	1.42	1.35	1.29	1.11
Revenues											
Sales			3.33	3.17	3.02	2.88	2.74	2.61	2.49	2.37	2.26	2.15
Liquidation											0.61
Total revenues	0		3.33	3.17	3.02	2.88	2.74	2.61	2.49	2.37	2.26	2.76
Net cash flow (Revenues-costs)	−1.00	−1.61	−1.36	1.29	1.23	0.39	1.12	1.07	1.02	0.97	1.65

tion rate of 5% per year. After estimating annual cash-flows, the next step in the analysis is to adjust these cash flows to account for the time value of money. This is achieved by dividing the net cash flow for each year (last line of Table 6.8) by the discount factor appropriate for both the year of the cash flow and the assumed interest rate (Table 6.9). In this example, a real constant-dollar discount rate of 8% is used. The resulting net present value of current and future cash inflows and outflows is 1.66 million US-$. The net present value is positive, implying that the project should be undertaken. But this calculation and conclusion implicitly assumes that all the input data are known with certainty, whereas in reality, none of the numbers are known with certainty. Thus, it is appropriate to test how sensitive this estimate of net present value is to changes in the underlying data, using one or more of the risk-analysis techniques described in Section 6.2.

6.3.3.5 The Prefeasibility Study

A prefeasibility study summarizes what is known at the end of initial target evaluation about a mineral deposit and how it might be developed. It analyzes both the technical and economic feasibilities, providing the basis for deciding whether to proceed to detailed target evaluation and the feasibility analysis, to abandon the project, or to conduct additional tests before making either of these decisions.

A prefeasibility study is based on the available geologic information, the preliminary engineering plans for the mine and any processing facilities, and initial estimates of revenues and costs. A typical study contains information and analysis in the following areas:

- *Project Description:* geographic area, existing access routes, topography, climate, project history, concessionary terms, schedule for development of mine and any processing facilities.
- *Geology:* regional geology, detailed description of the project area, preliminary reserve calculations, plans for detailed target evaluation.
- *Mining:* geometry of the ore body, proposed mining plan (and alternatives), required plant and equipment.
- *Processing:* technical descriptions of the ore and concentrate, processing facilities.

Table 6.9 Discount factors

Year	5%	8%	10%	15%	20%
1	0.9524	0.9259	0.9091	0.8696	0.8333
2	0.9070	0.8573	0.8264	0.7561	0.6944
3	0.8638	0.7938	0.7513	0.6575	0.5787
4	0.8227	0.7350	0.6830	0.5718	0.4823
5	0.7835	0.6806	0.6209	0.4972	0.4019
6	0.7462	0.6302	0.5645	0.4323	0.3349
7	0.7107	0.5835	0.5132	0.3759	0.2791
8	0.6768	0.5403	0.4665	0.3269	0.2326
9	0.6446	0.5002	0.4241	0.2843	0.1938
10	0.6139	0.4632	0.3855	0.2472	0.1615

- *Other Operating Needs:* availability of energy, water, spare parts and equipment (diesel oil, explosives, replacement parts, etc.).
- *Transportation:* description of the additional, necessary transportation facilities (roads, air strips, bridges, harbors, rail lines).
- *Towns and Related Facilities:* housing for workers, schools for children of workers, medical facilities, company offices.
- *Labor Requirements:* estimates of workforce broken down according to qualifications (skills) and local availability.
- *Environmental Protection:* plans to reduce or minimize environmental damage, description of relevant environmental legislation.
- *Legal Considerations:* review of mining law, taxation, foreign-investment regulations, political risk.
- *Economic Analysis:* cost estimates for plant and equipment, infrastructure, materials, labor, other factors; market analysis, including production, consumption, and price formation for the relevant minerals; revenue forecasts based expected production and mineral prices; cash-flow and net present value analysis; sensitivity analysis.

These 11 aspects of prefeasibility analysis usually are combined and reported in two analytical sections of the study: technical analysis and economic analysis. Tables 6.10 (technical analysis) and 6.11 (economic analysis) summarize the important aspects of the Metallgesellschaft's prefeasibility analysis of a lead-zinc deposit in the Northwest Territories of Canada. Lead-zinc-silver ore would be mined with room and pillar methods and then concentrated at a mill with a daily capacity of 1500–1600 metric tons of ore. The expected recovery rates are 96% for zinc and 80% for lead. Infrastructure requirements would be significant, including not only offices and other buildings directly associated with mineral production, but also a harbor, airport, and town.

The preliminary economic analysis includes a summary of the anticipated operating characteristics (for example, capacities of ore and concentrate production, personnel requirements, and minimum mine life); expected investment and operating costs, and sources of finance; and cash-flow and sensitivity analyses. The cash-flow analysis considers a period of 10 years. It begins at the top with estimates of revenues from production of lead and zinc concentrates (positive cash flows), and then subtracts operating costs, interest payments on debt, and taxes (negative cash flows) to arrive at estimates of net cash flows for each year.

The sensitivity analysis tests the sensitivity of the project's IRR to different assumptions about zinc prices, investment requirements, and the location of sales. The higher the zinc price, the greater the IRR, other factors remaining the same. Increasing the investment costs by 10% reduces the IRR, as expected. Increasing operating costs also lowers the project's IRR. Selling all lead and zinc concentrates in Europe raises the IRR, because it was assumed that European processors would be willing to pay a premium for lead and zinc concentrates.

Economic Evaluation of Mineral Deposits

Table 6.10 Technical elements of prefeasibility studies (example from the Strathcona Sound lead-zinc project in the Northwest Territories, Canada)

Locality	Strathcona Sound on the north coast of Baffin Island, Northwest Territories, Canada. 760 kilometers north of the Arctic Circle
Evaluation period	Early in 1973 to the middle of 1976
1. Deposit type	
Form	Horizontal, s-shaped sulfide lenses (length 3000 m width 80 m, average thickness 6 m) with vertical rootlike extensions
Ore distribution	Ore minerals are relatively evenly distributed throughout the ore body
Ore paragenesis	Sphalerite (with 3–5% iron), galena, with some pyrite and marcasite
Grain size	Sphalerite > 5 mm, galena averages 3.5 mm, pyrite somewhat smaller
2. Ore Description	
Main products	Sphalerite, galena
Byproducts	Silver in the galena
Geologic cutoff grade	7% zinc equivalent
Average grade	14% zinc, 1.4% lead, 1.77 ounces/ton silver
Tonnage	Geologic size of 6.97 million tons, mining size of 5.85 million tons
3. Extraction methods	
Mining	Room and pillar mining
Ore transport	LHD-Technik
Milling	Crushing, grinding, lead flotation, zinc flotation
Milling capacity	1500–1600 metric tons per day
Recovery	96% zinc, 80% lead
Concentrate quality	Zinc concentrate with 58–59% zinc, lead concentrate with 60% lead and 8–10 ounces per ton of silver
4. Infrastructure requirements	
Buildings	Milling facilities, service buildings, ancillary structures, offices
Energy	6000 kW, diesel generator
Transportation	Harbor for 50,000 ton ships, airport with 2000 m landing strip
Townsite	One- and two-family homes for 850 people, school, hospital, fire station, police station, restaurant, movie theater, gymnasium

Source: Metallgesellschaft AG, Frankfurt am Main, Federal Republic of Germany.

6.3.4 Detailed Target Evaluation and the Feasibility Study

The fourth phase – Detailed Target Evaluation – continues the work described above in even greater detail. The goal is to make a "go/no go" investment decision on deposit development. All of the initial estimates and preliminary plans from the preceding phase are revised and finalized in light of new and more detailed geologic, engineering, and economic information. The work focuses progressively less on geologic evaluation and progressively more on engineering and economic evaluation, broadly defined to include political, legal, and other factors that are not strictly economic in nature. This stage of exploration typically is much more expensive than earlier activities. Expenditures for detailed target evaluation tend to represent some 3–4% of the total investment in a project that proceeds to actual mining, whereas total expenditures for initial target evaluation and earlier reconnaissance exploration usually represent on the order of 1% of the total investment.

Table 6.11 Economic elements of prefeasibility studies (example see Table 6.10)

1. Projected mine capacity:

Run of mine:	525,000 tons of ore per year
Overburden:	400,000 tons per year
Lifetime (minimum):	13 years
Operating period:	350 days per year
Personnel requirements:	150–180 employees (including 40 eskimos)
Projected output:	approx. 125,000 tons of lead-zinc concentrate per year

2. Estimated costs:

Investment:	Exploration:	US-$	4.3 million
	Equipment:	US-$	36.9 million
	Interest:	US-$	3.1 million
	Current operating capital:	US-$	5.0 million
Financing:	Government subsidy (Canada)	US-$	8.5 million
	Company		
	Government loan (Canada)	US-$	10.0 million
	Bank loans	US-$	5.0 million
Operating costs (excluding depreciation and interest):		US-$/t	21.5 million 14.36

3. Sensitivity analysis on Internal Rate of Return (IRR):

zinc prices (cts/lb)	22.8	24.4	27.8
Example as above	15%	19%	26%
Without government subsidy	11%	14%	20%
Additional govt. subsidy of US-$ 5 m	20%	24%	31%
Increase of operating costs by 10%	13%	17%	24%
Increase of investment by 10%	12%	16%	22%
Concentrate sales to Europe	17%	21%	27%

4. Cash flow analysis (1976 figures)

	Unit	1977	1978	1979	1980	1981	1982	1983	1984	1985	1986	1987
Run of mine ore	tons	490000	490000	490000	490000	490000	490000	490000	490000	490000	490000	490000
Zn content	%	15.8	17.41	16.54	15.3	15.3	15.3	15.3	14.4	11.0	11.0	11.0
Zn concentrate	tons	121354	133875	127192	117669	117822	117822	117822	110449	84719	84719	84719
Pb concentrate	tons	3560	7640	12010	10700	9970	9970	9970	10480	12370	12370	12370
Zn concentrate revenue	1000 US-$	29989	33082	31431	29078	29175	29175	29175	27349	20465	20465	20465
Pb concentrate revenue	1000 US-$	319	685	1077	960	894	894	894	940	1110	1110	1110
Total revenue	1000 US-$	30308	33768	32077	30038	30070	30070	30070	28290	21576	21576	21576
Total cost	1000 US-$	16631	17356	17346	16901	17266	17036	16876	17096	16420	16190	16280
Operating profit	1000 US-$	13677	16411	15163	13136	12803	13033	13193	11193	5155	5385	5295
% interest	1000 US-$	6721	6035	5064	4121	3355	2532	1919	1919	1919	1919	1919
% taxes	1000 US-$	0	0	0	0	0	0	0	1912	151	90	50
% duties	1000 US-$	0	0	0	707	688	701	1828	591	237	249	244
Cash flow	1000 US-$	6955	10375	10098	8307	8760	9799	8734	6770	2846	3126	3081
Repayment of loan	1000 US-$	7885	11396	11223	8547	8760	9799	3109	–	–	–	–
Balance	1000 US-$	–930	–1023	–1125	–240	–	–	+5625	+6770	+2846	+3126	+3081

Detailed target evaluation concludes with the feasibility study, a comprehensive document assessing the technical and economic viability of the project. Typically included in the feasibility study are summaries of the project's geologic and mineral-processing characteristics, plans for mining and processing, schedules for construction and timing of investments, marketing plans, and estimates of all costs and cash flows, and financing (Petrick, 1985). The feasibility study contains the final analysis on which the decision on deposit development is based. It essentially is an enlarged and more detailed version of any earlier prefeasibility studies (see previous section). The engineering plans, in particular, are much more detailed and describe, for example, preproduction development work (such as overburden stripping, and construction of the mine, mill, and other facilities) and the plant and equipment necessary for mining and processing. The technical analysis also must provide for other required inputs such as energy, water, labor, roads or other transportation facilities, housing, and spare parts.

The technical analysis, in turn, is an important input into the economic analysis. The various technical aspects of the project must be converted into capital and operating costs, which then are combined with estimates of production and revenues to obtain a detailed description of the expected levels and timing of cash flows for the project. The cash flows determine the expected net present value of the project.

A typical feasibility study might have the following contents, although the exact form of report will vary from project to project (based on Crowther 1985; Lee 1984; Petrick 1985):

– Executive Summary

– Part I: Technical Analysis
  – Geology and reserves
  – Mining and processing
  – Infrastructure and services
  – Construction plans

– Part II: Economic Analysis
  – Cost estimates (capital and operating costs, contingency funds, inflation estimates, estimation methods, precision and accuracy of estimates)
  – Market analysis (general evaluation of industry structure, supply, demand, prices; project-specific evaluation of production, sales, timing and magnitude of revenues, potential sources of finance, legal factors, taxation, environmental requirements)
  – Cash-flow analysis (calculation of net present value or discounted cash flow rate of return; sensitivity to different assumptions about revenues, costs, discount rates, inflation, sources of finance, etc.)

– Part III: Appendices with Supporting Documentation

In addition to considering how these technical factors influence the economic viability of a project, the feasibility study also must evaluate a variety of other factors that affect a project's viability, including legal factors, fiscal regimes, and environmental regulations.

6.3.4.1 Legal Factors

Mining laws governing access to lands and ownership of mineral resources, described in Chapter 5, are not the only legal rules that importantly influence mineral development. A variety of other legal considerations must be reviewed prior to an investment in mineral-deposit development, especially when a private company is investing in a foreign country with legal traditions different from those in its home country. These laws, which vary substantially from country to country, of course need to be reviewed prior to detailed target evaluation for their general favorability toward mining. But the specific legal details influencing the timing and costs of investment need to be factored into the economic analysis of the feasibility study. The general laws pertaining to investment include:

- investment rules governing, for example, security of investment (including protection from expropriation), international contracts, repatriation of profits, taxes and tax holidays, customs duties, and government-equity participation in mineral development;
- banking laws governing access to local capital markets and other issues;
- foreign trade rules governing foreign exchange, export and import of funds and raw materials, export of ore or concentrate, licensing; and
- business and contract law.

6.3.4.2 Fiscal Regimes

For a private organization active in mineral development, the way in which governments collect their share of the proceeds from mineral production can significantly affect the economic viability of a project. Therefore, when evaluating investment alternatives, a private company must consider how fiscal measures alter expected cash flows from a project. Although, as noted by Kumar and Radetzki (1987), it may appear that the wide range of fiscal methods used around the world precludes any meaningful generalizations or comparisons, most regimes in fact reflect one or a combination of three types of fiscal measures.

The first method of government collection is a royalty based on total production. Royalties may be calculated as a percentage of the gross value of production, or may be based on the volume or tonnage of production (for example, 10 US-$ per ton of ore). This sort of government levy is advantageous from the government's perspective because it provides a fairly steady stream of tax revenues and is easy to assess. The disadvantage of a royalty, again from a government's perspective, is that it is not based on ability to pay and thus is likely to discourage investment in mines with low expected pretax profits. Consider, for example, two mines, one with an expected pretax profit of 1 US-$ per ton of ore and the other with expected pretax profits of 10 US-$ per ton. A royalty of 5 US-$ per ton will impose losses on the first mine and yet leave the second mine with a substantial profit.

The second type of fiscal regime is an income tax, calculated as a percentage of the difference between revenues and allowable costs. This sort of system is based more nearly on ability to pay, but is likely to result in a less stable stream of tax revenues for the government. Important differences exist among the income-tax codes of particular countries. The definition of allowable costs differs significantly among countries because of differences in, for example, the deductibility of royalties or export taxes, the speed of depreciation of capital costs, and depletion allowances. Differences also exist in the tax-rate structures; some are flat taxes (the same percentage applies to all levels of income), whereas others are progressive (higher levels of income are taxed at a higher rate than lower levels). A particular form of progressive income tax has gained in popularity during the 1980s: the additional-profits or resource-rent tax, with which a government attempts to capture a large portion of the very large profits earned during periods of high mineral prices or by particularly low-cost deposits. Perhaps the bestknown example of this type of tax is in Papua New Guinea, where – for large-scale mines – normal profits are taxed at the normal business rate of 35%, while excess profits are taxed at the higher rate of 70%. An excess profit is defined as a rate of return on invested funds greater than either 20% or the US prime interest rate plus 12%, as chosen once by the investor (Tilton et al. 1986).

Finally, the third type of fiscal regime is government-equity participation in a project. The extent to which a government actually benefits from equity participation depends on the terms of equity acquisition. When a government acquires equity in a project by paying the full market price, it earns nothing more per share than what any other owner of the project earns. But when government equity is acquired free of charge or at less than full market value, dividends paid to a government are equivalent to tax revenues. In some instances, governments pay for equity with payments in kind of, for example, permission to mine or infrastructure. In other cases, governments borrow money to acquire equity and then are required to repay the principal and interest out of dividend earnings. Some governments retain an option to purchase equity at less than full market value under certain conditions. The government of Papua New Guinea, for example, has the option to purchase up to 30% of a project's equity at concessionary rates at any time prior to actual deposit development (Tilton et al. 1986).

The terms of government-equity acquisition also influence incentives for private-sector investments in exploration and mining. When governments pay full market value, incentives should not be adversely affected because the former equity owners are compensated fully for their assets. But when governments acquire equity free of charge or at less than full market value, private companies are discouraged from investing in exploration and mining because they may be forced to sell their assets for less than what they are worth.

6.3.4.3 Environmental Regulations

Environmental regulations can influence the economic attractiveness of a mineral project at any stage of activity, from initial exploration to actual mining. But they often are of particular importance during detailed target evaluation and feasibility study. Analysis conducted at this stage determines whether or not to proceed with development of a deposit, mining, and processing, and it is at these later stages that most of the costs of complying with environmental regulations occur. Therefore, a feasibility study must determine the extent to which environmental regulations influence the economic attractiveness of developing a deposit.

Environmental regulations affect the economic viability of mineral projects in three different ways. First, they often increase the costs of mining and mineral processing by requiring, for example, scrubbers on smelter smokestacks that reduce the amount of sulfur dioxide emitted into the air, plastic liners at the base of tailings ponds that minimize the release of toxic heavy metals into adjoining ground and surface water, or reclamation of mined land. Second, environmental regulations often increase the time spent on nonmining activities, such as conducting environmental baseline studies, filing environmental impact statements, and applying for mining permits and waiting for their approval. Mine production thus is delayed. Third, regulations often increase the risks associated with an investment in mining, because of the discretionary authority that some regulations vest in government agencies to halt development or mining even after significant expenditures have been made. In all three cases, the net present value of an investment in mining or mineral processing is lower than in the absence of regulation.

What motivates environmental regulations that negatively influence the economic viability of some projects in mining and mineral processing? The answer is public concern about the potential impacts of the residuals of mineral production on human health, fish and wildlife and their habitats, and aesthetic values. Mining generates waste water, often acidic and containing significant amounts of heavy metals and organic contaminants. Waste water can affect both the quality and availability of surface and ground waters. Mining also generates waste rock, material that must be handled in addition to ore. Subsequent beneficiation or concentration of metallic ores generates large quantities of tailings, material that will not proceed to the next stages of processing. Tailings, as well as waste rock, are important sources for toxic metals and organic contaminants that escape and infiltrate surrounding surface and ground waters. The processing of metallic ores and concentrates generates significant quantities of sulfur dioxide and lead, which adversely affect air quality.

In response to these problems, governments in many countries have enacted rules and regulations governing the activities of mining and mineral processing as they affect the environment. Despite the many differences among the policies of various countries, there are similarities (MacDonnell 1987). Most policies require that the polluter pay the costs of mitigating the adverse effects of mining on the environment. Most policies also require environmental assessment of proposed mineral-development activities, as well as government approval for those activities significantly affecting the environment. Finally, most policies rely on monitoring systems to encourage compliance with the regulations.

6.4 Financing

This chapter, until now, has focused on the investment decision or, in other words, on how to evaluate the economic attractiveness of spending money on mineral exploration and development. Another important decision concerns financing: where to obtain the money to invest in these activities. A range of funding sources exists, and the choice of appropriate sources is the focus of the financing decision. This section of the chapter briefly reviews the range of available funding sources and then examines in greater detail the sources that are of particular importance to mineral exploration and development.

Economic Evaluation of Mineral Deposits

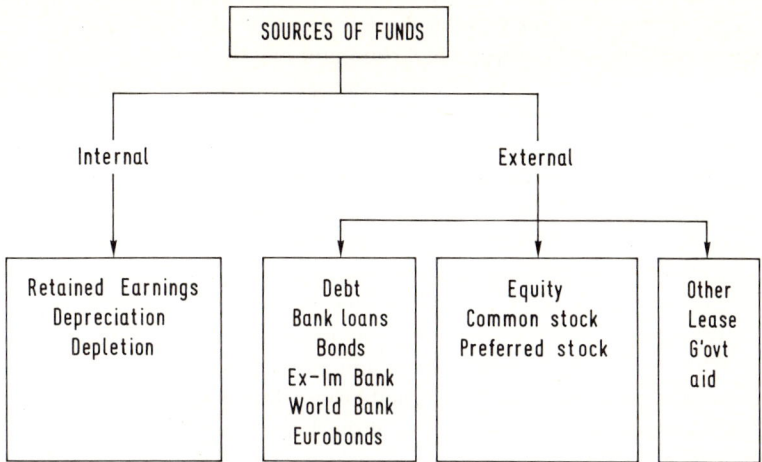

Fig. 6.6. Possible sources of investment funds

All sources of funds are either internal or external to the mineral-development organization (Fig. 6.6). Internal funds come from within the organization. For both private and state-owned organizations, the most important source of internal funds usually is retained earnings: revenues retained by an organization after all costs, taxes, and dividends to stockholders have been paid. For private companies in many countries, depreciation and depletion also are important. These are noncash costs that are deductible before income taxes are calculated. They are sources of funds in the sense that they are allowable costs for income tax purposes even though they are not incurred by the firm in the year claimed; rather, they represent reductions in the salvage value of equipment (in the case of depreciation) or of the deposit (in the case of depletion).

External funds originate from outside an organization. The most important external sources are debt and equity, although leasing, government appropriations to state-owned companies, and other forms of government assistance are also possible. Debt financing involves borrowing money from an outside party in order to pay for an investment. The borrower is required to repay the principal (the amount borrowed), as well as interest on this principal, usually over a specified period of time, but occasionally all at once in a lump-sum payment. Short-term debt requires repayment within a year or two, and thus is hardly ever used for financing mineral projects, which typically generate no revenue for several years following the development decision. Long-term debt, on the other hand, is repaid over much longer periods of time, sometimes up to 40 years. Important sources of long-term debt financing for mining projects are bonds and loans. A bond is a written agreement to pay the holder of the bond a certain amount of money at a specified time. Bonds typically are issued through a financial intermediary and sold to many types of investors. Many bonds are traded publicly (sold from one investor to another). A loan is an amount of money made available by one organization or individual to another with provisions for repayment of the principal and interest. Important sources of loans for mining projects include commercial banks, international organizations such as the World Bank's International Bank for Reconstruction and Development (see Sect. 8.3.3), and government lending agencies such as the US Export-Import Bank (the Ex-Im Bank) and Canadian Export Development Corporation (EDC).

The Ex-Im Bank and EDC actually provide what are known as *export credits,* loans to foreign buyers who will use the money to purchase goods and services from the United States (in the case of the Ex-Im Bank) and from Canada (in the case of EDC). Export-credit loans usually have a lower-than-market interest rate. The Ex-Im Bank and EDC also provide insurance or guarantees that US or Canadian exporters will be paid for goods or services sold to foreign buyers. Some western European countries have export-credit agencies similar to the Ex-Im Bank and EDC. All such agencies are designed to encourage exports from the home country.

Equity financing involves raising monies by selling stock in a company or project. In exchange for their money, stock purchasers become part owners of the company or project. Stockholders are residual owners in the sense that they are entitled to the earnings or assets of a company or project only after debt payments have been paid. Equity, as well as debt, financings actually are much more complicated in their implementation than suggested by this overview. There are many variations on these general themes, and new types of equity and debt arrangements appear constantly. Moreover, many differences exist among countries in what types of debt and equity are permitted.

In summary, three broad categories of funding sources are important for mineral exploration and development: internal funds, debt, and equity. The relative importance of these sources varies from one country to another, from exploration to development, from one mineral or deposit type to another, and over time.

6.4.1 Exploration Projects

As a general rule, internal funds dominate exploration financing. Most large, multinational mining companies rely almost solely on internal funds to finance exploration in foreign countries and at home. Debt is almost never used by any type of organization to finance exploration, presumably because (1) potential lenders will not lend to enterprises as risky as exploration (risky in the sense that the payoffs to exploration are highly skewed: most individual projects do not find minable deposits, do not recover their costs, and thus would not be able to repay the principal and interest on a loan), and (2) exploration organizations are reluctant to hurt their competitiveness relative to other organizations by revealing information on exploration strategies and programs that potential lenders would want to see when considering a loan request.

Equity financing is used in several countries to fund exploration, typically by small companies with only one or a few small operating mines – if any – and thus with little or no retained earnings to devote to exploration. These companies often sell any discoveries to larger organizations for detailed exploration or development. Small stock companies are particularly common in Canada, where funds are raised on, for example, the Vancouver and Toronto stock exchanges. To a lesser extent, equity funds for exploration are raised by small companies in the United States (on the Spokane, Denver, and several other stock exchanges) and in Australia.

A particular type of equity financing for exploration that became common in Canada in the 1980s is the flow-through share. Investors purchase shares in one of several investment funds that then directs the money to the individual mining companies participating in the fund. Investors, therefore, own shares in a fund rather than in a particular mining company or group of companies. Shares are flow-through in the sense that tax benefits flow-through from the company to the purchaser of shares: investors are permitted to deduct more than 100% of their investment for income-tax purposes. To qualify, money must be spent on exploration in Canada. The Prospectors and Developers Association of Canada estimates that some 40% of the money spent on metallic mineral exploration in Canada in 1983, 1984, and 1985 was raised using flow-through shares: 38 of 308 million US-$ in 1983, 145 of 420 million US-$ in 1984, and 300 of 500 million US-$ in 1985 (constant 1985 Canadian dollars; *Northern Miner,* March 2, 1987).

Prior to 1983, the tax benefits associated with flow-through shares were not nearly as attractive.

In developing countries, exploration is financed in a variety of ways. Foreign companies, domestic private companies, and state-owned companies tend to rely on internal funds. In addition, several multilateral agencies support exploration. Multilateral aid has been motivated by desires to promote economic development in potentially mineral-rich areas that lack the necessary technical expertise or financial capital, as well as by concern that inadequate exploration by private companies has occurred in some developing countries because of political risk. The two largest programs of multilateral aid are the United Nations Development Program (UNDP) and the United Nations Revolving Fund for Natural Resources Exploration (see Sect. 8.3.2). The UNDP supports both exploration and development projects, and works to improve institutions relating to minerals, such as geological surveys and university geology departments. In contrast, the Revolving Fund, established in 1973, has a more limited range of activities. It chooses projects based on their commercial viability and conducts exploration only up to the point of detailed deposit delineation (stopping short of the feasibility study). It does not support deposit development, but instead makes exploration information available to potential investors. It undertakes no institution-building such as described above for UNDP. The Revolving Fund is intended to be self-supporting, with contributions from successful projects funding all exploration (see Crane-Engel and Schanze 1988).

6.4.2 Mine Development

A wider range of funding sources is used to finance development compared to exploration. The data in Table 6.12 for 72 development projects in 1984 and 1985 – although not necessarily representative of all development projects during this period – illustrate this range. Bank loans were the most important source in terms of amount of money, followed by internal funds, export credits, stock offerings, government assistance, gold loans, and other debt. Although export credits represent some 18% of the money included in the survey, they were used in only 2 of the 72 projects (Ok Tedi in Papua New Guinea, and Cerro Colorado in Chile).

Government assistance (other than export credits) most frequently was in the form of loans but also as grants and direct expenditures, often for mining infrastructure such as roads, power lines, or port facilities. An important element of the other-

Table 6.12 Sources of 116 mine financings, 1984–1985

Source	Amount financed (million US$)	Percent of total	Number of financings	Percent of total
Bank loans	1300	38.3	25	21.5
Gold loans	90	2.7	10	8.6
Export credits	600	17.7	6	5.2
Other debts	70	2.1	3	2.6
Government assistance	312	9.2	18	15.5
Stock offerings	405	11.9	35	30.2
Internal funds	614	18.1	19	16.4
Total	3391	100.0	116	100.0

Source: *Mine Development Bimonthly,* vol III, no 3 (December 31, 1985), pp. 1–7.
Notes: 72 individual mines are represented in these 116 financings (24 in the United States, 20 in Australia, 17 in Canada, 2 in Peru, and 1 each in Burkino Faso, Chile, Costa Rica, Fiji, Ghana, Papua New Guinea, the Philippines, South Africa, Zambia). These include 61 gold or gold/silver projects; 5 zinc or lead/zinc projects; 4 copper, copper/gold, or copper/cobalt projects; 1 tin project; 1 diamond project.

debt category is the consumer credit, a variant of the export credit. Future consumers of the metals mined loan money to a project in exchange for a guarantee of supply; the future consumer might, for example, be a smelter in need of concentrate.

Gold loans are a relatively new financing instrument, introduced in the 1980s and used almost exclusively to finance gold mines, particularly in Australia and to a lesser extent in the United States and Canada. Although accounting for only 2.1% of the money in the survey (Table 6.12), gold loans were important sources of development funds for 10 of the 72 projects surveyed (some 14%). The lender lends physical gold to the borrower, who then sells the gold to fund deposit development. The borrower repays the loan in physical gold out of mine production. Typically the interest rate is significantly below the rate on a commercial bank loan. To protect the lender against falling gold prices, the borrower generally is required to enter into a forward-selling or price-hedging agreement for the duration of the loan, guaranteeing the price to be received for an agreed-upon level of production.

For the 72 development projects included in Table 6.12, debt financing represents over 60% of all funds. Both the importance of debt financing for mine development *and* the range of debt-financing instruments for mineral development have increased dramatically over the years. Prior to the middle 1960s, mines typically were financed with internal funds (retained earnings) and, to a lesser extent, equity issues. Debt financing was extremely rare. The rationale for relying heavily on internal funds was that external financing was viewed as higher cost than internal financing; external investors, in other words, required a higher-than-normal rate of return on their debt or equity investments to compensate for the riskier-than-normal nature of mining investments. In developing countries, where development of new mines and expansions of old ones were dominated by multinational companies from North America and Europe, debt financing is estimated to have accounted for no more than 10–20% of the monies raised for mine development (Radetzki 1980). In 1950, for example, US investments in mining and mineral processing were dominated by shareholders' assets (Table 6.13). These data are estimated to represent about one half of US total foreign

Table 6.13 US investments in mining and mineral processing in selected developing countries, 1950 (million US-$)

	Chile	Mexico	Peru	Sterling area developing countries	Total for countries listed
Total assets	402	198	68	162	830
Shareholder assets:					
US-owned	351	126	55	67	599
Non-US-owned	18	7	—	74	99
Total	369	133	55	141	698
Debt:					
US-owned	—	–6	—	13	7
Non-US-owned	33	71	13	8	125
Total	33	65	13	21	129

Source: Radetzki (1980).

investment in mining and processing in 1950. The data are not a perfect measure of new mine financings because they include old or established projects.

Between the middle 1960s and early 1980s, the role of debt in mine financing increased considerably, as illustrated by the data in Table 6.14 from the 1970s in several developing countries. (These data may overstate the role of debt financing because these projects are well-known examples of very expensive projects that relied heavily on debt funds). There are at least four reasons for this shift to predominantly debt financing. First, individual mining projects generally became larger and more expensive, and thus were too expensive to be financed solely out of internal funds. Technological changes in mining and mineral processing encouraged larger and more expensive operations that took advantage of scale economies (that is, costs per unit of output were lower for large-scale mines and processing facilities than for small-scale operations). Moreover, in several countries, lower-grade ores and a variety of regulations governing the environment and health and safety also contributed to rising financial requirements per project, as did the fact that more remote mines required larger investments in infrastructure, such as roads, townsites, and power sources.

Second, during the same period, a variety of debt-financing alternatives emerged and generally made debt financing more available. These alternatives included

Table 6.14 Equity finance for selected mineral projects nin developing countries, 1970s

Country and project	Mineral	Start-up year	Total finance (million US-$)	Equity finance	Equity's share (%)
Papua New Guinea (Bougainville)	Copper	1972	479	166	35
Guinea (Boke)	Bauxite	1973	224	17	8
Botswana (Selebi-Pikwe)	Copper/nickel	1974	204	47	23
Peru (Cuajone)	Copper	1976	726	264	36
Brazil (Samarco)	Iron ore	1977	405	110	27

Source: Radetzki (1980).

financing from the eurodollar market, export credits, and multilateral lending agencies.

Third, for mine development in industrialized countries, perceptions of risk were low during much of the 1960s when mining companies incurred substantial amounts of debt. Overall economic growth had been brisk for much of the period since World War II, and as a result demand for most metals had grown steadily and considerably. Inflation was low. Interest rates were low and steady. Both lenders and borrowers (mining companies), therefore, were confident that there was little risk of default on loans for mine development.

Fourth, for mine development in developing countries, mining company perceptions of risk were generally high following the political and economic emancipation of these countries after World War II (Radetzki 1980). Most developing countries desired and, in fact, demanded greater control over the development and production of their mineral resources. Government actions varied considerably from country to country and from project to project. Some governments raised tax rates on foreign-owned businesses. Other governments required greater control over development and production decisions in the mineral sector or demanded equity participation in mineral projects. Still others nationalized the assets of foreign companies. As a result, multinational mining companies became more cautious about investing internal funds in developing countries.

Instead, these companies reduced their exposure to political risks by relying more heavily on debt financing, an important element of a new type of financing called project financing, which differs in several important respects from the traditional method of financing new mines and processing facilities out of internal funds. (The term "project financing", referring to a particular type of financing for mine development, described below, should not be confused with the general issue of financing exploration and mine development, a broader topic which is the focus of this entire Sect. 6.4). First, it relies on funds raised from a variety of sources – usually in the form of debt – rather than on internal funds from a single investor. Although a foreign multinational company may provide a small portion of the required financial capital from internal funds or perhaps new share issues, the majority of funds typically come as loans from a consortium of commercial banks and other investors, sometimes including the World Bank group, export-credit agencies, or consumers. Second, these lenders have little or no recourse to the assets of the foreign multinational, but rather accept the project itself as collateral. In other words, lenders rely on the expected revenues of the project, rather than on the revenues of the parent company. But to compensate for lack of recourse to the parent company, lenders charge a higher rate of interest than would be charged if the parent company, rather than the project, were responsible for repaying the debt. Lenders also typically impose additional requirements on a project. The parent company, for example, usually guarantees that the project will be completed and brought into production; up until that point the parent company is responsible for repaying the debt. Normally lenders require that the project commits a significant portion of the expected production to specific buyers through long-term purchase agreements. The project may be required to purchase political risk insurance from an organization such as the US Overseas Private Investment Corporation (OPIC), which would compensate lenders in the event of nationalization. Export-credit agencies, such as the US Ex-Im Bank, may guarantee that suppliers receive payment for goods and services exported to the project. Third, as a consequence of the first two points, project financing places little, if any, of the political and other project risks on the multinational mining company. Rather, the risks are spread over a large number of organizations, including commercial banks, export-credit agencies, multilateral lending agencies, and consumers. Furthermore, the overall risks of government actions detrimental to a project – such as nationalization – may actually be reduced because the costs to the host government of undertaking such actions are raised by the nature of the agreements among the organizations sharing the project risks. (An interesting example of project financing is the Ok Tedi project; see McGill 1985.)

What is the likely future course of financing for investments in metal mines and processing facilities? The era of such heavy reliance on debt financing is undoubtedly over. Several factors suggest that equity financing will be of greater importance in the late 1980s and early 1990s than in the previous 20 years. First, real inflation-adjusted interest rates are higher in the middle 1980s than during much of the period from the late 1960s to the early 1980s, increasing the costs of debt financing. Second, many of the developing countries and mining companies that relied on debt financing for mine development are heavily in debt and are working hard to reduce this debt. These countries and companies will be cautious about incurring substantial amounts of new debt to finance new mines and expansions, and commercial banks will be wary of lending them these monies. Finally, in many developing countries the risks of nationalization now are much lower than in the late 1960s and 1970s, and thus multinational mining companies are less apt to use debt and project-financing arrangements to reduce exposure to political risks. Although some nationalizations are likely to occur in the future, nationalizations were essentially a one-time phenomenon associated with the political and economic emancipation of many developing countries. Most host governments and foreign investors have become less adversarial and more willing to compromise and work toward mutually acceptable agreements. Host governments are generally more eager to promote foreign investment, and foreign investors are more willing to accept the legitimate rights of host governments to participate in the development and extraction of mineral resources.

Notes on the Literature

Those interested in more detail on investment analysis should consult Stermole (1987) *Economic Evaluation and Investment Decision Methods*, Schenck (1985) "Methods of Investment Analysis for the Mineral Industries", and Rudawsky (1986) *Mineral Economics: Development and Management of Natural Resources*. Benefit-cost analysis and other aspects of government policy analysis are covered by Stokey and Zeckhauser (1978) *A Primer for Policy Analysis*. Newendorp (1975) discusses risk analysis – as applied to petroleum exploration – in *Decision Analysis for Petroleum Exploration*. Those interested in financing are referred to Brealey and Myers (1984) *Principles of Corporate Finance;* Tinsley, Emerson, and Eppler (eds) (1985) *Finance for the Minerals Industry;* and Radetzki and Zorn (1979) *Financing Mining Projects in Developing Countries*. Those interested in project finance should read Nevitt (1984) *Project Financing*. Feasibility studies are the subject of Northwest Mining Association (1984) *Mine Feasibility – Concept to Completion* and Crowther (1985) "Feasibility Studies."

Chapter 7 Mineral Markets

A "market" is a hypothetical place where buyers and sellers of a given commodity meet to determine price. Mineral markets are material goods markets, and many are regarded as world markets because of the easy negotiability of the traded commodities.

Strictly speaking, markets for any particular mineral may exist at several stages of production and for several levels of quality. The copper market for example is divided into:

- a market for copper ore (today limited to local markets for rich ores),
- a market for copper concentrate (today mainly at a national level),
- a market for crude copper (as a product of smelting),
- a market for electrolytic copper (as a product of refining), which is further divided into markets for wirebars, cathodes and fire-refined copper.

The petroleum market includes, among others:

- a market for light petroleum (34°– 40° API)
- a market for heavy petroleum (27°–31° API)
- a market for low-sulfur petroleum and
- a market for sulfurous petroleum.

These subdivisions are especially important for detailed market analysis. In practice, however, the market usually referred to is that of the standard trade quality for the mineral commodities most important in world trade: today, for copper, electrolytic copper (cathodes) with a minimum copper content of 99.9%, and for crude oil, Arabian Light (34° API). These are homogeneous, fungible commodities negotiable on exchanges at an approximately uniform world market price.

The importance of various types of products in international trade, however, varies continuously. Now that the oil-exporting countries are increasingly building up their own processing facilities for instance, crude oil is less important in international trade and is being replaced by petroleum products. Examples also exist for other mineral commodities. Thirty years ago, trade in tin concentrates was still substantial; today world trade is limited essentially to refined tin. Iron ore or bauxite, in contrast, is mainly exported as a mine product, often unconcentrated. Antimony and wolframite are largely traded as ore concentrates and tantalum in the form of tin slag, making up the major part of world trade.

7.1 Market Structures

The underlying features of supply and demand determine the structure of any particular market. A structural analysis must thus consider the number, geographic distribution, and market participation of producers and consumers as well as the communications and competition between them. The production of minerals is influenced by geologic conditions and therefore market analysis needs to be conducted by mineral economists with good geologic inside knowledge.

The form of market is defined by the extent of free competition. It constitutes the sum of all factors influencing competition and hence pricing. Empirical investigations on market structure are usually confined to determining the number, size, and market position of the buyers and sellers. It is assumed that a large number of producers and consumers implies competition resulting in an economically efficient price; a small number, imperfect competition, where the price can be influenced; and one single producer or consumer, the absence of competition and the possibility of price fixing.

Market theory thus differentiates according to the number of participants as follows:

Competition: markets with a large number of small suppliers in atomistic competition. Each supplier sells the same product and has no influence on price.

Oligopoly: markets with a small number of large suppliers in imperfect competition, in which the members have a certain influence on the pricing. If the market has a small number of large buyers, it is referred to as an oligopsony.

Monopoly: markets with only one supplier in which competition is absent and the monopolist exercises considerable influence on prices. If the market has only one buyer, it is known as a monopsony.

Another important factor when analyzing market forms is the behavior of the members, which has a decisive influence on pricing. This can take the form of peaceful adjustment to market conditions or aggressive strategies. Examples of the former are quantity adjustment (typical in competitive markets), price or quantity fixing (typical in monopolistic markets and in oligopolistic markets). Agreements to restrict competition (cartels) or to integrate competitors (trusts) are instances of aggressive strategies.

The crucial question for a structural analysis of markets is, of course, the size of the share needed to influence the market and hence prices, or to dominate the market and thereby fix prices. This share is particular to the market and depends on the competitive conditions and substitution possibilities for the mineral. In the short run, mineral supply typically is inelastic with respect to the price, which means that a change in price results in a proportionally smaller change in mine output. It also means that a relatively small share is sufficient to dominate the market, often as little as 30% to 40% of overall production. The presence therefore of a large producer accounting for roughly 35% of supply is enough to create a monopolistic mineral market.

7.1.1 Determination and Alteration of Market Structures

World trade in mineral commodities is dominated by petroleum, natural gas, coal, metals and some industrial minerals. The reason for the trade is that only few countries are self-sufficient in these commodities. In the course of the last few decades, the separation between consumer countries dependent on mineral imports and producer countries exporting their surplus has become increasingly distinct. In the latter half of the nineteenth century, the young industrial countries of Western Europe, headed by Great Britain, were also the largest mining countries, joined by the United States at the turn of the century. During this period 50 to 100 years ago, oil and mining companies were established in Western Europe and the United States, which continually expanded and soon began to extend their activities beyond their national borders. This consistent expansion of business activities, combined with a process of concentration, led to the emergence of international mining companies. These companies accumulated power over the market, which resulted in reduced competition in many mineral markets and increasingly in oligopolistic or even monopolistic markets.

Any market analysis must account for the international character of the mining industry. The usual national statistics on production and consumption are of little value in determining the market form. The structural relationships on the supply and demand sides of a market are much more complicated than country statistics show. Market structures change significantly over time, due for example to:

- nationalizations in the mining sector, as occurred in several developing countries in the 1960s and 1970s;
- extensive changes in ownership relations, as occurred in the early 1980s, when oil companies acquired a number of mining enterprises;
- the discovery and development of new mineral deposits.

A market analysis must first of all take the following into account:

- The production statistics available are generally unsuited to ascertain the supply structure. They are confined to individual countries and reveal little about the actual participants in the market, which complicates the analysis considerably, since detailed research on the enterprises, their business activities, and ownership are needed. The most useful sources of information here are the annual reports of the mining and oil companies.
- On the supply side of a mineral market, the mining and processing sectors are quite often separate. As a rule, the market analysis has to concentrate on the sector producing the mineral in that form in which it is traded on the world market. Both sides nevertheless influence one another and in many cases their structures interlock.

The complexity of market structures and forms and the changes to which they are subject are illustrated with the following examples of three specific markets:

- The tin market, which has undergone profound structural changes in recent years.
- The aluminum market, characterized by an oligopoly and vertical concentration of supply.
- The gold market, in which supply is affected by exceptional factors of influence.

Mineral Markets

7.1.1.1 The Tin Market

Now that tin smelters have been put into operation in all major tin-mining countries, the world trade in tin is limited entirely to refined tin.

The usual national statistics (Table 7.1) show that the smelting sector underwent a structural transformation between 1972 and 1982, brought about in particular by the expansion of capacity in Vinto/Bolivia (ENAF plant) and production increases in Volta Redonda/Brazil (CESBRA plant) on the one hand and the closure of the Kirkby smelter in Britain and the decline in production in Nigeria and Malaysia on the other. The statistics provide hardly any information, however, on the market

Table 7.1 Tin output (western world, in tons)

Country	Tin-in-concentrate output		
	1972	1982	1986
Malaysia	76 830	52 342	29 134
Indonesia	21 766	33 800	24 634
Thailand	22 072	26 207	15 992
Bolivia	32 405	26 713	10 824
Australia	12 081	12 308	8 669
Zaire	5 892	2 483	1 889
United Kingdom	3 327	4 175	4 345
Brazil	2 813	8 297	25 449
Peru	20	1 700	4 000
Nigeria	6 731	1 822	72
Canada	159	206	2 500
South Africa	2 216	3 035	2 055
Other	10 113	21 782	7 686
Total	196 425	194 870	137 249

Country	Primary refined output		
	1972	1982	1986
Malaysia	91 001	62 836	43 788
Thailand	22 281	25 479	19 744
Indonesia	12 010	29 755	22 080
Bolivia	6 528	18 980	9 000
Brazil	3 583	9 298	25 069
United Kingdom	21 333	10 211	11 582
Australia	7 025	3 105	1 302
Nigeria	6 744	1 808	91
Netherlands	—	2 754	5 114
Singapore	—	4 000	500
Spain	4 100	3 400	1 365
USA	960	400	2 000
South Africa	1 491	2 725	1 816
Other	12 944	9 414	4 842
Total	190 000	184 165	148 293

Sources: International Tin Council, Monthly Statistical Bulletin and Quarterly Reports, London 1975 – 1987; – Tin International, London 1987.

shares of suppliers of standard refined tin. Market positions become clearer if we analyze the ownership structures in the tin industry.

In 1972, roughly 40% of refined tin was produced by Consolidated Tin Smelters Ltd. (CTS) in London. The CTS emerged in 1929 from an integration process in the tin industry of that time; the Bolivian "tin king", Patino, acquired complete or partial ownership of various smelting firms in Great Britain, Malaysia, and the Netherlands. CTS was also interlocked with tin-mining enterprises, partly due to Patino's ownership of mines in Bolivia but also to a minority interest in the London Tin Corporation, founded in 1925 with mines in Malaysia and Nigeria.

In 1972, CTS owned the large tin-smelting plants of the Sharikat Eastern Smelting Bhd. in Penang, Malaysia, of Williams, Harvey & Co. Ltd. in Kirkby, Britain, the Makeri Smelting Co. in Jos, Nigeria and shares in the Associated Tin Smelters Ltd. in Alexandria, Australia. Four other medium-sized producers, however, weakened CTS' exposed monopolistic position on the market: at that time, the Straits Trading Co. in Butterworth, Malaysia, with more than 15% of the market in tin metal, the THAISARCO in Phuket, Thailand, with roughly 12%, the P.T. Timah in Mentok, Indonesia, with roughly 8% and the newly established ENAF in Vinto, Bolivia, with 4%. The supply structure in 1972 could thus be described as a weak monopoly. The market structure in the mining sector was substantially less concentrated. A few government enterprises (P.T. Tambang Timah, Indonesia; COMIBOL, Bolivia) and private enterprises (London Tin Corp. and Charter Consolidated in Malaysia, Thailand and Nigeria) formed a weak oligopoly, each with about 10% of the world market, the rest being shared by a large number of smaller and medium-sized producers.

Within 10 years, the ownership structure in the tin-smelting industry had changed considerably. In 1975, CTS' parent company, Patino N.V., amalgamated CTS in London with the Amalgamated Metal Corporation (AMC), founded in 1968. Then in 1978, Preussag AG of Germany (F.R.) acquired 79.3% of the AMC from the Patino N.V.; the AMC still controls to a large degree the smelting companies in Penang, Malaysia (Datuk Keramat Smelting Bhd., 50.5%), and in Jos, Nigeria (58%), whose share of the market, however, has declined to less than 20% in 1986.

The Straits Trading Company Ltd. was also affected by this regrouping of capital. In 1981, the newly established Malaysia Mining Corporation (MMC) acquired 42% of the shares, thus attaining substantial influence over the Malaysia Smelting Corp. Sdn. Bhd. in Butterworth, Malaysia, and its 16% share of the world market. The MMC also represented a concentration in tin mining, since this conglomerate, controlled by the Malaysian Government parent enterprise PERNAS, has progressively taken over the shares of the London Tin Corp. and the Charter Consolidated.

THAISARCO has now been wholly taken over by Billiton N.V. (Shell), its shares having been sold by Union Carbide, USA in 1975. THAISARCO's share of the world market in primary refined tin metal is now 11%.

After the considerable extension of the plant in Vinto near Oruro, the Bolivian state-owned company ENAF (1986 incorporated into COMIBOL) has also been able to increase its share of the market. Approximately 7% of world production derived from there in 1985 (but only 5% in 1987).

The market share of P.T. Timah, Indonesia, has increased even more since the expansion of smelting capacity in Mentok on the island of Bangka; in 1985 it accounted for 13% of world production.

The most impressive increase in tin production, however, was achieved by the Paranapanema SA, Brazil, after the construction of the Mamoré tin smelter in Sao Paulo, producing 16% of the world tin market.

In 10 years the structure of tin supply thus changed radically from a weak monopoly to a clear oligopoly with six main producers.

The demand side of the tin market does not display this degree of concentration. A few large tinplate manufacturers (US Steel Corp., Nippon Steel Corp., British Steel Corp., Rasselstein AG) each account for 3% to 7% of world tin consumption at most, so the demand structure is slightly tending toward oligopsony.

7.1.1.2 The Aluminum Market

The aluminum market displays a strongly marked vertical concentration. Although there are three basic distinct sectors – bauxite mining, aluminum oxide plants (refineries), and aluminum smelters – a characteristic of the ownership structure is that the smelting companies also control the other sectors. This vertical amalgamation quite often extends to the production of semifinished products and even to final consumption goods. This facilitates an analysis of the market structure, because the shares of the processing firms correspond to their share of the world markets in bauxite and/or aluminum oxide. The national statistics, on the other hand (Table 7.2), reveal very little.

The primary aluminum industry in the Western World has a capacity of roughly 14 million t.p.a., over half of which is produced by six multinational enterprises:
- The Aluminum Company of America (Alcoa), headquartered in Pittsburgh, USA, with production plants in 16 countries, global production capacity of 2,000 million tons per year (production in 1986: 1.67 million t) and activities in bauxite mining (Australia, Brazil, Guinea, Jamaica, Suriname and the USA), aluminum oxide

Table 7.2 World bauxite and aluminum production 1986

Country	Mine production (1000 tons bauxite)	Country	Smelter production (1000 tons aluminum)
Australia	32 432	USA	3037
Guinea	14 656	Canada	1355
Jamaica	6964	Germany (W)	764
Brazil	6446	Brazil	757
Suriname	3731	Norway	729
Yugoslavia	3459	Venezuela	423
Greece	2225	Spain	355
India	2338	France	322
Guyana	2074	Gt. Britain	276
France	1379	India	257
USSR	6275	USSR	2350
Hungary	3022	China	480
China	2200	Romania	260
Total	91 205	Total	15 550

Source: Metallgesellschaft, Metallstatistik 1976–1986, Frankfurt a.M. 1987.

production (Australia, Brazil, Jamaica, Japan, Suriname, USA) and aluminum smelting (Australia, Brazil, Mexico, Norway, Suriname, USA).
- The Alcan Aluminium Ltd., in Montreal, Canada, with production plants in 31 countries, a production capacity worldwide of 1.84 million tons per year (production in 1986: 1.65 million t) and activities in bauxite mining (7 countries), in aluminum oxide production (7 countries) and smelting (9 countries).
- Reynolds Metals Co. in Richmond, USA, with production plants in 17 countries and a global production capacity of 1.2 million tons per year (production in 1986: 1.1 million t) and activities in bauxite mining in aluminum oxide production, and in aluminum smelting.
- Pechiney Ugine Kuhlmann S.A. in Paris, France, with production plants for aluminum in 11 countries, a worldwide production capacity of 1.0 million tons per year (production in 1986: 0.8 million t).
- Kaiser Aluminum Chemical Corp. in Oakland, USA, with production plants in 22 countries, a global production capacity of 1.0 million tons per year (production in 1986: 0.64 million tons) and activities in bauxite mining and in aluminum smelting.
- Alusuisse (Schweizerische Aluminium AG) in Zürich, Switzerland, with production plants in at least 12 countries, a worldwide production capacity of 0.9 million tons per year (production in 1986: 0.75 million t) and activities in bauxite mining, aluminum oxide production and aluminum smelting.

These six conglomerates constitute a clear oligopolistic market for aluminum and bauxite; they are, however, characterized by activities in bauxite mining, alumina production, and aluminum smelting which means a vertical concentration of supply. The rest of the market is made up of a number of small and medium sized producers, mostly European.

The demand side, at least for final consumption of aluminum, is competitive.

7.1.1.3 The Gold Market

On the world market for the precious metal gold, this commodity is traded in the form of fine gold bars of 99.5% purity. Since 1974, there has also been a demand for coins such as Krugerrands.

Trade in this coinage has increased considerably: in 1978, over 6 million ounces (187 t or 26.5% of South African production) was sold, in 1983, 4.7 million ounces (16%), in 1984 however, only 3.9 million ounces (12%). Since economic sanctions were introduced in 1983 by countries in Western Europe and North America against South Africa, the Krugerrand sales have decreased. In 1985, the Royal Leaf of the Royal Canadian Mint in Ottawa was the best selling coinage (66.4 t in 1985).

The national statistics clearly show South Africa's leading position in primary gold production (Table 7.3). Thirty-four gold mining enterprises (cf. Table 7.4) contributed to the 1985 production figure of 671.7 t (55.5% of non-communist world production). An analysis of the structures of ownership, however, shows that practically all these companies are controlled directly or indirectly by six larger mining groups (cf. Table 7.4). Worthy of particular mention are three multinational mining enterprises:

Table 7.3 Gold-mining production, western world, in tons

Country	Production 1980	Production 1986
South Africa	675.1	640.0
USA	30.5	108.0
Canada	51.6	107.5
Australia	17.0	75.0
Brazil	35.0	67.4
Philippines	22.0	39.9
Papua-New Guinea	14.3	36.1
Colombia	17.0	27.1
Chile	6.5	19.2
Zimbabwe	11.4	14.9
Japan	6.7	14.0
Ghana	10.8	11.5
Other	61.1	120.2
Total w. world	959.0	1280.8

Source: Chamber of Mines of South Africa, Statistical Tables 1986, Johannesburg 1987.

- The Anglo American Corporation of South Africa Ltd. (AAC), which accounted for 36% of South African production in 1986;
- The Gold Fields of South Africa (GFSA), accounting for 24%;
- The General Mining Union Corp. (Gencor), with 18%.

Of importance too are the Charter Consolidated, the Johannesburg Consolidated Investment (JCI) and the Anglovaal Ltd. with production shares of between 6% and 8% each.

A further concentration on the supply side can be deduced from the interlocking of some of these mining houses. The Anglo American Corp., for example, is the connecting link to Gold Fields S.A. via a direct participation (9%) and an indirect participation (via Anglo American Gold Investment, 11% and Mineral Resources Corp. 29%) as well as to Gencor (6%), JCI (40%) and Charter Cons. (via Minerals and Resources Corp. Ltd. 36%).

This interlocking implies that more than 90% (1986:94%) of South African gold production and thus just about half of the production in the Western World is influenced by one international mining company, Anglo American Corp. of South Africa Ltd. This also means that the market for primary, newly mined gold on the supply side is monopolistic.

The supply of gold on the market, however, is not determined by mine production alone. Three other factors play a role, although they are particularly difficult to quantify:

- The net sales of the Soviet Union, which fluctuate greatly (1970: 3t, 1976: 412t, 1980: 90t, 1981: 280t, 1983: 93t, 1986: 420t) and depend on its foreign exchange requirements.
- The sales and purchases of the International Monetary Fund (IMF) and government-controlled investment and monetary agencies, which are geared to the finan-

Table 7.4 Gold-mining companies in South Africa

Mining company	Group*	Production 1986 (kg gold)
Anglo American O.F.S.	AAC	2 743.3
Barberton	Gencor	1 671.5
Blyvooruitzicht	RM	13 238.2
Bracken	Gencor	3 070.0
Buffelsfontein	Gencor	35 112.4
Deelkraal	GFSA	7 587.6
Doornfontein	GFSA	8 661.2
Driefontein Cons.	GFSA	59 765.5
Durban Deep	RM	7 458.3
East Rand P.M.	RM	9 223.4
Elandsrand	AAC	11 836.0
Free State Geduld	AAC	106 488.9
Grootvlei	Gencor	5 735.0
Harmony	RM	28 730.0
Hartebeestfontein	AV	30 912.4
Kinross	Gencor	12 355.5
Kloof	GFSA	28 841.0
Leslie	Gencor	3 493.0
Libanon	GFSA	8 931.6
Loraine	AV	8 734.0
Marievale	Gencor	1 006.0
Randfontein	JCI	27 059.0
St. Helena	Gencor	9 743.0
Stilfontein	Gencor	8 555.7
Unisel	Gencor	9 407.0
Vaal Reefs	AAC	81 501.4
Venterspost	GFSA	5 902.9
Western Areas	JCI	16 074.0
Western Deep Levels	AAC	37 200.1
West Rand Cons.	Gencor	3 893.0
Winkelhaak	Gencor	13 760.8
Other		6 615.0
Total		615 306.7

*AAC = Anglo American Corp. of South Africa Ltd.
Gencor = General Mining Union Corp. Ltd.
GFSA = Gold Fields of South Africa Ltd.
RM = Barlow Rand Ltd.
JCI = Johannesburg Consolidated Investment Co. Ltd.
AV = Anglovaal Ltd.

Source: Chamber of Mines of South Africa, Statistical Tables 1986, Johannesburg 1987.

cial requirements of these institutions, amounting to as much as 544 t net sales in 1979 or 276 t net purchases in 1981 and 181 t net purchases in 1986.
- The sales of the United States Treasury Department to redress the balance of payments, which attained a record level of 365 t gold in 1979. Although these sales have been suspended in the 1980s they are a potential source of supply.

Although Anglo American Corp. and its affiliates account for a large share of the primary gold production, it has relatively little impact on world gold prices because these are importantly influenced by the existing stocks of gold.

7.1.2 Market Shares of International Mining and Oil Companies

The industry analyses given above clearly demonstrate the international nature of mining and mineral trade. In order to give an impression of the market positions of the companies involved (Tables 7.5 and 7.6), some statistics on the largest international oil companies and the major mining companies will be presented. A comparison of earnings shows most clearly the significance of the oil firms, the largest of which, EXXON, headed the list of largest industrial corporations between 1980 and 1984 (1985/86 second), published in the American business journal Fortune. The indicators employed for the sequence of companies were their sales (revenues). Annual net income (net profit) and expenditures are also given (cf. Table 7.5).

In recent years, the international oil companies have diversified their activities substantially. After losing direct access to the oil fields in OPEC and other developing countries, they started to focus on oil refining and the distribution of petroleum products and petrochemicals. They also extended their activities to other energy subsectors (coal, electricity) and to the mining sector (copper, gold).

Some large mining enterprises have also pursued a similar strategy and have diversified above all into the smelting industry.

The increased involvement of oil companies in mining, through direct investment or equity participation, has led to an interlocking of the two major sectors of the private mineral economy: Exxon (Exxon Minerals), British Petroleum (BP Minerals), Shell (Billiton), Sohio (Kennecott), Atlantic Richfield (Anaconda) and Union Oil (Molycorp) acquired or established mining companies (not included in Table 7.6). By the mid-1980s, however, many of these oil companies had reduced or eliminated their activities in mining.

As can also be seen from Tables 7.5 and 7.6, the leading oil and mining companies differ greatly in terms of size: the oil companies' earnings and profits are larger by at least one order of magnitude. Five of oil's "Seven Sisters" are American oil firms (Table 7.5). The largest mining empire emerged in South Africa, where the Anglo American Corporation of South Africa Ltd. acquired direct and indirect participation

Table 7.5 Operation and financial data of the major oil groups 1986

	Exxon (1986)	Shell (1986)	Amoco (1986)	Chevron (1986)	BP (1986)	Mobil (1986)	Texaco (1986)
Crude oil supply (1000 BOPD)	1796	5029	810	1989	2738	1306	1146
Refinery runs (1000 BOPD)	3032	3220	914	1800	1895	1545	1934
Natural gas sales (million cu. ft/day)	5329	6205	2819	2329	637	3659	2705
Net income (million US-$)	5360	3714*	747	747	734*	1407	725
Expenditure (million US-$)	7200	6816*	3181	3018	4855*	3046	2369
Fixed assets value (million US-$)	49289	23706*	23706	22746	22356*	24304	21661

* 1 L = 1.45 US-$ (as of December 1986).
Source: Annual company reports 1987.

Table 7.6 Major international mining companies

	Company	Revenue 1986 (million US-$)	Net profit (loss) 1986 (million US-$)
1.	Anglo American Corp. of S. Africa Ltd. (AAC)*	n.a.	287 (1985)
2.	Broken Hill Pty Co. Ltd. (BHP)	8503	1009
3.	Barlow Rand Ltd.*	6580	301
4.	Rio Tinto-Zinc Corp. plc (RTZ)*	6298	474
5.	Alcan Aluminium Ltd.	6056	244
6.	Aluminum Co. of America (Alcoa)	5751	256
7.	Metallgesellschaft AG (MG)*	4973	16
8.	Fluor Corp. (St. Joe Minerals)	4660	(60)
9.	Preussag	4125	0.4
10.	Noranda Mines Ltd.	3545	43
11.	Alusuisse*	3446	(420)
12.	General Mining Union Corp. Ltd. (Gencor)	n.a.	236
13.	Kaiser Aluminum & Chemical Corp.	2221	(33)
14.	Consolidated Gold Fields plc*	1617	99
15.	Amalgamated Metal Corp. plc (AMC)	1543	1.5
16.	Inco Ltd.	1452	0.2
17.	Cominco Ltd.	1329	(152)
18.	AMAX Inc.	1307	14
19.	Asarco Inc.	1034	9.1

* US-$ exchange rate in June 1986: R = 0.40; L = 1.44; DM = 0.41; sfr = 0.48; can$ = 0.72.
Source: Annual company reports 1986.

in well over 100 enterprises, including mining enterprises, industrial enterprises, financing companies, insurance firms and business corporations. The market value of the R 30 million shares amounted to 204 billion Rand (approx. 100 billion US-$ in 1987). Anglo American's most significant subsidiaries are De Beers Consolidated Mines Ltd. (34.2%), Charter Consolidated Ltd. (29%), Mineral and Resources Corporation Ltd. (39%) and Johannesburg Consolidated Investment (40%).

7.2 Market Organizations

The markets for mineral commodities are characterized by the fixed locality of the deposits, low elasticity of supply with respect to price in the short run (as discussed earlier), frequently a high degree of vertical concentration, and price fluctuation. As early as the end of the nineteenth century, companies in several mineral industries started to make efforts to reach agreements at the international level in order to influence prices. Such measures were taken at an early stage in the case of copper, lead, zinc, tin, aluminum and nickel:

- for copper, the Sécrétan-Syndicate (1887–1889), the Amalgamated Copper Co. (1899–1901), the Copper Export Association (1919–1923), the Copper Exporters' Incorp. (1926–1932) and the International Copper Cartel (1935–1939);
- for zinc, the European Smelter Convention (1909–1914) and the International Zinc Cartel (1931–1934);
- for lead, the Lead Smelters' Association (1909–1914), the Lead Producers' Reporting Association (1931–1932) and the Lead Producers' Association (1938–1939);
- for tin, the Bandoeng Pool (1921–1925), the Malayan Tin Trust (1928), the Tin Producers' Association (1929) and the International Tin Control Scheme (1931–1946) with a Tin Quota Scheme and a Tin Buffer Pool;
- for aluminum, the European Aluminium Cartel (1901–1907; 1912–1915; 1923–1925; 1926–1931) and the Alliance Aluminium Compagnie (1931–1939);
- for nickel, the Entente du Nickel (1900–1914).

Prior to the Second World War, it was large companies that organized themselves; after 1945 agreements were reached at an intergovernmental level.

The following types of organizations can be distinguished:

- Producer organizations in the form of trade associations or intergovernmental agreements for the purposes of informing their members, conducting public relations and representing members' interests in UNCTAD commodity negotiations.
- Producer cartels, here defined as international organizations attempting to regulate competition on the supply side of the market and thus influence prices (this differs from the more restricted definition of cartels as organizations that actually have the power to limit competition).
- International study groups as a common discussion forum for producer and consumer countries.
- International commodity agreements in the form of contracts between producer and consumer countries to regulate the market.

An industry often has different types of market organizations at different points in time, as the history of the tin market shows. Between February 1921 and the beginning of 1925, the major producers, the Netherlands East Indies and the Federation of the States of Malaya, maintained a joint buffer stock, the Bandoeng Pool, in order to remove surplusses from the market to prevent a fall in prices. Then, at a conference in London in 1929, the Anglo-Oriental Mining Corp. established the Tin Producers' Association, which imposed "voluntary" production cuts on its members to maintain prices. In 1931, representatives of the major tin-producing countries, the States of Malaya, Bolivia, the Netherlands East Indies, Nigeria, and Siam signed the International Tin Control Scheme, which remained in force at least theoretically until the end of 1946. Finally in 1947, an International Tin Study Group based in The Hague, Netherlands, was set up, which also drew up a draft for an international tin agreement which was negotiated at a tin conference in 1953. Since 1 July 1956, the International Tin Agreement (cf. Sect. 7.2.2) has been in existence, the first commodity agreement on a metal market. Thus, three organizational forms emerged consecutively on the tin market: producer cartel, study group, and commodity agreement. After the termination of the present tin agreement (30.6.1989), however, the Association of Tin Producing Countries as a cartel will continue to exist and the establishment of a new Tin Study Group is expected.

7.2.1 Producer Associations

As already mentioned, in the initial phase of cartel formations on mineral markets, it was exclusively private mining companies that pursued a common price policy. Today, it is largely the government of the mineral-exporting countries that desire or reach agreements on international cooperation. Experience has shown that a prerequisite of effective participation in international mineral cartels is the control of the mining industry on the part of the governments concerned. This need not imply complete nationalization.

The requisite influence can be obtained by the following:

- Granting of export licenses for mining products, as control is only required for that portion of production that is exported.
- Compulsory sale of mining products to government establishments such as "mining banks".
- Government participation in mining companies.
- Mining legislation with provisions on government control in export and mineral trade.

The commodity cartels pursue active price strategies, which depending on the market conditions aim at:

- reaching agreements on minimum prices, which can for example be secured by means of control measures;
- publishing posted prices to be taken as recommended prices and as a basis for taxation;
- fixing producer prices.

Not all commodity markets, however, have proved suitable for the establishment of cartels, which presuppose a number of conditions as regards market structures, geological factors, and the areas of use of a given mineral, particularly:

- The mineral exporting countries must be to a large extent independent from the importing industrial countries.
- The major portion of supply must be concentrated in a few producing countries in order to minimize the problem of outsiders, e.g. producers in non-member countries.
- The producer countries must be strong enough economically to withstand possible trade sanctions by consumers.
- The potential mineral resources in other countries which could become minable if prices rise cannot be extensive.
- The technological possibilities for recycling must be limited, as the recovery of secondary metals is normally performed in consumer countries.

Mineral Markets

- The commodity in question must be the preferred one for certain areas of application in order to restrict substitution possibilities.

In view of the above listed prerequisites for the establishment of commodity cartels, it is understandable that only a few markets have as yet been affected. The main ones are crude oil, copper, bauxite, iron ore, wolframite, and uranium.

In response to the UN Resolution on the New International Economic Order (cf. Sect. 8.2.2), the producer cartels even want to establish an umbrella organization. On the instigation of OPEC, representatives of eight producer organizations met in Georgetown, Guyana in August 1977, to discuss the establishment of a joint secretariat and a council of producer organizations. In April 1978, the Council of Association of Developing Countries, Producers-Exporters of Raw Material (APEC) of 32 developing countries was formally established; it has not as yet, however, made its presence felt.

7.2.1.1 Petroleum Associations (OPEC, OAPEC)

The cartel of oil-exporting countries has been the subject of much discussion and has been viewed by many other mineral commodity-exporting countries as a prototype for their own attempts at regulating their respective markets. It is thus instructive to examine this cartel and to analyze its strategies and successes.

The Organization of Petroleum Exporting Countries (OPEC) was founded in 1960 as an emergency measure of the oil-exporting countries, which were facing a price collapse due to the oil glut. The multinational oil companies, above all today's EXXON, repeatedly lowered the posted price (from 2.08 US-$ to 1.76 US-$ per barrel petrol for Arabian Light, 34°API), thus curtailing government revenues of the leading exporting countries. During the First Arab Petroleum Congress of 1959 in Cairo, the oil ministers of Venezuela and Saudi Arabia proposed the establishment of a producer organization. In August 1960, the international oil companies demanded further price concessions, which induced the oil ministers of Iraq, Iran, Kuwait, Saudi Arabia, and Venezuela to arrange a spontaneous meeting in Baghdad from 10 to 14 September 1960. The original intention was merely to agree on a trading strategy toward the oil companies; instead the ministers surprisingly decided to sign the original agreement for the establishment of the OPEC.

The five Founding Members, which were accorded special status in the OPEC Statute, were soon joined by other exporting countries (Full Members): Qatar (1961), Indonesia (1962), Libya (1962), Abu Dhabi (1967, United Arab Emirates replacing Abu Dhabi in 1974), Algeria (1969), Nigeria (1971), Ecuador (1973), and Gabon (1975), raising the membership to 13.

According to Article 2 of the Statute, the principal aim of the OPEC shall be the coordination and unification of the petroleum policies of member countries and the determination of the best means for safeguarding their interests.

OPEC established the following bodies:

- The Conference as supreme authority, consisting of delegations representing the member countries normally headed by the oil minister concerned. It holds at least two ordinary meetings a year and its decisions on oil policy are formulated in the

form of resolutions. The Conference decides on the general policy of OPEC and determines the appropriate ways and means of its implementation.
- The Board of Governors, composed of Governors nominated by member countries and responsible for the implementation of Conference resolutions. It submits reports and recommendations to the Conference.
- The Secretariat with a Secretary General to conduct routine business, prepare the Conferences, and assist the OPEC Economic Commission. Owing to Switzerland's refusal to recognize it as an international institution, the Secretariat shifted its Headquarters from Geneva, where it was established in 1961, to Vienna, Austria, on 1 September 1965. Since 1 July 1978, the Secretariat has had a new structure: a Division of Research (with an Energy Studies Department, Economic and Finance Department, Data Service Department), a Personnel and Administration Department, a Legal Office, and a Public Information Department. The Secretariat is composed of about 60 officers appointed from member countries.
- The Economic Commission as a specialized body assisting OPEC in promoting stability of international oil prices at equitable levels.

Over the years, OPEC's policy has shifted focus. Since 1960 it has pursued the following consecutive strategies:

- Harmonization of national legislation for oil production, especially to standardize the franchise system and tax regulations (compilation of a Code of Uniform Petroleum Laws).
- Establishment of national petroleum companies to offset the power of the multinational oil companies.
- National participation in foreign oil companies by acquiring equity via government corporations.
- Raising taxes such as royalty, profit tax, or export duty for foreign enterprises on the basis of fixed posted prices.
- Nationalization of foreign oil companies by taking over their entire stock.
- Concluding service contracts (e.g. PERTAMINA model) and supply contracts with international oil companies.
- Administering oil prices for international trade.
- Setting production quotas for member countries to avoid excess supply.

The greatest source of dispute between the member countries, whose interests diverge widely, has always been OPEC's oil-price policy. Serious differences exist with regard to political systems, economies, and the population densities in the 13 OPEC countries.

Although they cooperated during the two price hikes of 1973/74 (1st oil-price crisis) and 1979 (2nd oil-price crisis), various groups disagreed, for example in December 1976, when Kuwait, Saudi Arabia, and the United Arab Emirates rejected a further 10% price rise and the price system was split until 1 July 1977. This led OPEC to form a Long-Term Strategy Committee, which in 1978 submitted proposals for a long-term price policy. Prices were to be adjusted according to three basic market features: rates of inflation, the dollar exchange rate, and growth rates in gross domestic products in Western industrialized countries. Algeria, Libya, and Iran refused to comply with this price adjustment mechanism as early as mid-1979. This refusal resulted in a second wave of substantial price rises, particularly due to the change of government in Iran.

In order to reduce the obvious oversupply of crude oil on the world market after 1981, OPEC attempted to reach several agreements of production quotas. An extraordinary meeting in Vienna (19–20 March 1982) decided to curtail the production of all OPEC countries as of 1 April 1982 to 17.5 million barrels oil per day (BOPD) (OPEC production in 1979: 32 million BOPD – December 1981: 21.5 million BOPD). The agreement on establishing national quotas required long consultations. In 1983, no quota was allocated to Saudi Arabia, who agreed to act as a swing producer. As of 1 November 1984 OPEC reduced the production ceiling to 16 million BOPD but the solidarity needed to sustain such production cuts proved inadequate. Despite repeated efforts OPEC could not prevail upon all its members to observe the quotas; particularly Iran and Libya did not abide by their promises. Instead, countries competed for shares of the market by granting price concessions. In 1983, the "marker" price for Arabian Light 34° API f.o.b. Ras Tanura was lowered from 34 US-/bbl to 29 US-$/bbl. At the end of January 1985 the OPEC Conference meeting in Geneva decided to abandon the once sacrosanct marker crude oil price and agreed on a set of official prices. In 1987 OPEC used a basket of seven crude oil prices to set its official $ 18 benchmark. In view of the gloomy demand prospects, OPEC continued with the quota system in 1986 and 1987 (16.8 million BOPD, set in August 1986 and 15.8 million BOPD from 1 January 1987, see Table 7.7).

By exploiting the dependence of the importing countries, however, the cartel was able to achieve considerable success, especially between 1973 and 1980:

- A continuous rise in income for the member countries (cf. Table 7.8), providing the financial means for rapid economic development. Saudi Arabia earned the major part, however, and the income of certain OPEC countries declined considerably after 1981, a trend which continued until 1986.
- Achievement of a unified petroleum export policy until 1982 when some disagreements with the production quota scheme emerged.
- Harmonization of national legislation with regard to oil production, particularly licensing contracts and tax provisions.

Table 7.7 OPEC national quotas as of February 1987 (in million BOPD)

Saudi Arabia	4.133
Iran	2.255
Venezuela	1.495
Iraq	1.466
Nigeria	1.238
Indonesia	1.133
Kuwait	0.948
Libya	0.948
United Arab Emirates	0.902
Algeria	0.635
Qatar	0.285
Ecuador	0.210
Gabun	0.152
Total	15.800

Table 7.8 OPEC government revenues (in billion US-$)

Country	1966	1971	1974	1980	1982	1984	1986	1987
Saudi Arabia	0.78	2.15	22.6	102.0	76.0	43.7	20	21.5
Iran	0.59	1.94	17.5	13.5	19.0	16.7	5	9.5
United Arab Emirates	0.10	0.43	5.5	19.5	16.0	13.0	7	8.8
Kuwait	0.71	1.40	7.0	17.9	10.0	10.8	6	5.4
Iraq	0.39	0.84	5.7	26.0	9.5	10.4	7	11.4
Qatar	0.09	0.20	1.6	5.4	4.2	4.4	1	1.9
Libya	0.48	1.77	6.0	22.6	14.0	10.4	5	5.7
Algeria	–	0.35	3.7	12.5	8.5	9.7	4	3.4
Nigeria	–	0.92	8.9	25.6	14.0	12.4	7	7.4
Gabon	–	–	0.7	1.8	1.5	1.4	1	0.9
Venezuela	1.11	1.70	8.7	17.6	16.5	13.7	7	7.4
Ecuador	–	–	0.7	1.4	1.2	1.6	1	0.6
Indonesia	0.24	0.28	3.3	12.9	11.5	11.2	4	4.5
Total OPEC	4.95	11.98	90.5	278.8	201.9	159.4	75	88.4

Sources: Petroleum Economist, London, March 1988; – Shell International Petroleum, London.

- Progressive nationalization of the petroleum sectors in the member states by continually increasing participation in the franchise areas of foreign oil companies up to complete takeover.
- Autonomous fixing of export prices for petroleum. The last time international oil companies were officially involved in price negotiations with OPEC countries was on 12 October 1973.

These successes of the OPEC countries were offset by a number of negative impacts on the world economy:

- The balance of payments situation of the oil-importing countries was affected critically, at least temporarily. The recycling of OPEC foreign exchange earnings functioned relatively smoothly with the extensive help of European banks after 1974, but was less smooth after 1979.
- The volume of world trade 1980–1983 contracted as a result of rocketing interest rates and global recession.
- The economic situation of developing countries dependent upon the imports of energy worsened dramatically, particularly from 1980 to 1984. Especially newly industrializing countries ran up enormous debts (Brazil, Argentina, South Korea).

In 1986, in response to an oversupply, the oil prices went down to 12 US-$ per barrel and less on the spot market (see Fig. 7.1). Several OPEC meetings were organized in order to solve the crisis. The main problem was the outsider production, and efforts were made to convince countries like Mexico, Oman, Egypt, Malaysia, Brunei, and even Norway to join the cartel. In the longer run, however, OPEC prospects are promising because more than 75% of all proven reserves are located in member countries.

In January 1976, OPEC initiated the establishment of a special fund for development aid (OPEC Special Fund) to issue loans to so-called non-oil-exporting countries

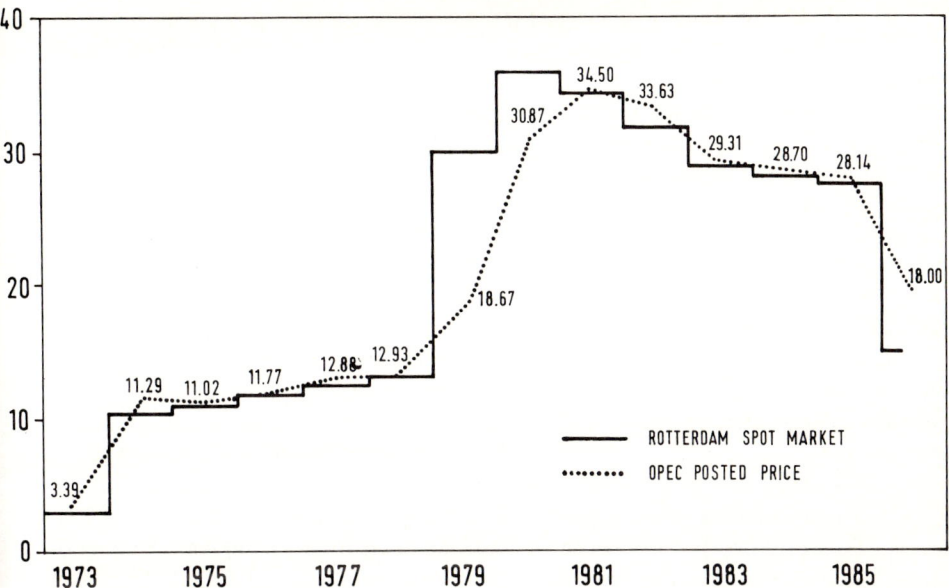

Fig. 7.1. Development of crude oil prices

(NOEC) for development projects (initial resources: 800 million US-$). After the 1979 wave of price rises, the Fund was increased in May 1980 from 2.4 billion US-$ to 4 billion US-$ and it was agreed that repayment should be direct to the Fund and not to individual OPEC countries. The Special Fund, initially to be administered by a Board of Governors in Vienna, was converted by statutory amendment in 1980 into an international corporation in its own right, the OPEC Fund for International Development, headquartered in Vienna. It grants balance of payments support, loans on concessional terms to finance projects and programs as well as grants for technical assistance. Since the Fund's operations were launched in August 1976, up to December 1987 it had implemented seven lending programs. By the end of 1986, the Fund approved 419 loans amounting to 2028 million US-$ for 44 African, 21 Asian, and 20 Latin American and Caribbean countries. However, owing to a decline in petroleum export revenues after 1982, the OPEC countries increasingly defaulted on their payments to the Fund, which in turn hampered the issuing of loans (by the end of 1986 the commitments stood at 3.2 billion US-$, but the disbursements reached 2.48 billion US-$ only).

In addition to the OPEC Fund, some OPEC countries grant substantial amounts of bilateral aid, though mainly to Arab (or Muslim) "brother states". The Kuwaiti Fund for Development, the Saudi Fund for Development and the Abu Dhabi Fund are particularly active. Of all bilateral Arab/OPEC development assistance in 1984, 77.8% was conferred to other Arab countries, mainly to Syria and Jordan. Other important recipients were Kenya, Senegal, and Bangladesh.

Besides OPEC, a second international oil cartel has been formed, based in Kuwait, the Organization of Arab Petroleum Exporting Countries (OAPEC). Founded at the beginning of 1968 as a loose association of the Arab OPEC members Saudi Arabia, Kuwait, Iraq, and Libya, it was joined by Algeria, Abu Dhabi, Dubai (the latter two to form the United Arab Emirates in 1974), and Qatar in May 1970. After an amendment of the statute, which provided for membership of non-OPEC Arab countries, Egypt (membership suspended in April 1979), Tunisia, Syria, and Bahrain joined at the end of 1971. Its supreme authority is a Council of Ministers, below this an Executive Bureau and a Secretary General in Safat (Kuwait), which oversees departments for petroleum projects, economics, exploration, training and international relations. Apart from purely economic goals, its intention has been since 1968 to apply oil exports as a political tool to strengthen the Arab world. Its political motivation became particularly clear in 1973/74 during the October War in the Middle East: while OPEC confined itself to pushing through price increases, the OAPEC countries cut production and imposed a politically directed embargo on supplies to the USA (October 1973–March 1974) and the Netherlands (October 1973–July 1974) in response to these countries' pro-Israeli position.

In 1974, the OAPEC announced the creation of a joint fleet of tankers, which in 1978 had eight tankers with a capacity of 2.1 million tdw, the Arab Maritime Petroleum Transport Co. In 1975, seven OAPEC members founded the Arab Shipbuilding and Repair Yard Co. which put Bahrain's first large-scale shipyard into operation at the end of 1977. The Arab Petroleum Investment Co. (1976), the Arab Petroleum Services Co. (1977), and the Arab Engineering Co. (1981) were founded to promote the autonomy of OAPEC countries in project financing and exploration.

In 1986, eight African oil states established the African Petroleum Producers Association (APPA), aimed at gaining national control of the continent's oil resources and to help non-oil African states to cut their expenditure on oil. The first ministerial meeting of the APPA was conducted in July 1987 in Algiers, Algeria. Founding members of the APPA are Algeria, Angola, Benin, Cameroon, Congo, Gabon, Libya, and Nigeria.

7.2.1.2 Copper Associations (CIPEC and Others)

After World War I, US copper mining dominated world trade, accounting for two-thirds of all exports. Only when new producers in South America and Africa entered the market and raised supply did it become necessary to support the price for copper by means of international cartel agreements. For this purpose, the Copper Exporters Inc. (CEI) was founded in 1926, a cartel consisting of mining companies from the USA, Chile, the Belgian Congo, and Great Britain.

Of interest is the strategy with which CEI intended to raise copper prices: explicitly rejecting production cuts, it set up a sales office in Brussels to circumvent middlemen (including the London Metal Exchange) by fixing prices in direct contractual agreements with the European consumers. At the outset of the Great Depression of mid-1929, it was quite successful and managed occasionally to monopolize copper prices. These efforts were, however, thwarted both by the Great Depression itself and the agreements reached at that time by outsiders, mainly Canadian and Rhodesian firms, who were able to compete ever more effectively with CEI. Between 1930 and 1935, the price of copper declined continually, ultimately leading to the establishment of the International Copper Cartel, which agreed on production and export quotas to curtail supply. The substantial price rise until 1938 was only in part attributable to the cartel, because demand also expanded due to armament programs, decisively improving their market position.

Although after World War II attempts were made at cooperation between copper producers as well as between producer and consumer countries, it was not until the beginning of June 1967 that concrete steps were taken by the governments of Chile, Peru, Zambia, and Zaire at a conference in Lusaka: they founded the Conseil Intergouvernemental des Pays Exportateurs de Cuivre (CIPEC). The four founding members were particularly important copper-exporting countries, whose economies were highly dependent on the copper trade. It was extended in November 1975, when, at the eighth Ministerial Conference in Lima (Peru), Indonesia was made a full member, Australia and Papua New Guinea being accorded associate membership. In 1976, Yugoslavia and Mauritania also decided to become associate members (though the latter withdrew again in August 1977).

Since 1977, there have been two kinds of membership:

- Full members, i.e. at present the founding members (Chile, Peru, Zaire, Zambia, and until 1986 Indonesia). They must be net exporters of domestic copper.
- Associate members, i.e. at present Australia, Indonesia (since 1986), Yugoslavia, and Papua New Guinea. They must be net exporters – or likely to be in the near future and accepted by the full members. Their budget contributions are only half those of full members.

CIPEC formed the following bodies:

- The Conference of Ministers, consisting of one minister responsible for mining from each member country, and taking place once annually. Here, the guidelines for future policy are laid down and the work of the Executive Committee is assessed. Motions relating to matters of overriding importance must be passed unanimously, others require a two-thirds majority.
- The Executive Committee (called Governing Board until 1977) consisting of at least one permanent representative from each member country, to meet four to five times a year and take decisions with two-thirds majority on current price policy measures, technical cooperation, coordination of training measures and the harmonization of marketing.
- The Secretariat, based in Paris and headed by a Secretary General to perform administrative functions, particularly the compilation of statistics, public relations work, publication of literature (Quarterly Review, Statistical Bulletin) and preparation of the Conferences. The Secretariat consists of four divisions dealing with marketing, research and development, economic problems, and administration.

In mid-July 1982, the eighteenth Ministerial Conference also created a Special Committee for cooperation with UNCTAD and the Integrated Program for Commodities (cf. Sect. 8.2.2.2) and abolition of trade constraints for copper and semi-finished goods.

The CIPEC defines its goals as follows:

Coordinate measure to increase earnings from copper exports.

- Harmonize cartel policy on problems relating to the production of copper.
- Improve the exchange of information and advisory facilities for members.
- Improve the socio-economic development of the member countries.
- Promote solidarity amongst the member countries.
- Promote the coordination of CIPEC policy with similar efforts by international organizations.

The measures taken to achieve these goals can be summarized as follows.

In order to avoid a fall in prices, an effort was made at the end of 1974 to curtail supply artificially by cutting production and exports. First, export quotas, 10% below those for the previous year, were imposed on each member country, followed by production quotas from April 1975 to the end of June 1976, lowering production by 15% compared to 1974. These initiatives were modestly successful; the price for copper recovered, which then, however, induced Chile to refuse further production cuts from mid-1976. In order to raise earnings for exporting countries, the CIPEC submitted proposals to UNCTAD in February 1977 to fix CIPEC producer prices based on LME quotations and US producer prices. Supply contracts concluded by the government marketing organizations of CIPEC members as of 1975 stipulated an agio or premium based on cartel recommendations to be added to the London Metal Exchange prices. Reasons given by CIPEC for the premium are that producers provide special delivery, e.g. copper supply at required places and at required quality. These surcharges for direct contracts reached sizeable proportions (10–20 L/t, in 1980 even more). Discussions at Ministerial Conferences on other measures, such as intervention on the LME or the setting up of a buffer stock have not as yet led to any concrete agreements. Finally in 1982, CIPEC recommended that members undertake individual price support measures, to which above all Chile and Peru agreed in principle, without, however, taking concrete steps. The CIPEC countries favor longer-term direct contracts with consumers, by-passing the metal exchanges, thus stabilizing prices and maintaining the system of producer premiums. Though CIPEC claims that it does not have the power to act as a cartel, the copper-consuming countries, especially the USA, consider CIPEC as a cartel, which means that no formal contacts have been established to US-based organizations like the International Copper Research Association because of the strict US antitrust laws.

CIPEC's efforts to coordinate marketing have been particularly successful. It has also improved market transparency by compiling and publishing statistics. Another important activity is CIPEC's close cooperation with UNCTAD; it has for years been conducting negotiations on an international copper agreement, vigorously representing the interests of producer countries. CIPEC wants to arrive at such an agreement while at the same time retaining the cartel.

7.2.1.3 Bauxite Association (IBA)

The mining and processing sectors for aluminum are for the most part geographically remote from one another. Although the aluminum industry disposes of many of its own deposits, these are in the mining countries, so that a large portion of bauxite production is traded on the world market: the exporting countries recognized the potential for a joint market strategy. At a conference in Belgrade in 1973 therefore, it was agreed to establish an association of bauxite exporting countries, which then took place in March 1974 in Conakry, Guinea: the International Bauxite Association (IBA) based in Kingston, Jamaica. In November of the same year, the seven founding members, Australia, Guinea, Guyana, Jamaica, Yugoslavia, Sierra Leone, and Surinam, were joined by the Dominican Republic, Ghana, Haiti (withdrawal in 1982 after suspending production), and Indonesia. In 1984 India joined the association. The IBA countries account for roughly 85% (1985) of western world mining production. Not including Australia, however, this share in world production has decreased by 13% to 44.5% (1985) since the establishment of the IBA. The members dispose of about 70% of the workable deposits.

The successes of OPEC in 1973/74 formed the precedent for the aims formulated by the bauxite association. After the founding members ratified the agreement establishing the IBA, they officially took up their work on 27 July 1975 with the following bodies:

- The Council of Ministers as supreme authority, consisting of the mining ministers responsible in the member states, to determine policy.
- An Executive Board comprising two representatives of each member country to meet at least three times a year and concern itself with the implementation of the Council's resolutions.
- The Secretariat in Kingston, Jamaica, headed by a Secretary General and divided into four departments for administration, economics, statistics and technical information to manage the Association's everyday affairs.

Article III stipulates the major objectives of the IBA:

- Promote the bauxite industry.
- Secure fair and reasonable returns for member countries from the exploitation, processing and marketing of bauxite.
- Safeguard the interests of member countries in relation to the bauxite industry.

As is the case for CIPEC members, the economies and above all the foreign trade earnings of many IBA countries depend to a decisive extent on bauxite exports. Only seven members have processing plants to produce aluminum oxide. The Guyana plant, however, was closed down in 1982. Aluminum plants are operating in six member countries. Unlike the majority of other minerals, bauxite cannot be processed economically everywhere, the production of aluminum being highly energy-intensive (13,000 to 15,000 kWh/t Al in modern plants). Numerous IBA countries, however, do not have access to cheap energy or to the financial means to develop it. It is true that since the establishment of IBA the share of its member countries in world aluminum production has more than doubled to just under 14% in 1985, yet this increase can almost exclusively be attributed to the expansion of Australian plant capacity. In spite of great efforts on the part of IBA and its member countries, the only new processing country within the cartel is Indonesia. In view of the present excess capacity in the field of aluminum oxide production, IBA does not expect an additional expansion of the capacity of its member countries in the near future.

The policy of the IBA also focuses on raising government revenues by imposing taxes and duties on the mining and exporting of bauxite. By means of a joint initiative by West African and Caribbean member countries in 1975, royalties and special export duties in most member states were raised considerably, in Jamaica, Surinam, Guyana, Haiti and in the Dominican Republic for example, from roughly 2 US-$ per ton in 1974 to 12 US-$ per ton in 1975. This tax burden on bauxite mining, however, pushed the margin of profitability to its limit. Between 1975 and 1985, production in these countries fell by as much as 42%. Prior to the economic difficulties of bauxite production in the 1980s, tax reforms had led to a substantial increase of government revenue. There is a growing pressure on the governments of the Caribbean, South American, and West African member countries to abolish export taxes or to reduce them drastically. Australia's decision not to levy such taxes has led to a decisive improvement of her competitive position and to an increase of production by 30% from the middle of the 1970s onwards, to the detriment of the other cartel members (cf. Table 7.9).

Since 1979, IBA has been recommending minimum cif prices for bauxite to its members and since 1980 for aluminum oxide. As well in the past, these recommended prices had been geared to the American posted prices for aluminum bars (99.5% Al). Though during the initial years the recommended minimum price had usually been realized, the situation changed in the middle of the 1980s, especially where aluminum oxide was concerned. In order to find a more realistic basis for its minimum price recommendations, IBA took into consideration the spot market price of aluminum in the USA and in Europe.

Additional emphasis was put on IBA participation policy. Since the establishment of IBA, national participation in the mining companies operating in IBA member countries has risen substantially.

The very participation of Australia in this commodity cartel is interesting. It views its membership as a means to counter radical trends and work towards harmonizing the interests of consumer and producer countries. Not only does it contribute the most in terms of technical know-how to IBA, it also accounts for the largest share of the 1.2 million US-$ budget.

Table 7.9 Tax burden on bauxite mining in 1985 (US-$/t)

Australia	1.20	Haiti	23.69*
Brazil	1.00	India	0.49
Dom. Republic	17.00–23.20*	Indonesia	1.05
France	0	Jamaica	13.80
Ghana	3.42	Sierra Leone	0.27
Greece	0	Surinam	15.00
Guinea	13.0	USA	0
Guyana	0.50	Venezuela	0

* 1982 before mining was stopped.

The Integrated Program for Commodities of UNCTAD mentions bauxite in its "Second Window", e.g. measures for stabilizing the market in the long term (cf. Sect. 8.2.2.2). Though skeptical in the beginning, IBA is now of the opinion that such a program would enable extensive cooperation between producer and consumer countries, provided that an agreement is signed.

The IBA also maintains contact with the trade association for primary aluminum producers, the International Primary Aluminium Institute (IPAI). Founded on 28 April 1972 and based in London, the IPAI is an international association of 49 smelting companies engaged in the production of primary aluminum from 24 countries. Its aims are to promote broader use of its products, study environmental health and safety, facilitate the exchange of information, support research, publish literature, and represent the aluminum industry in international negotiations. The Institute cannot therefore be classified as a cartel, its influence on prices being at best indirect.

7.2.1.4 Iron Ore Association (APEF)

In 1968, on the occasion of UNCTAD II in New Delhi (cf. Sect. 8.2.2.1), the Indian Government instigated the formation of an informal group aimed to establish an organization of iron ore exporting countries. After a preparatory meeting at ministerial level in November 1974 in Geneva, the Association of Iron Ore Exporting Countries (Association des Pays Exportateurs de Minerai de Fer, APEF) was formally established in April 1975 and its Secretariat in Geneva officially commenced business in October 1975.

APEF comprises exclusively founding members, which together account for roughly 50% of world trade in iron ore: Algeria, Australia, India, Liberia, Mauritania, Peru, Sierra Leone, Sweden, and Venezuela.

According to Article 2 of the Statute, membership is also open to Brazil, Canada, Chile, the Philippines, Swaziland, and Tunisia, none of whom, however, has as yet taken up the offer.

APEF's supreme authority is the Conference of Ministers, to meet once in 2 years. Its decisions require unanimous agreement. The Board, made up of representatives of member countries, meets twice a year and is concerned with the management of APEF and the implementation of the Conference's decisions. The Secretariat in Geneva deals with everyday business matters and particularly the procurement of information for member countries and the preparation of technical and economic studies.

The APEF has repeatedly denied being a cartel. It wishes to demonstrate that the legitimate interests of export countries can be safeguarded without jeopardizing those of importing countries. The main objectives of APEF are therefore to promote the iron ore export industry, to ensure the healthy growth of export trade in iron ore and to assist member countries to secure their export earnings and terms of trade. The agreement recognizes officially that the interest of importing countries need to be borne in mind. This is based on an awareness that active price controls on the world market for iron ore are hardly practicable. Physical scarcity of iron ore is unlikely in the foreseeable future. Even relatively small price rises can result in a number of deposits in various countries becoming minable. In addition, numerous developing countries

Table 7.10 Changes in world steel market, 1973 – 2000. Annual steel consumption, million tons

1973		2000	
1. USA	149.6	1. China (PR)	134
2. Japan	89.3	2. USA	94
3. Germany (W)	40.3	3. Japan	72
4. China (PR)	33.6	4. India	32
5. France	25.3	5. Germany (W)	31
6. United Kingdom	24.9	6. Brazil	29
7. Italy	22.6	7. Korea (S)	25
8. Canada	14.2	8. Italy	22

Source: Metal Bulletin, Oct. 1986.

accord priority to industrialization projects to set up steel plants supplied from domestic ore deposits, which curbs demand for iron ore on the world market. APEF has thus enjoyed little success so far. It has remained an information and consultation agency and now hopes during UNCTAD negotiations in the context of the Integrated Program for Commodities (cf. Sect. 8.2.2.2) to represent the interests of the exporting countries as effectively as possible.

Iron ore prices are usually fixed by annual negotiations between mining companies and steelmakers. For historical reasons, the European buyers settle earlier each year than the Japanese purchasers, resulting in some crucial differences (in 1986, the mines conceded a 2–3% price reduction for Japanese steelmakers) and tough competition. The miners and the APEF are aware of the problems the different settlement prices can create and try to solve them by closer consultations.

The world steel crisis has, however, exacerbated the position of iron ore-exporting countries even further, but the steel market is also in a period of structural change which will lead to major shifts in the trade of iron ore (see Table 7.10).

7.2.1.5 Tungsten Association (PTA)

At a meeting of representatives from various mining enterprises in La Paz, Bolivia, in April 1975, the Primary Tungsten Association (PTA) was formed. In 1987, it had an official membership of 14, both government and private enterprises domiciled in Australia (2), Bolivia (3), Brazil (1), France (1), Peru (1), Portugal (1), Rwanda (1), Spain (1), Sweden (1), Thailand (1) and Zaire (1). The Chinese trade corporation, Minmetals, is accorded observer status. PTA's base was originally to be in Brussels but a British consulting firm conducted its business until an independent Secretariat was established in London on 1 January 1982.

The organs of PTA consist of the General Assembly as supreme authority, the Executive Committee as management body, and the Secretariat.

Its activities cover the following areas:

– To represent tungsten industries in regional or international organizations or agencies.
– To advise and support UNCTAD in negotiations for an International Tungsten Agreement (cf. Sect. 7.2.2).

Mineral Markets

- To prepare market studies and statistics on the production and consumption of primary tungsten.
- To organize symposia to promote the tungsten industry (Stockholm September 1979; San Francisco, June 1982; Madrid, May 1985; Vancouver, September 1987).
- To cooperate with the tungsten consumers in the Consumer Reporting Group to introduce price indices. In July 1978 it was agreed to establish an International Tungsten Indicator (ITI) to serve as a guideline for world trade in tungsten concentrates in the hope of stabilizing prices and rendering them more transparent.
- To maintain contacts with the US stockpile agency GSA so as to influence the US stockpile disposal policy, especially surplus sales at "uncompetitive" prices.

In 1987, tungsten mines worldwide have been shutting down at such a rate that the PTA has decided to close at the end of 1987, at least temporarily.

7.2.1.6 Uranium Association (UI)

On 12 June 1975, 16 leading uranium mining companies established the Uranium Institute in London. The motivating force for establishing this association on the lines of a cartel was the uncertain constitution of the uranium market in the aftermath of the 1973/74 rise in energy prices.

The Uranium Institute now has 69 members from 18 countries and the European Community, both private and government enterprises as well as agencies of the electricity sector and the uranium mining sector. The members are from Australia, Brazil, Canada, China (Taipei), Japan, Belgium, France, Germany (F.R.), Finland, Gt. Britain, Sweden, Italy, Switzerland, Spain, South Africa, Gabon, South West Africa/Namibia, and the United States of America.

Interestingly, the Uranium Institute extended its membership as early as 1976 to include both producers and consumers of nuclear fuel. Since then the producer and consumer groups each have 16 seats in the Council of Management, which meets twice a year. The resolutions passed by the Council are carried out by an Executive Committee.

The Institute is not a price cartel; it is a forum for members of the industry to exchange market information. Its working program is wholly geared to obtaining information and is conducted by three specialized committees which have been producing studies since 1978:

- Committee on Supply and Demand, which prepares market studies and forecasts.
- Committee on International Trade in Uranium, which cooperates closely with atomic energy organizations like the International Nuclear Fuel Cycle Evaluation (INFCE) and International Atomic Energy Agency (IAEA).
- Committee on Nuclear Energy and Public Acceptance, which mainly researches into environmental problems associated with nuclear power and performs public relations work.

The Uranium Institute maintains a well-furnished specialist library in London and its Council of Experts helps to solve technological problems and provides advice on financing projects.

7.2.1.7 Other Producer Associations

There are producer associations in numerous metal markets which cannot be designated as cartels, since they do not aim at directly influencing prices by restricting competition, but are concerned rather with researching into application possibilities and disseminating technical information to stimulate demand. Though raising demand influences prices in the producers' favor, it is perfectly in keeping with the market and does not restrict any of its mechanisms.

Such producer associations for base metals have already been in existence for decades and in the last 10 years many more have been formed for minor metals. They are useful when preparing market studies, not only because they dispose of specialist libraries but also because they provide statistics on supply and demand.

Producer associations are of significance in the following markets:

Copper: the Copper Development Association (CDA) based in Potters Bar near London, England, has been in existence since 1903. In 1986, 38 important multinational copper mining and smelting companies from Australia, Britain, Japan, Canada, and the United States belonged to the CDA. A major aim is to support the development of the copper market by organizing the exchange of technical information, which it does through publications, trade journals, and conferences.

Lead, zinc, cadmium: since all three metals occur simultaneously in many types of deposits, their markets are also closely linked and the producer associations concerned interlocked. A clear example is the International Lead-Zinc Research Organization Inc., whose foundation in 1958 was largely instigated by the Lead Industries Association and the American Zinc Institute. It conducts basic research on the application of lead and zinc in order to develop new products.

The Zinc Institute Inc. (called the American Zinc Institute until 1968), a platform for American zinc producers and its European counterpart, the European Zinc Institute, pursue similar activities.

Lastly, worthy of mention are:
– The Zinc Development Association.
– The Lead Development Association.
– The Cadmium Association.

Based in the same office in London, these three organizations are largely concerned with disseminating technical know-how and promoting market transparency by organizing seminars, publishing technical reports, and compiling statistics.

To further cooperation between producers and consumers, UNCTAD set up the London-based Lead and Zinc Study Group in 1960.

Nickel: in November 1985, UNCTAD organized a meeting of some 30 nickel producers and consumers in Geneva to confer on the establishment of an international study group to monitor the world nickel market.

Aluminum: for details on the International Primary Aluminium Institute (IPAI) see Section 7.2.1.3.

Magnesium: the International Magnesium Association (IMA) based in McLean, Virginia, USA (previously in Dayton, Ohio), was founded in 1943. Its regular members are enterprises producing, trading, or consuming magnesium. It also accepts those enterprises supplying the magnesium industry as associate members. It holds an annual Magnesium Conference, publishes a journal, Magnesium, and answers members' specialized inquiries.

Mercury: the desire voiced in 1974 by major mercury-producing countries for supraregional cooperation led to the establishment of the Association Internationale des Producteurs des Mercure (Assimer) on 16 April 1975 in Geneva. The members, Algeria, Italy, Yugoslavia, Mexico, Spain, and Turkey, who control roughly 70% of the market, are aware of its weakness due to the availability of substitutes and environmental legislation limiting its use. Its activities and achievements are therefore extremely limited.

Bismuth: founded at the end of 1972 by six large producer countries in La Paz, Bolivia, the Bismuth Institute compiles and disseminates technical information.

Cobalt: the leading cobalt-producing countries decided on 9 November 1981 to establish a "world cobalt institute", the Centre d'Information de Métaux Non Ferreux (CIMNF) based in Brussels and allotted the task of researching into new areas of application and preparing market analyses. It was initiated by Zaire, Zambia, the Philippines, Morocco, and Finland and aims at close links between producers as well as consumers of cobalt.

It succeeded the "Centre d'Information du Cobalt", which terminated its activities on 20 May 1976 in Brussels after a period of 20 years.

Tantalum, niobium (columbium): the Tantalum-Niobium International Study Center (TIC), largely instigated by tantalum and niobium producers, was established on 24 October 1974. The 67 members (as of 1983) account for over 90% of noncommunist world production. Since its foundation, a sizeable number of consumer enterprises have become members. It has set itself the task of disseminating statistics and technical information on the tantalum and niobium markets and organizing congresses. It explicitly refrains from influencing prices. The TIC in Brussels is under the direction of a Secretary, who receives his directives from the Executive Committee and the General Assembly. Until May 1986, the TIC was named Tantalum Producers' International Study Center.

Selenium, tellurium: since 1963, the primary producers of tellurium and selenium have maintained a Selenium-Tellurium Development Association, Inc. (STDA) with an information agency in Darien, Connecticut, USA (predecessor: Selenium Development Committee, 1938–1963). Its membership includes such important mining companies as AMAX, Asarco, Inco, Phelps Dodge, Noranda, Centromin/Peru, Boliden, Mitsubishi, and Nippon Mining. Via its Market Development Committee it concentrates on promoting consumption, including sponsoring research projects which are conducted in university laboratories in order to extend the areas of application for selenium and tellurium.

Gold, silver: The Silver Institute Inc. was established in New York City in January 1971 and has now (1987) 86 members from 15 countries. The institute advances the interests of the silver industry by developing and distributing information on silver uses, research, technology, and market conditions. In 1976, The Gold Institute was incorporated in Canada and has today (1987) 64 members from gold-mining companies and gold manufacturers in eight countries. It collects and publishes statistics and other information on production, marketing and uses of gold. The secretariat of both The Silver Institute and The Gold Institute is located in Washington, D.C., Connecticut Avenue.

Finally, one more international association founded in London by traders dealing in minor metals should be mentioned: the Minor Metals Traders Association (MMTA). The MMTA was created in 1973 to encourage and foster the activities of traders in minor metals with a view to maintaining fair trading practices and healthy competition. The Association comprises about 60 members, most of whom are London-based companies trading in metals such as antimony, mercury, cadmium, gallium, germanium, or silicon.

7.2.2 International Commodity Agreements

International commodity agreements are signed between countries producing and consuming a particular commodity for the purpose of regulating the market. They differ primarily from cartels in that producers and consumers dispose of parity voting rights in the decision-making bodies. The instigation of such agreements after World War II was initially the concern of the Interim Coordinating Committee for International Commodity Agreements of the United Nations Economic and Social Council and then of UNCTAD (cf. Sect. 8.2.2.1).

Such activities on mineral markets have for 30 years been concentrated on tin, copper, and tungsten; through the Manila Declaration (1979) they now also encompass bauxite and iron ore.

Concluded between the governments of producing and consuming countries, international commodity agreements invariably represent a compromise between the commodity policy goals of both parties, in which international development assistance also plays a part.

The major aim of the producing countries (export countries), the large majority of which are developing countries, is to secure foreign exchange revenue, i.e. to attain the highest and most stable world market price. The major aim of the consuming countries (importing countries), mainly the industrialized nations, is to ensure that the market is adequately supplied at reasonable prices.

The contracting parties thus concur as to improving market transparency and stabilizing prices, disagree, however, on price fixing and above all production quotas.

The measures agreed upon can be classified as follows:

– Compiling detailed statistics and comprehensive market studies to improve market transparency.
– Promoting research and development in the mining and use of the minerals concerned, including the organization of trade conferences and the publication of research findings.
– Minimizing price fluctuations by manipulating supply and demand via a buffer stock.

– Guaranteeing minimum prices by fixing supply quotas or by agreeing on production and/or export quotas for the producing countries.

The only concrete agreement hitherto reached is the International Tin Agreement, which was for a long time viewed as a prototype for future contracts. Its instruments and mode of operation will thus be examined.

a) The International Tin Agreement. The first Tin Conference took place with the assistance of the UNO in Geneva from 25 Oct. to 12 Nov. 1950. The Tin Study Group submitted a draft for an International Tin Agreement, which, however, required revision before the 30 participants of the second Tin Conference in Nov./Dec. 1953 could agree on a text. After much discussion it was signed and ratified and the 1st International Tin Agreement (ITA) came into force on 1 July 1956 for a period of 5 years. The following Agreements have since come into being:

1.7.1961–30.6.1966: 2nd International Tin Agreement
1.7.1966–30.6.1971: 3rd International Tin Agreement
1.7.1971–30.6.1976: 4th International Tin Agreement
1.7.1976–30.6.1981: 5th International Tin Agreement
1.7.1981–30.6.1982: Extension of the 5th ITA
1.7.1982–30.6.1987: 6th International Tin Agreement
1.7.1987–30.6.1989: Extension of the 6th ITA.

The members of the Agreement are grouped into producing countries comprising Malaysia, Bolivia (until 1982), Thailand, Indonesia, Nigeria, and Zaire, who were joined in 1971 by Australia, and consuming countries, whose membership constantly fluctuates. The 5th ITA enjoyed the largest participation with 23 consuming countries. Significantly, membership included not only all leading Western industrialized nations, such as the USA, Japan, and the EC countries but also numerous Eastern Bloc (COMECON) countries, such as the Soviet Union, Hungary, Poland, Bulgaria, and the CSSR. Even India joined the 5th Agreement as a consumer country. The Federal Republic of Germany was not ratified as a member of ITA until July 1971, the United States not until as late as July 1976. The USA remained a member only for the 6-year duration of the 5th ITA and withdrew in 1982 because it opposed import controls.

The Agreement's main objectives are to prevent excessive fluctuations in the price of tin, to help increase the export earnings from tin, and to secure an adequate supply of tin at prices fair to consumers and remunerative to producers.

Its bodies are:

– The International Tin Council (ITC), in which the producer group and the consumer group dispose of 1000 votes each. It consists of one delegate per member and holds four sessions a year. The membership of Producing Members and Consuming Members is based respectively on their domestic mine production and their consumption of tin metal. The Council appoints an independent Executive Chairman, who presides over meetings but has no vote.
– The Secretary of the Council, with a permanent secretariat in London, whose task is to implement the Council's resolutions and publish statistics.
– The Buffer Stock Manager to manage the buffer stock, who, like the Secretary, is appointed by the ITC.

The ITC's most important instruments to regulate the market are the buffer stock and export controls.

The *buffer stock* is a stockpile to stabilize prices by manipulating supply and demand. Such an instrument alters the quantities of minerals available for purchase without completely neutralizing market mechanisms.

The volume of the buffer stock was the subject of frequent discussion; it was to be extensive enough to be effective for as long as possible and at the same time it was limited by the financial resources available to sustain it. Both considerations must be viewed in relation to one another. At first, the buffer stock was provided with 25,000 t of tin or the equivalent in money (at that time L 16 million), contributed exclusively by the producing countries. The 2nd ITA's buffer stock contained 20,000 t (L 14.6 million) the 3rd and 4th ITA Pools the same volume, which, however, meant an increase in value to L 20 million and L 27 million respectively. During the 5-year period of an agreement, however, the purchasing power of the Pool Manager declined as tin prices rose.

For the first time, the Consuming Members France and the Netherlands made voluntary contributions towards financing the buffer stock in 1971/72. In the 5th ITA a volume of 20,000 t for compulsory contributions from the Producing Members and the same amount for voluntary contributions from the Consuming Members was agreed upon. Voluntary contributions, however, were made only by Great Britain, France, the Netherlands, and Belgium. The 6th ITA finally raised the Pool to a nominal 50,000 t of tin, 30,000 t of a "normal stock" to be furnished by compulsory member contributions, 20,000 t of an "additional stock" to be financed from borrowing, using as security stock warrants and government guarantees.

The management of buffer stock operations is subject to directives stipulated in Articles 28–30 of the Agreement. Floor and ceiling prices for tin metal are fixed so as to keep prices stable. The level to be set has always been heavily disputed and in the history of the Agreement often revised. In 1956 the floor price was fixed at L 640 per ton, the ceiling at L 880 per ton. After the oil price increase and the subsequent mineral boom in the second half of the 1970s, the floor price was raised so rapidly that it doubled over the 6-year period 1975–1981 (31.1.75: 14.89 Malaysian /kg; 17.10.81: 29.15 M$/kg). In the 6th ITA the floor price is 29.15 Malaysian Ringgit (M$) per kg ex-works Penang (equivalent of approx. 5.60 US-$/lb), the ceiling price 37.89 M$ per kg. The range between the floor and ceiling prices is 30% of the floor price and is divided into three equal sectors determining the scope of the Buffer Stock Manager's actions.

Export quotas are a very effective instrument for reducing supply and raising prices, representing, however, a form of intervention not in keeping with the market mechanism. They are applied when the tin price threatens to fall through the floor. When 70–80% of the maximum volume is held in the buffer stock, the ITC declares a 3-month control period, in which each producing country is subjected to an export quota.

The ITC has repeatedly had to resort to export controls to guarantee the floor price. The first time was between 15 December 1957 and 30 September 1960, when export quotas were imposed, then again from 19 September 1968 to 31 December 1969 and most recently from 27 April 1982 to 31 March 1987. This latest period is indicative of a market imbalance which could not be corrected, despite a buffer stock of 50,000 t.

Above all, the limited elasticity of supply with respect to price in the short run causes substantial short-term price fluctuations. The buffer stock has proved effective in minimizing such price fluctuations and since market mechanisms are not neutralized, only market figures altered, this instrument can be viewed as conforming to the market. The efficency of the stock depends on its size. A volume of 20,000 t tin proved inadequate and was therefore raised to 50,000 t in mid-1982 (roughly 25 % of annual world production). The 70,000 t proposed by the USA could not be financed. The buffer stock is, however, ineffective when the market is subject to structural imbalance, as occured in 1982, when it had to be stockpiled with 50,000 t surplus supply in 6 months without achieving the intended effect.

On 24 October 1985 the ITC collapsed with buffer stock holdings of approximately 62,000 t and further 50,000 t or so of forward contracts with LME brokers. The roots of the problem which led to the demise of the International Tin Council and the subsequent closure of the LME tin market go back to at least 1982 or even 1972.

The following main factors of influence should be mentioned:

- The continuous decrease of world total primary tin consumption from 173,220 t in 1979 to 140,625 t in 1983, which led to a substantial oversupply.
- The revision of buffer stock price limits between 1975 and 1981. In this period the ITC raised the floor price nine times from 14.89 M /kg to 29.15 M /kg but rejected all proposals for an adjustment after 1982, when the market changed from shortage to oversupply.
- The increasing outsider production caused by Bolivia (resigned ITA membership in 1982) and countries with new tin mining capacities like Brazil (Pitinja), Canada (East Kemptville), Great Britain (Cornwall), Peru (San Rafael) and China. This development reduced the share of the ITC producing members in the world tin market from 71 % in 1981 to 54 % in 1985.
- The fluctuations on the international money markets because in 1972 the ITC decided to denominate the buffer stock price ranges in Malaysian Ringgit (M) rather than in L sterling, but the buffer stock funding remained in L and the Buffer Stock Manager (BSM) continued to operate on the London Metal Exchange (LME). The M is linked to the US-$ and since 1980 the M$ tin price in Kuala Lumpur (KLTM) has fallen but the LME tin price has risen. The BSM realized paper profits which he used for additional borrowing power. The fall in US-$ and M$ in 1985 provoked nervous reactions on the part of the London brokers and bankers.
- The US Stockpile influence on the tin market because the disposals of surplus tin continued over the years despite the clear indications of surplus and increased the pressure on the tin price.

The consequences of the tin crisis are rather complex; three main facts are of special interest:

- The future of tin mining in traditional producing countries. The price collapse forced many mines in Malaysia, Thailand, and Bolivia to close down. In Malaysia, the number of operating tin mines decreased from 460 in September 1985 to only 175 at the end of June 1986. In Thailand, 274 out of the 580 registered tin mines were closed in April 1986 and the remaining were operating with reduced capacity. In Bolivia, the state-owned COMIBOL reduced the labor force from 27,000 to

11,000 and ceased operations in well-known tin mines like Catavi-Siglo XX or San José/Oruro. In fact, the tin mining industry is undergoing a major structural change. The exploitation of large-scale primary tin deposits will replace small-scale alluvial mining.
- The future of the London Metal Exchange. The suspension of tin trading at the LME led to a loss of confidence which resulted in a reduction of turnovers of no less than 50%. All other metals traded on the LME such as copper, lead, zinc, aluminum, nickel, and silver are affected in a similar way. As the mining industry in many developing countries is nationalized, the governments will attempt to negotiate with consumers independently.
- The future of commodity agreements in general. The failure of the International Tin Agreement will lead to the resumption of discussions about the necessity of commodity agreements. The opponents will use the behavior of the ITC and the BSM as an argument both for objecting to the establishment of future agreements and for the funding of the Integrated Program for Commodities.

The LME tin price, which had exceeded L 10,000/ton in February 1985 decreased by over L 2000/ton by 24 October 1985 and fell as low as L 3500/ton (14 M$/kg) in May 1986. Every L 1000/ton drop in the tin price will cost the ITC about L 112 million. The total loss to the ITC is estimated at L 350–400 million (optimistic sources estimate L 200 million only) and it will have to pay interest to both banks and brokers which could be L 30 million per annum. The 22 governments that signed the 6th ITA have to settle the debt problem. Experts are of the opinion that this may take many years. As a result, the 6th ITA might be the last international commodity agreement in metals for a long time. It seems possible, however, that an International Tin Study Group will be established under the auspices of the United Nations as a forum for discussion between tin-producing and consumer countries.

Another of the Tin Agreement's aims was to encourage exploration and protect tin deposits from destructive exploitation, such as highgrading (selectively mining the high grade ore only). A number of approaches have been adopted to attain these goals. In 1974, the leading producing countries of SE Asia, Malaysia, Thailand, and Indonesia, decided with the support of ESCAP to found the South East Asia Research Development Centre in Ipoh, Malaysia (cf. Sect. 8.1.1). The Technical Conferences organized by the ITC succeeded in making important contributions to disseminating the latest results of scientific research in all areas of tin mining and use.

As yet, international commodity agreements contain no provisions for an equitable distribution of raw materials in the case of an acute scarcity of supply. A system of quotas to ensure a balanced supply for consumers is still needed.

The recession in the industrialized countries between 1980 and 1983 had also curbed the consumption of tin world-wide and put producers under heavy pressure. Malaysia, Indonesia, Thailand and above all Bolivia were dissatisfied with the results of the negotiations for the 6th ITA and the USA's refusal to take part. This has induced the producing countries to form a new association on the lines of a cartel. On 2 April 1983 in London, Malaysia, Indonesia, Thailand, Bolivia, Australia, Nigeria and Zaire signed an agreement for the formation of an Association of Tin Producing Countries (ATPC). After the termination of the extended 6th ITA on 30 June 1989, the ATPC might be the only international organization on the tin market. The first substantial market intervention of the ATPC was the introduction of export quotas for member countries as of April 1987. The total export of ATPC members should not exceed 96,000 t tin per year, and in September 1987 Brazil has agreed to maintain its exports at 21,000 tonnes per year in line with the ATPC supply rationalization scheme.

b) Preparatory Meetings for Other Agreements. In the context of the UNCTAD commodity talks, negotiations on further international agreements for mineral commodities are being conducted.

Since spring 1976, a copper agreement has been under discussion in 17 UNCTAD intergovernmental meetings (Preparatory Meetings on Copper); no concrete result has as yet (1987) been reached. In 1978, the producer and consumer countries almost agreed to the creation of a forum for consultations between the groups to improve market transparency, but the developing countries feared that such an institutionalized consultation mechanism might permanently replace an agreement and rejected the proposal. The situation on the copper market is difficult: on the one hand, the developing countries' share of the world market is not sufficient to effectively manipulate supply, and on the other, a buffer stock to stabilize prices is too costly. According to expert opinion, a copper buffer stock would have to contain 2 million tons, the equivalent of approx. 3 billion US-$. A pool of this size would seem prohibitively expensive. Nevertheless, efforts are still being made to find a solution in the context of the Integrated Program for Commodities.

So-called Preparatory Meetings were also instigated for bauxite (1982), manganese ore (1977, 1980), and phosphate (1977, 1978). Leading producers did not, however, attend, thus indicating a clear lack of interest.

Preparatory Meetings (first in October 1977) and two sessions (1978) of an Ad-hoc Intergovernmental Group of Experts on Iron Ore were also organized for iron ore under the auspices of UNCTAD, during which various measures were agreed upon to promote market transparency, compile statistics, prepare studies, and discuss questions of sea transport. The latest meeting took place in April 1984 and discussed only in general the need for a regular dialog between iron ore exporting and importing countries.

Negotiations on tungsten have a particularly long tradition. After the establishment of a UN Committee on Tungsten in January 1963, UNCTAD took over responsibility in May 1965 and formed the UNCTAD Committee on Tungsten, which since that time has convened almost annually; it publishes the journal "Tungsten Statistics", organizes conferences for exporting countries, and has formulated a draft text for an international tungsten agreement. The Tungsten Committee comprises at present 29 countries, on the production side for instance, Australia, Bolivia, the People's Republic of China, Portugal, South Korea and Thailand, and on the consumer side, the USA, the Federal Republic of Germany, Japan, and Great Britain.

The draft of the agreement provides for floor and ceiling prices, a buffer stock and possible export control: instruments already employed by the International Tin Agreement. In addition, an UNCTAD Preparatory Working Group was established in 1978 to prepare a negotiating conference for the International Tungsten Agreement. The rift between consumers, demanding above all greater market transparency, and the producers, who regard an exchange of information as feasible only within the context of an agreement, could still not be overcome.

In reaction to the sluggish negotiations, the producers have initiated the Meetings of Tungsten Producer Countries, which may lead to a pure producer association parallel to the Primary Tungsten Association. Until now, Australia, Bolivia, South Korea, Peru, Portugal, Rwanda, Spain and Thailand have been attending.

7.3 Mineral Pricing

A price is established on the markets when goods or services are exchanged. Pricing takes various forms, depending upon the form of market, such as for example,

- a competitive price formed under conditions of perfect or nearly perfect competition;
- a monopoly price determined by a sole supplier or major supplier according to his price-demand function – it is usually higher than the competitive price;
- an oligopoly price fixed by the oligopolists according to their own profit functions and the estimated parameter of action of the competitors. It usually ranges between competitive and monopoly prices.

Pricing can also result from cartel agreements and it can be on an organized (exchanges) or informal basis.

Most minerals in standard brands are fungible goods, i.e. goods whose number and weight are stipulated in the trade and are thus exchangeable and negotiable worldwide, hence the existence of a number of metal exchanges where pricing occurs in an organized way. Private controlled price formation is also common, particularly on commodity markets with limited competition, where monopolistic or oligopolistic price setters lead the market or producer prices are fixed.

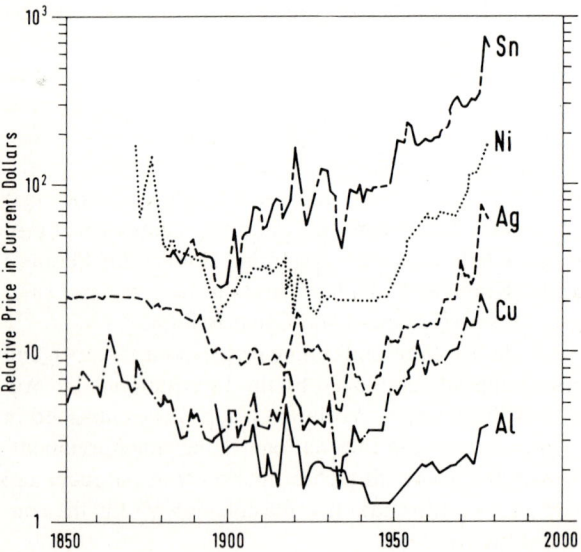

Fig. 7.2. Relative prices in current dollars for major metals. (After Petersen and Maxwell 1979)

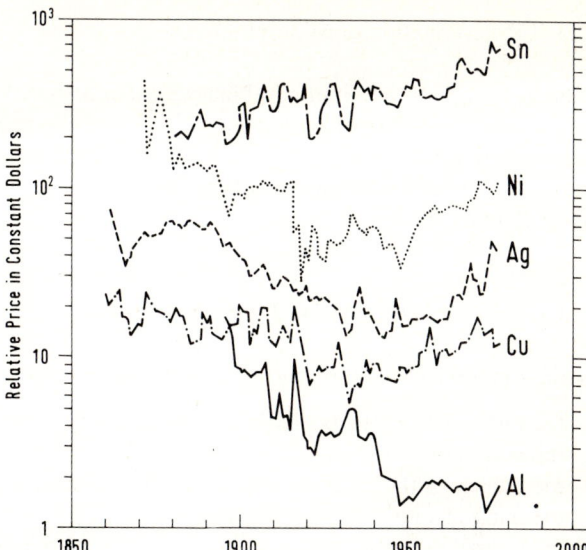

Fig. 7.3. Relative prices in constant dollars for major metals. (After Petersen and Maxwell 1979)

The pricing of mineral commodities can thus occur in four different ways:

- established on an exchange according to competitive supply and demand;
- regulated by international cartel or commodity agreement;
- negotiated between producers and consumers;
- fixed by monopolistic or oligopolistic producers as posted prices.

The long-term price movements of five selected non-ferrous metals is presented in Figs. 7.2 and 7.3.

Strictly speaking, however, there is no perfectly competitive pricing on mineral commodity markets. The reasons for this have to do with both geology and the structure of the markets. The margin of adjustment of mineral production to changes in demand is very narrow; mining reacts to increases in demand, for example by exploring for and opening up new deposits, which involves a delay of 5 to 10 years. Supply on mineral commodity markets in relation to price is thus inelastic, at least in the short term.

In addition to short-term price fluctuations, a further issue on mineral markets is the general price trend. For developing countries exporting minerals, the absolute rise or decline of prices is not the only important consideration: the ratio between mineral prices and those for industrial goods is equally significant. For decades, the terms of trade of countries producing and exporting commodities worsened until the period 1973–1981, when the situation altered, at least for the oil-exporting countries. The OPEC countries thus demand a parallel price trend for petroleum and industrial goods.

The quantities offered and demanded are determined by a number of factors, which will be dealt with in the following.

7.3.1 Determination of Mineral Supply

The quantitiy offered in mineral markets is influenced by costs, which in turn are determined by a variety of geologic, technological and economic factors, usually having a long-term effect on the production trends. There are also a number of special short-term determinants which can often destabilize the market and induce substantial price fluctuations.

7.3.1.1 Factors Determining Production Trends

The essential factors of importance for the long-term supply trend are:

- the mineral reserves available, i.e. for the long-term supply of minable deposits of the mineral;
- the quality of the reserves, the ore grade, the associated minerals, the grain size of the minerals, etc.;
- the type of deposits, above all the geometry, depth, and size of the ore body;
- the paragenetic association of minerals, such as the simultaneous occurrence of copper and cobalt, copper and molybdenum, lead and silver, or zinc and cadmium;
- the mining and processing technology available, including the efficiency of extraction of the valuable minerals, the relative importance of capital and labor, and the operational capability;
- the possibilities for financing new mineral exploration and mining projects;
- the infrastructural conditions in the regions in which the deposits are located, i.e. the accessibility for mines and processing plants;
- the overall legal and economic conditions for mining in mineral-producing countries.

7.3.1.2 Special Supply Factors

A number of market conditions can, in the short term, increase or reduce supply, thus affecting prices. Although these impacts are especially difficult to forecast, they are indispensable for any market analysis.

a) *Stocks*

The increase and decrease of stocks can have a substantial effect on short-term supply in mineral commodity markets. Stockpiling takes two basic forms depending on functions and aims:

- commercial stocks of industry to furnish the production process with material;
- government strategic stockpiles to ensure the supply of vital sectors of industry in the case of scarce supply resulting from political instability.

Commercial Stocks. At al levels of production and trade, the industry maintains stockpiles. For metals, commercial stockpiles are usual

- in mines, which "hoard" concentrates for reasons of transport or earnings expectations. Here, small and large enterprises differ greatly: only the latter can afford to maintain stockpiles for speculative purposes, while the former depend on immediate receipt of income. This is one point where oligopolists and monopolists can influence the market;
- in smelting plants, which can store concentrates as well as the metals after smelting and refinement. Possibilities for speculation are, as a rule, greater for smelting than mining companies;
- in the processing industry, which varies its stocks simply for reasons of supply but also in anticipation of price movements;
- in the licensed warehouses of the metal exchanges. As opposed to commercial stockpiles, reliable information is available on these stocks in government statistics. Market analyses should therefore take into account metal exchange stocks at least (see Table 7.11). Price analyses have demonstrated that changes in stocks at the LME and COMEX can have an impact on the market and that they are also capable of reacting to it (Müller-Ohlsen 1981).

As mentioned, commercial stocks are also influenced by the price expectations of large producers and manufacturers. Only changes in the quantity of mineral commodities stored is of importance for pricing, as these can have a stabilizing as well as destabilizing effect on prices (see Table 7.11). Stockpiling can stabilize prices if stocks are reduced when prices rise, because the additional supply will limit the price increase. The opposite can also be the case, however, if further price rises are anticipated and stocks are built up. The size of commercial stocks is also affected by the rate of interest. High rates of interest lead to reductions in stocks, particularly in the case of expensive mineral commodities such as tin, silver, or tantalum. It is important to point out once again that a part of consumers' commercial stocks serve as minimum reserves that are technically necessary and cannot be disposed of at will. Only the quantity exceeding this minimum can influence pricing. The size of commercial stocks depends therefore on the price of the mineral commodity, the demand trend, the price expectations and the procurement risk, the latter being determined by the safety of transportation routes or the political stability in the exporting countries.

Table 7.11 Stocks in London Metal Exchange warehouses and various other commercial stocks of tin metal (except US Stockpile and ITC Buffer Stock); in tons

Location	1983	1984	1985	1986
LME: warehouses	42750	22520	57080	51610
in transit to USA	1316	1010	1281	921
USA: importers	608	761	1642	222
USA: consumers	7740	8206	9440	3810
Japan: consumers/dealers	2980	2495	2947	3268
Germany: smelter	49	26	85	68
Germany: consumers	1566	1331	1642	1503
Thailand: smelter	193	960	962	2722
Brazil: smelter	114	212	502	1148
Malaysia: smelter and port	15005	21951	9006	12180

Source: International Tin Council, Quarterly Statistical Bulletin, London, June 1986.

Strategic Stockpiles. Government stockpiles are built up to ensure material supplies in times of national emergency and are orientated above all to the degree of dependence on imports and the (political) supply risk. According to the United States Strategic and Critical Materials Stock Piling Act, a stock of strategic and critical materials is to be maintained to decrease dependence upon foreign sources of supply in times of national emergency (cf. Sect. 9.1.4). Here, only those influences exercised by strategic stockpiles on pricing will be considered.

The USA began setting up a strategic stockpile in 1939 and intensified purchases after World War II and the Korean War (1950/51), which led in certain commodity markets to "stockpile booms". The authority responsible at that time, the Reconstruction Finance Corp. (R.F.C.) was able to dictate prices in bilateral supply contracts in 1953 and 1954.

The stockpile purchase on the tin market (1950–1956) contributed decisively to the excessive price rises in boom periods; they helped to stabilize prices, however, on falling markets. At the end of the 1950s, the build-up of the US Stockpile led to overproduction on the tungsten market, which resulted in the closure of nearly 700 US mines in the early 1960s.

After stocktaking in 1962, the General Services Administration (GSA) was officially instructed to conduct purchases without influencing commodity prices; this however proved impossible on some markets. During periods of scarce supply of antimony and tungsten, due to declining Chinese exports for example, the price for both metals was determined by the fixed "Stockpile prices" of GSA sales. Reportedly, the market price of tin between 1966 to 1973 was depressed by an average of 5.3% due to GSA sales.

In 1976, surpluses had almost disappeared, when the US Congress decided on a new build-up on 1 October 1978, for which requisite administrative preparations were complete in 1979. For the first time in 20 years, the US administration began to make large purchases in 1981, starting with cobalt for 78 million US-$. These purchases had a large influence on the market.

Government stockpiles are also maintained for petroleum and coal. In the Federal Republic of Germany, the oil companies were obliged to store enough petroleum to provide for the needs of the country for 90 days. The OECD's International Energy Program also includes provisions for stockpile quotas, which, however, are hardly subject to change and thus only affect prices when stocks are raised.

b) *East-West Trade*

The East-West trade in minerals is a special market factor because the volume of exports and imports fluctuates from year to year. Trade relations with the COMECON countries have altered since the end of World War II, including mineral commodity trade. The leading exporting countries of the Eastern Bloc are the Soviet Union (Table 7.12) and the People's Republic of China; the East European countries are the major importers.

East-West trade is subject to pronounced fluctuations as to quantity and, owing to lack of information, difficult to predict. This is true especially of some metals, though long-term contracts have been signed for supplies of oil and natural gas. The Soviet Union's actions in trading some mineral commodities (tin, gold) have aroused suspi-

Mineral Markets

Table 7.12 USSR trade in selected base metals (tons)

Metal	1976	1980	1983	1986
Copper				
import	11 689	9 974	—	264
export	105 589	44 834	57 209	26 079
Zinc (slab)				
import	37 985	53 805	29 490	21 551
export	40 940	24 774	35 095	6 429
Tin				
import	9 403	15 149	11 308	14 274
export	—	—	—	—

Source: Metallgesellschaft AG, Metallstatistik, Frankfurt/Main.

cions that it occasionally attempts to disrupt world markets or at least to engage in speculation. China's foreign trade is also erratic and almost impossible to forecast in the long term (cf. Table 7.13). Certain markets (antimony, tungsten, tin) are thus particularly unpredictable because China is a major world market supplier of antimony, tungsten, tin, barite, and fluorspar. On the other hand, the Chinese Government plans to allow imports of mineral commodities to meet growing domestic requirements (aluminum, copper, steel, phosphate rock according to a recent projection).

c) *Recycling*

The recovery of metals from scrap, also known as secondary production, has increased only slowly in the last two decades. Recycling affects metal markets as follows:

- Supply is further increased.
- The new suppliers can weaken oligopolistic or monopolistic market positions.
- World trade diminishes, because recycling takes place directly in the consuming countries (with few exceptions, like a substantial amount of trade in ferrous scrap).
- Mineral resources are preserved.
- Recycling of waste contributes to protecting the environment.

Table 7.13 Tin exports of the PR of China (tonnes)

Exports to:	1978	1980	1985	1986
W Europe	1 485	930	571	211
E Europe	1 939	50	—	—
Japan	154	84	4 848	981
Hongkong	81	572	3 577	2 344
USA	1 571	858	4 514	2 259
USSR	175	—	—	—
Others	10	30	173	84
Total	5 414	2 524	13 683	5 879

Sources: International Tin Council, London; – Tin International, London.
Note: Tin export in 1987 is estimated at 18 000 t.

Table 7.14 Energy consumption in metallurgical processes

Metal	Raw material	Process	Energy consumption (millions Joule/kg)	Energy savings in recycling (%)
Aluminum	Bauxite	Electrolysis	220	—
	Alloyed scrap	Remelt	17	92
	Used beverage cans	Remelt	10	95
Copper	Ore concentrate	Outokumpu	98	—
	Copper scrap	Reverberatory furnace	20	80
Lead	Ore concentrate	Blast furnace	16	—
	Battery scrap	Vessel	10	38
	Scrap lead	Vessel	8	50

Source: Pawlek and Fischer, 1982.

- Energy can be saved, because the specific energy consumption in recycling is generally lower (cf. Table 7.14).

Metal scrap and residue are not reprocessed primarily for ecological but rather for economic reasons. Recycling is quite profitable, particularly when energy prices are high, because secondary smelting is much less energy-intensive. A cost-benefit analysis on recycling of metals also has to take costs for the dumping of scrap into account.

Metals are recycled from the following sources:

- Metal-containing sludge, such as iron-containing red sludge from bauxite treatment, jarosite sludge from zinc recovery, galvanic mud or sludge from gas purification in smelting.
- Metal-containing dust and ash gathered in the filters of metal smelters.
- Metal-containing slag or refining residue.
- Residue and manufacturing scrap from the metal-processing industry during the manufacture of semifinished products (e.g. zinc dross during galvanizing) or end usage (soldering residue in the electronics industry; scrap during the manufacture of tinplate beverage cans).
- Old scrap and waste; house waste makes up over 50% with 0.67% nonferrous metals (of which approx. 60% is aluminum), industrial waste having a higher percentage of metal content. A special area in scrap processing is the recycling of wrecked cars. A car in the medium range (unloaded weight 900 kg) contains roughly 400 kg carbon steel, 100 kg cast-iron, 100 kg alloy steel with soldering elements nickel, manganese, molybdenum, vanadium, cobalt, and niobium, 25 kg galvanized steel with aluminum, zinc, chromium, nickel, tin, or cadmium coating, up to 60 kg aluminum (or up to 20 kg magnesium), 30 kg copper, 20 kg lead, 3 kg tin, and 2 kg zinc.

The proportion of recovery depends on the metal, largely on its areas of application and the rate of increase in consumption (cf. Table 7.15). Lead, for example, can be recovered relatively easily from battery scrap; tin, on the other hand is used for such thin electrolytic coatings in tinplate that recycling is technically difficult and uneconomic. The rate of consumption of aluminum, in contrast, is increasing by some

Table 7.15 Recycling of metals: percentage of apparent consumption in the USA

	1978	1981	1986	1987
Aluminum	27	37	34	32
Antimony	69	47	35	n.a.
Chromium	8	10	18	25
Copper	51	59	54	53
Lead	51	44	41	49
Nickel	17	23	21	26
Silver	26	37	39	48
Tin	33	33	25	27
Tungsten	15	15	25	20
Zinc	4	5	37	35

Source: US Bureau of Mines: Mineral Commodity Summaries.

per cent annually, which means that although a substantial portion of old aluminum scrap can be recycled, the ratio of these quantities to present production remains relatively small. Some metals can be "co-recycled", such as antimony, which is yielded when recovering lead from battery scrap.

In addition to metals, certain industrial minerals can also be recovered, such as asbestos or bentonite; glass is also recovered from domestic garbage.

Fuels are not suited to recycling; after combustion or conversion, they are irretrievable for today's technology. Efforts are, however, being made to reprocess radioactive waste from nuclear reactors, and old lubricating oil can be re-used after cleaning.

Developments in recycling are the improvement of scrap and residue collection, the processing of domestic waste and the adaptation of the areas of application to the specific products (e.g. alloys) of raw material recovery. The rates of increase in the recycling of base metals have diminished markedly (Table 7.15), the possibilities in this "traditional" area being almost exhausted. The influence of secondary metals on market supply is thus easier to quantify.

7.3.2 Determination of Mineral Demand

The consumption of energy, metals and industrial minerals is closely linked to the industrial production of goods. Forecasts of future demand trends therefore make use of a number of indicators:

- The ratio of energy consumption or metal consumption to the gross national product of a country. Industrialized countries with a high GNP tend to consume more than developing countries with a low GNP. Neither overall nor in relation to minerals can a direct, let alone linear, function be established.
- The ratio of mineral consumption to the proportion of industrial production in relation to the GNP. This specific relationship is even more significant than the ratio of mineral consumption to the GNP.
- The ratio of mineral consumption to that of mass-consumption goods or to the "living standard" in a country. Statistical data on this is, however, incomplete.

Investigations on the per capita consumption of nonferrous metals and the degree of industrialization have produced clear correlations between the two (Müller-Ohlsen 1981).

Market studies of mineral demand trends will have to consider both final consumption and the manufacture of semifinished products. Here, the following determinants are of importance:

- The results of research into new areas of application for specific mineral commodities (e.g. new metal alloys or chemicals), leading to an incresae in demand.
- Technological innovations in the processing of minerals to reduce the specific consumption (e.g. the introduction of electrolytic tinplating, which uses a mere fraction of the tin metal needed in hot tinning), leading to a decrease of demand.
- The substitution of expensive or rare minerals by cheap ones that are more easily available (e.g. substituting the tinplate cans by aluminum cans for drinks). Substitution processes must be studied carefully, since they are reversible, as the example of aluminum cans has shown: after the rise in energy prices, production became more expensive than that of tin cans. Substitution also proceeds in stages, when a large manufacturer decides to re-equip with new production technology.
- Changes in end users' habits of consumption, which can cause the market for certain consumer goods to decline (e.g. when lipstick and nail varnish are no longer used, which reduces the demand for bismuth used in large quantities in the form of bismuth oxychloride for the production of cosmetics).
- Laws and regulations on the processing of certain metals for ecological reasons (e.g. the use of unleaded gasoline or less mercury and cadmium in areas of danger to health), leading to a decrease in demand.

As this list clearly shows, the analysis of demand for mineral commodities cannot be based exclusively on economic and industrial growth.

Consumption forecasts produce projections on overall consumption of a commodity (global analysis) and also conduct investigations on a sectoral level (sectoral analysis) for transport, industry, agriculture, etc., if adequate statistics are available. The methods employed are econometric models, trend extrapolation and also single and multiple regressions (cf. Sect. 7.4).

7.3.3 Competitive Prices

Mineral commodities, particularly metals, can be standardized according to quality and quantity and are thus negotiable on exchanges. Here, prices are subject to competitive supply and demand.

At present there are two important metal exchanges in the world, where competitive prices are fixed:

- The London Metal Exchange in London, Great Britain
- The New York Commodity Exchange in New York City, USA.

The metal trade in London has a long tradition. In the nineteenth century, import requirements for metals from abroad increased as a result of incipient industrialization in Europe. Merchants started to meet informally in coffee houses and then organized themselves into an association for the purpose of conducting an exchange. In 1876, leading metal merchants agreed to establish the London Metal Exchange Company, which then opened the London Metal Exchange (LME) in Lombard Court in January 1877, whence it moved to Whittington Avenue and finally (on 29 September 1980) to its present premises, Plantation House, Fenchurch Street.

The Exchange is today owned by the Metal Market and Exchange Company, whose shareholders are the "subscribers" to the market. Twenty-eight out of the 36 seats in the Ring are occupied by subscribers (1986). Its organs are the Committee of Subscribers and the Board of Directors, which are responsible for the orderly conduct of business. The Quotations Committee oversees the registration of daily price quotations, which it fixes at the end of the morning trading session.

Although membership is permitted for individual persons, all subscribers are now trading firms or mining companies registered in London.

The following metals are officially listed on the LME:

- Copper (since 1883) divided into wirebars (refined) and cathodes (electrolytic copper with 99.90% Cu + Ag).
- Tin (since 1883) as standard tin in ingots or slabs (99.75% Sn) and high grade tin in slabs or ingots (99.85% Sn), both suspended on 24 October 1985.
- Lead (since 1903) as pig lead (99.9% Pb).
- Zinc (since 1915) as standard zinc in bars (98.00% Zn).
- Silver (1897–1911, 1935–1939, since February 1968) as fine silver in small bars (99.90% Ag).
- Aluminum (since October 1978) as pure aluminum in ingots (99.50% Al).
- Nickel (since April 1979) as pure nickel in cathodes (99.80% Ni).

Trading takes place on the basis of standard contracts stipulating quality, quantity, form (ingots, bars, etc.), delivery dates and terms of payment. Only specific "brands" (trademarks) of metal are traded and delivery is via licensed warehouses in London, Liverpool, Hull, and Manchester and on the continent since 1963 in Antwerp, Rotterdam, and Hamburg. Customs law differentiates between "free port", "transit" and "inland" commodities. Silver is stored in the Westminster Bank, London, the Amro Bank, Amsterdam, and the Commerzbank, Hamburg.

On the LME, two sorts of contract are concluded, depending on date of payment: cash transactions for prompt delivery and immediate payment (spot market) and forward pricing (3 months) for delivery and payment after 3 months (future market or terminal market). For silver, forward contracts of 7 months' duration are also possible, as this metal is used as a safeguard against currency fluctuations.

Forward transactions serve as important indicators of the market and can also prevent erratic price fluctuations. The futures price points to the price expectations of consumers. If the 3-month price exceeds the cash price, this is termed "contango"; if it is lower than the cash price, it is termed "backwardation". On a contango market, supply bottlenecks and high prices are expected, in the case of backwardation, excess supply at reduced prices. A balanced stable market should continually display low backwardation, because a number of producers usually sell their anticipated production forward, thus offsetting possible loss due to falling prices. For this coverage they are prepared to accept a limited backwardation as a kind of "insurance premium".

Consumers also make use of forward transactions to insure against upward price movements (hedging).

Hedging keeps price fluctuations within bounds. There are also speculative purchases seeking to profit from rising metal prices, so-called short buying, which theoretically also contribute to minimizing price fluctuation. Speculation on commodity markets involves risk as an inherent part of forward trading; it is concentrated on precious metals, largely silver, and occasionally, tin.

The importance of the London Metal Exchange is reflected in its volume of transactions. London is still the major world trading center for copper and a substantial portion of trade in lead is conducted there. Competition from Kuala Lumpur tended to reduce trade in tin until 24 October 1985, when tin trade was suspended at the LME. For the foreseeable future, the Kuala Lumpur Tin Market will be the only official tin metal exchange. Despite widespread skepticism, trade in silver has developed well, and nickel and aluminum, only recently listed, have held position.

The LME performs two essential functions on the metal markets:
– It serves as a price indicator on the world markets. The quotations are based directly and visibly on real trade transactions, thus rendering the market more transparent and price trends easier to discern. Many producers include price clauses in their sales contracts that are orientated to LME quotations.
– It helps to stabilize prices on the world markets; hedging prevents erratic price fluctuations.

On 1 January 1982, gold forward-trading was commenced under the aegis of the LME. The London Gold Futures Market Ltd. contracts are for the current month and the 6 subsequent months. This trade supplements the London gold market located in the premises of N.M. Rothschild & Sons since 1919. Zurich is also a gold market with a long tradition. New exchanges emerged at the beginning of the 1970s, largely as a result of the demonetization of gold, for example:

– Winnipeg, Canada, the Winnipeg Commodity Exchange (since 1972);
– Sydney, Australia, the Sydney Future Exchange (since April 1978);
– Singapore, the Gold Exchange of Singapore (since November 1978);
– Hongkong, the Hongkong Commodity Exchange (since 1981).

In the USA, on 5 July 1933, the New York Commodity Exchange Inc. (COMEX) was established. Its more than 400 members are also shareholders. The metals traded are copper, gold, zinc, and silver, only copper transactions and somewhat fewer silver transactions being of significance; for long periods, no trade in zinc takes place at all (it was suspended from 15 October 1970 to 8 February 1978). The rules governing trade and market times are by no means as strict as those of the LME. In general, trade is largely confined to the American market and focused on metal imports into the USA, as American producers prefer to trade directly with consumers. COMEX deals mainly in long-term contracts of as much as 14 months' duration, which also contain provisions on quality standards as well as stipulations limiting price fluctuations in relation to the previous day. It maintains licensed warehouses for copper delivery in New York and Chicago.

Another metal exchange has been in existence since 1889 in Penang (occasionally in Singapore and now in Kuala Lumpur), dealing exclusively in physical tin metal. It

Mineral Markets

is run by the large Malaysian smelting company Datuk Keramat Smelting Bhd., Penang, in cooperation with the Malaysian Smelting Corporation Snd. Bhd., Butterworth. The price is determined by the amount of tin concentrate offered by miners or individual traders and the offers of consumers of melted tin. Purchase offers are registered and the market price is the lowest price offered at which the remaining tin concentrate can be sold, being also the highest price at which the entire quantity of tin on offer can be traded. The terms of delivery stipulate a maximum delivery period of 60 days for fine tin purchased from smelting plants (ex-works). Until the end of 1980, quotations were in Ringgit (Malaysian dollars – M$) per picul (equals roughly 60 kg) and since then prices have been listed in M$ per kg (ex-works).

On the Penang (now Kuala Lumpur Tin Market) tin exchange therefore, there is no hedging or speculation. An important function remains that of price indicator for the regional market in SE Asia.

After the opening of the Kuala Lumpur Commodity Exchange (KLCE) in October 1980, the Penang tin exchange was transferred there in 1984. On 25 October 1985 the Kuala Lumpur Tin Market (KLTM) suspended the tin trade but reopened on 3 February 1986. In May 1986, the Government of Malaysia set up a task force to consider the possibility of trading nondomestic tin, at least from Thailand, Indonesia, and Australia. In October 1987, the KLCE started trading in tin futures. The future contract is denominated in US dollars and there are two trading sessions, the second coinciding with the LME morning session.

The Chicago Board of Trade (CBT) also lists gold and silver. Transactions in 1000 and 5000 fine ounces are conducted, as well as forward business (futures market).

7.3.4 Producer Prices

When the market structure permits, the producers of mineral commodities also pursue an active price policy. Owing to the high level of concentration in the mining sector (cf. Sect. 7.1.1), i.e. the predominance of oligopolistic and monopolistic markets, producer price fixing is common.

These producer prices are not determined by competitive price mechanisms and usually exceed a free market price, generally resulting in additional producer's profit. In some markets, however, such as copper in the 1980s, producer prices closely follow changes in LME or COMEX prices.

The price fixers on mineral commodity markets are either suppliers who dominate the market, or powerful oligopolists, usually alternating with one another.

Here are some examples of metals whose prices are fixed as so-called producer prices. In addition to these posted prices, so-called spot market prices can exist:

– aluminum with posted prices of US producers;
– copper with posted prices of US producers for wirebars;
– zinc with posted prices of US and West European producers;
– lead with posted prices of US producers;
– nickel with a "world producer price" for cathodes;
– cobalt with producer prices of the government corporations Gecamines, Zaire and Sozacom, Zambia;

- cadmium with a posted price of US producers;
- platinum with a producer price of the major supplier Rustenburg Platinum Holding Ltd., South Africa;
- rare earths (especially RE oxides) with posted prices of the monopolist supplier Molycorp, USA.

The posted prices of the major producers are standard for direct contracts between small or medium suppliers and their customers. In fixing prices, the major producers often lead the market, with leadership often rotating among oligopolistic producers. For cadmium for example (cf. Fig. 7.4), the large US producers Asarco, AMAX, and St. Joe alternately take the lead. The remaining oligopolists promptly follow any changes in leading prices.

The aluminum and nickel markets have clearly demonstrated the influence of an exchange on producer prices. After aluminum was officially listed on the London

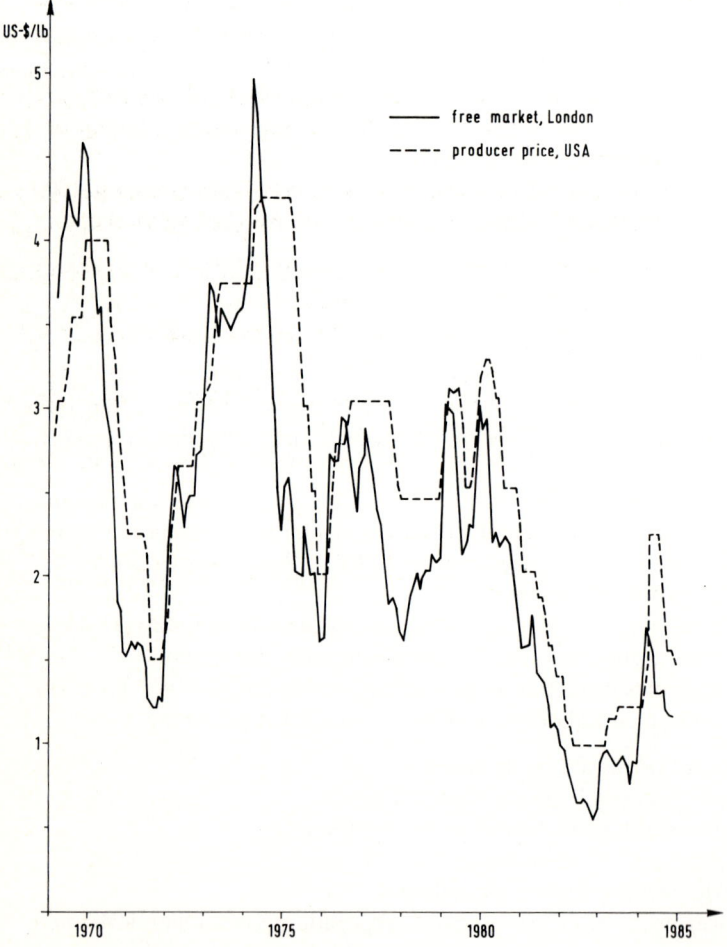

Fig. 7.4. Prices for cadmium (monthly average)

Metal Exchange in October 1978 and nickel in April 1979, the producers aligned their posted prices with those of the LME.

Differences in producer prices can also indicate differences in quality when they differ as to region. Low sulfur petroleum from Libya, for instance, can claim a higher price than sulfurous petroleum from Venezuela; chromite concentrate from South Africa, Turkey, and the Soviet Union sell for different producer prices, depending on the grade of concentration and impurities.

Usually, the producers themselves benefit from relatively high producer prices, which provide them with additional profits. It is argued that consumers benefit from the absence of price fluctuations and the existence of long-term supply contracts at agreed prices, although this argument is questionable.

In a number of cases, however, this price policy has proved disadvantageous to producers. Artificially high prices induce consumers to search increasingly for substitutes and obscure the emergence of structural imbalances on the market, which can lead to a critical decline in demand.

7.4 Market Models and Forecasts

A mineral-market model describes the relationship between the attributes of a particular market (such as production, consumption, prices, or inventories) and the determinants of these attributes (such as resource endowments, market structure, government policies, technologies of production and use, income, and many others). Factors that are determinants in some models are described by other models; for example, price is a determinant in many production models and yet is itself modeled in models of price formation. Some models only partially describe the market under study (such as a model of copper production in the United States), whereas others more completely describe a market (such as a model of the world copper industry). Some models are formal in the sense that they explicitly and quantitatively describe the important relationships in a particular market, whereas other models are largely informal, intuitive, or judgmental. This chapter focuses largely on formal models.

Models are used by a variety of people and organizations for a variety of purposes. Private companies active in mineral development, for example, use models as one of several tools for making decisions concerning mineral production, marketing, pricing, exploration, and investments in new mines and processing facilities; often these models are used to forecast or predict various factors that influence the economic attractiveness of a project, including prices, capital and operating costs, world production, and consumption. Government organizations active in mineral development use models for these same purposes. Government organizations – whether or not active in mineral development – also use models for other purposes, such as assessing the potential impact of different fiscal or concessionary regimes on investment, and evaluating the likely impact of a supply disruption in an important producing country on the availability of a particular mineral in the home country.

More generally, mineral-market models have at least three important purposes: (1) analysis, to understand the relationships between the attributes of a market (for example, the price of a mineral) and the important determinants of these attributes (for example, factors of supply and demand, including raw material costs, technology, income, government policies, and consumer tastes and preferences), (2) simulation, to assess the potential market impact of a change in one of the underlying determinants, and (3) forecasting, to predict future values of, for example, prices, production, or consumption.

Given this multiplicity of purposes for mineral-market models, it is not surprising that many different types of models exist. The next section reviews several of the important types.

7.4.1 Types of Mineral-Market Models

Most mineral-market models belong to one of two families of model types (adapted from Labys et al. 1985). Some models incorporate characteristics of both model types. The first family of mineral-market models is econometric models. This type of model describes the relationships among supply, demand, prices, and inventories, as well as the determinants of these market attributes (cf. Fig. 7.5). Prices adjust to changes in supply, demand, and inventories, which in turn respond to the change in prices. A typical econometric model consists of equations for supply, demand, inventories, and prices, as shown in the illustrative example below:

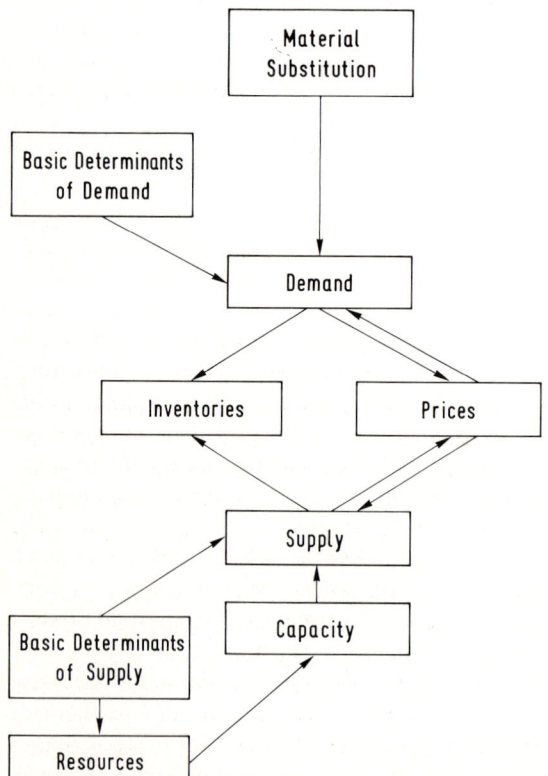

Fig. 7.5. General form of econometric mineral models. (Based on Labys et al. 1985)

$S(t) = s[P,G,GP,MS]$
$D(t) = d[P,PSC,Y]$
$I(t) = i[I(t-1) + S(t) - D(t)]$
$P(t) = p[I(t)/D(t)]$.

Four important attributes of a market are explained in terms of a number of explanatory variables. The quantity of mineral supplied in any year t, S(t), is a function of four explanatory variables: the mineral's price (P), geologic factors (G), government policies affecting mineral production (GP), and market structure (MS). The rationale for including price in the model is that an increase (decrease) in price provides an incentive for a producer to increase (decrease) production, other factors remaining the same. Geologic factors importantly influence the availability of a mineral resource for mining, and production is likely to rise or fall with changes in the geologic favorability for mining. Government policies influence the level of mineral production by encouraging or discouraging mining. Market structure also influences the level of production; monopolistic markets tend to supply smaller quantities at higher prices than competitive markets, whereas oligopolistic markets tend to have quantities and prices intermediate between those of monopoly and competition (see Sect. 7.1).

The quantity demanded by consumers depends, in this example, on three explanatory variables: the mineral's own price (P), the prices of substitute and complement materials (PSC), and income or industrial activity (Y). The quantity demanded is expected to be inversely related to a mineral's price; the higher the price, the less consumers will demand, other factors remaining the same. The relative price of a substitute material reflects the attractiveness of substituting one material or mineral for another in a particular use (such as aluminum for copper in electric-power transmission). As the price of a substitute falls relative to the mineral in question, consumers will demand smaller quantities. A complement is a good or service that is used along with the mineral in question (for example, copper and zinc in brass). As the price of a complement rises (falls), it is expected that consumers will consume less (more) of the mineral in question (again, other factors remaining the same). Income, sometimes measured as industrial activity, is also expected to importantly influence the quantities demanded by consumers: as income or industrial activity rise and fall, so, too, is the demand for minerals used in this industrial activity likely to rise or fall. Inventories in any year are simply the inventories from the previous year, I(t-1), plus current supply minus current demand. Finally, the system is closed by relating prices to inventories and demand. The relative importance of each explanatory variable in each equation, or in other words the sensitivity of the attribute being modeled to changes in particular explanatory variables, is determined by regression analysis using historical data. All four equations then are solved simultaneously to determine market-equilibrium values for each attribute.

As noted earlier, this example is merely illustrative. Not all models include exactly the same equations or variables, and some models are much more complicated, while others are simpler. The wide variety of mineral-market econometric models should not be surprising. First of all, mineral markets differ in a number of important respects, including the extent of competition among producers, the nature of government policies influencing costs of production, the relative importance of the various sources of supply (main product, by-product and co-product, and scrap), and the nature of end uses. Second, there are different models for different purposes; for instance, a model designed for policy analysis will not necessarily be appropriate for a more purely academic study striving to understand the past.

There are many examples of econometric models of mineral markets, illustrating the differences among models (although it is beyond the scope of this study to compare and contrast models). There are numerous models for the copper market, for instance. Fisher et al. (1972) is the seminal work in copper. Their model, estimated for the period 1948–1968, divides the world into two separate, but linked, markets. The first market is the United States, where administered or producer prices dominated during the period under study. The second market is the rest of the world, governed by free-market prices based largely on the prices of the London Metal Exchange. Seven equations describe copper supply: five equations for primary production (in four important producing countries, and the rest of the world) and two equations for secondary – or scrap – production (the United States and the rest of the world). Four equations describe copper demand in the United States, Europe, Japan, and the rest of the world. Although an important purpose of the model was to assess the impact of potential increases in Chilean output on copper prices and Chilean revenues, the model has perhaps been more influential as a standard against which to compare subsequent models of the copper industry. Charles River Associates (1978), in one subsequent model, incorporate exploration and discovery of new reserves into a copper model. Lasaga (1981) examines the role of the copper industry in the Chilean economy. Takeuchi et al. (1987) use an econometric model to forecast future consumption, mine capacity and production, production costs, and prices. For additional copper models, as well as models for other mineral markets, see Labys et al. (1985), Mikesell (1979), and the references cited therein.

The second important family of mineral-market models is engineering models. This class of model includes a wider variety of models than the econometric family and thus can only be defined loosely. What distinguishes engineering models is their mathematical description of the production process largely in technical, rather than in more purely economic or behavioral, terms.

Engineering models vary significantly in the extent to which they incorporate the economic determinants of the attributes of a market. The general form of this type of model is illustrated in Figure 7.6. National economic activity determines the demand for final products, which in turn determines the derived demands for inputs of raw materials (including minerals), energy, and labor. An important strength of an engineering model is its technical description of how inputs are transformed into outputs of final products. The description typically is disaggregated into several transformations intermediate between the initial use of raw materials and the production of final goods.

Most engineering models are designed either to determine the optimal combination of inputs given an initial set of constraints (technical, economic, or other) or to simulate what would occur under a given set of assumptions. One specific purpose of optimization techniques is to determine the best location for a new plant or processing facility given what is known about the costs, technologies, and location of existing and expected future plants or processing facilities. Another purpose is to choose the optimal combination of inputs for an assumed mix of final products. Important examples of optimization techniques include linear programming, process optimiza-

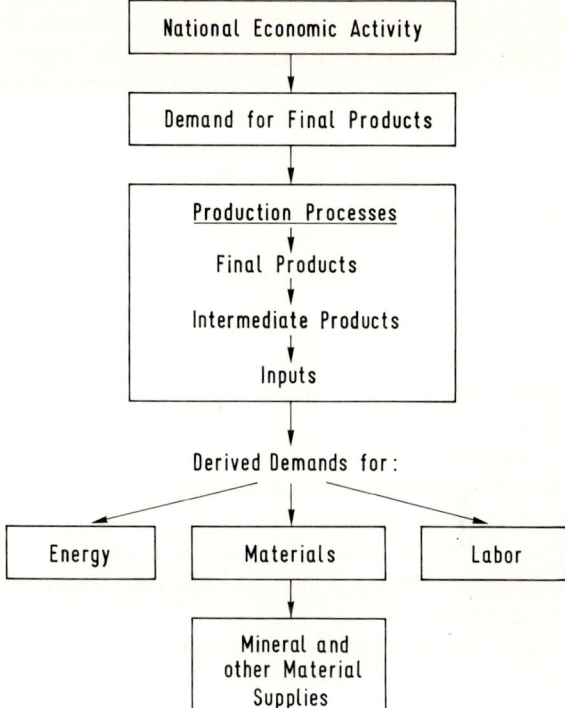

Fig. 7.6. General form of engineering mineral models. (Based on Labys et al. 1985)

tion, and spatial optimization. Simulation or system techniques, the second broad class of engineering models and including system-simulation and system-dynamic models, can also be used for a variety of purposes, including: to estimate the effects of changes in costs, government policies, or technologies on supply and demand; to assess the impact of mineral demands on national or regional economic activity; and to forecast a variety of factors such as future production, consumption, and the geographic distribution of production and consumption.

Engineering models typically are best suited for analyzing the important technological and to a lesser extent policy determinants of mineral supply and demand. Because engineering models emphasize the technologies of mineral production and use, they are often useful for long-term forecasts, which depend critically on technological change. But engineering models tend to be less useful for predicting the short and medium term, periods generally short enough so that major unanticipated changes in technolgy do not occur. (Over the short and medium term, mineral markets tend to be more strongly influenced by changes in the level of industrial activity and other largely economic factors.) Similarly, engineering models tend to account better for material substitution than econometric models. Engineering models can also capture the impact of policy changes to the extent that they properly describe the effects of policy changes on costs.

Econometric models, on the other hand, tend to be strong in exactly those areas where engineering models are weak, namely in accounting for the important economic determinants of mineral supply and demand. This type of model is valuable for assessing the responsiveness of production and consumption to changes in economic variables such as price and income, a concept known to economists as elasticity; for instance, a mining company may want to know whether a given percentage change in price will cause demand for its product to change by a greater, lesser, or the same percentage. Econometric models are often used for short and medium term forecasts, periods in which technological change is more likely to be smaller and more gradual than in the longer term. But because econometric models are based largely on historical data reflecting old and existing technologies, they are not able to capture significant changes in production and consumption technologies. Similarly, econometric models are less able to account for significant changes in government policies and other institutional factors that importantly influence mineral markets. Given these complementary strengths and weaknesses of econometric and engineering models, a number of researchers have developed hybrid models that attempt to take advantage of the strengths of each type of model.

An important lesson from this brief review of mineral-market models is that there are different types of models for different purposes. The appropriateness of any particular type of model depends on the purpose of the modeling exercise. There is no all-purpose type of model. Finally, a word of caution is in order: a model is only as good as the input data and, perhaps more importantly, the underlying rationale or theory behind the model. Bad data fed into a theoretically sound model and good data used to test a model based on bad economic or engineering thinking both yield bad results.

7.4.2 Forecasting

An important use of mineral-market models is forecasting. Although many forecasts are derived from formal or quantitative models similar to those described above, others are based on less formal and more qualitative methods. Regardless of the method, forecasting is indispensable for any individual or organization that must make decisions today whose correctness will be determined by future events. For mining companies considering an investment in a new mine or processing facility, or banks considering a loan to the project, estimates of future profits hinge critically on estimates of future demand, prices, and production costs. For government organizations, the impacts of stockpiling, land-use, environmental, and other policies need to be considered with some sort of forecast.

Most forecasts belong to one or – more frequently – a combination of three families of forecasting techniques (Tilton 1983). The first family includes statistical methods, which essentially extrapolate from past trends of prices, production, consumption, or other market attributes. These methods range from simply projecting a straight-line trend of past data into the future to sophisticated statistical methods of extrapolation (such as the Box-Jenkins technique). Compared to other forecasting methods, statistical methods tend to be simple and inexpensive, and require little data. They are often good enough for short-term forecasts. However, these methods implicitly assume that there are no significant changes in the fundamental determinants of price, demand, supply, or whatever factor is being predicted. In other words, these methods by their very nature will not foresee major turning points in the mar-

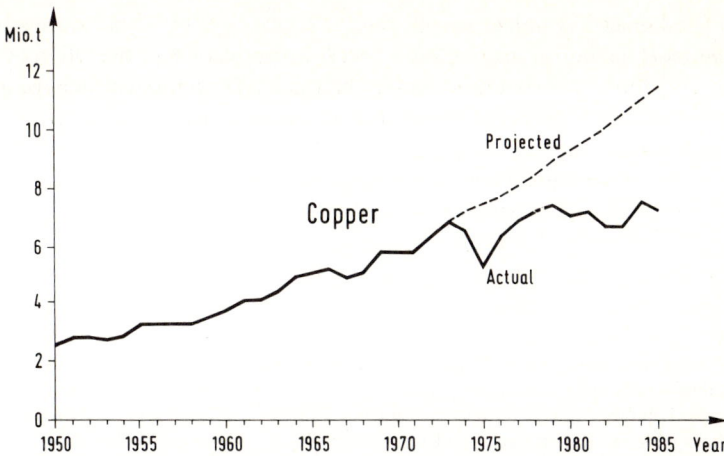

Fig. 7.7. Western World consumption of copper, actual and projected trends, 1950–1985. The projected trend assumes that consumption had continued to grow since 1973 at the annual average growth rate of the period 1950–1973. (After Tilton 1987, based on data from the Metallgesellschaft)

ket. For example, many mines and processing facilities were built in the middle 1970s largely on the basis of projecting historical trends in metal consumption (compare, for instance, actual copper consumption since 1973 with a simple projection of earlier trends in Figure 7.7).

The second type of forecasting method is causal or behavioral. These methods identify cause and effect, and predict the future on this basis. Important examples include the econometric and engineering models described previously, as well as several others. Causal methods can better predict the future, especially major turning points, if they correctly identify cause and effect. However, they are usually more time consuming and expensive and tend to require larger amounts of data, compared to statistical methods.

The third type of forecasting method is judgmental. In contrast with the first two types of methods, judgmental methods are not explicit in their analysis of either cause and effect or of historical data. The methods are largely qualitative, usually based on the informed judgments of experts, who presumably have some sort of implicit cause-and-effect model in their minds. If the experts have correctly identified the cause-and-effect relationships, they may be better able to foresee major changes in mineral markets than statistical and causal methods of forecasting. Judgmental methods can adjust current trends for anticipated changes in policies and technologies, changes which in many cases are likely to be unstatable in an explicit or formal forecasting model. Judgmental methods suffer, however, from their lack of quantitative rigor.

The general problems of forecasting can be appreciated by considering two types of forecasts: short-term forecasts of this year and perhaps the next several years, and long-term forecasts of the next several decades. Over the short term the major influence on mineral markets is the general state of the economy, which in turn determines the demand for final goods and services and thus the derived demands for minerals, as well as mineral prices and production. Thus, short-term projections require a good macroeconomic model. Over the longer term, changes in technology – including material substitution – are at least as important as overall economic acitivity as driving forces for change in mineral markets. Technological change and material substitution are both exceedingly hard to predict. Thus, an important message is that forecasting is inherently difficult. Nevertheless, those who make decisions concerning mineral development and policy have no choice but to use forecasts – whether formal or informal, quantitative or qualitative – in making their decisions.

Notes on the Literature

Those interested in more details on methods of mineral modeling are referred a review article by Labys, Field, and Clark (1985) "Mineral models" (an important source for section 7.4.1). Those particularly interested in econometric modeling and forecasting should read Kennedy (1985) *A Guide to Econometrics,* and Pindyck and Rubinfeld (1981) *Econometric Models and Economic Forecasts.*

Part III International Mineral Policies

Chapter 8 Policies and Cooperation Programs of International Organizations

8.1 General Problems and Objectives

Mineral policies significantly influence the environment in which mineral production occurs, the level of distribution of international mineral trade, and the optimum use of scarce mineral resources.

A realistic mineral policy must take the following basic geological considerations into account:

- The quantity of mineral resources currently accessible in the earth's crust is limited.
- Minerals are nonrenewable natural resources.
- The regional distribution of the known mineral deposits is uneven (cf. Fig. 8.1, Table 8.1).
- A number of minerals occur as associated products, which means that supply is highly inelastic (for example cobalt is a byproduct of copper smelting in Zaire and Zambia and the amount of cobalt thus depends on the copper production).

Viewed almost without exception by planners as the source and means of economic prosperity, industrialization and industrial production still primarily determines economic policy, irrespective of the socio-economic system adopted. The way to maintain the standard of living in industrialized nations and to enhance socio-economic development in Third World countries is to increase the production of industrial goods. In terms of mineral policy, this means:

- Total world demand for minerals tends to increase.
- Some minerals become relatively scarce, leading to the exploration and development of new deposits.
- Mineral exploitation costs tend to rise due to more remote and lower grade deposits being opened up, costly measures to protect the environment and increasing taxation imposed by mineral exporting countries.
- The exploitation and utilization of minerals increasingly endangers the environment.
- An increasing rift develops between the mineral importing countries and the mineral exporting countries.

These developments have on the one hand led to an expansion in world trade but on the other caused problems for the world markets in numerous minerals, in particular:

- Strong, usually short-term price fluctuations;
- Substantial fluctuations in the export earnings of producing countries, which are

Table 8.1 Distribution of mineral resources (1986) (in % of the total identified reserves)

Mineral	OECD*	Country Group Eastern Bloc	LDC
Energy resources:			
Petroleum	9.9	11.5	78.6
Natural gas	15.8	44.6	39.6
Coal	44.5	46.3	9.2
Uranium**	60.0	20.0	20.0
Iron and steel alloys:			
Iron ore	46.5	29.3	24.2
Manganese	47.2	37.2	15.6
Nickel	33.9	24.5	41.5
Cobalt	3.6	31.8	64.5
Chromite	80.0	13.1	6.9
Base metals:			
Bauxite	25.8	3.8	70.4
Copper	26.3	16.8	57.0
Zinc	64.4	13.7	22.0
Tin	10.3	17.1	72.6
Antimony	14.5	59.4	26.1
Tantalum	27.3	13.0	59.7
Precious metals:			
Gold	71.9	16.4	11.7
Silver	40.4	23.6	36.0
Platinum	90.7	9.3	0.0
Industrial minerals:			
Fluorspar	54.3	15.6	30.1
Phosphate rocks	90.9	9.1	0.0
Kyanite	92.9	0.0	7.1
Diamonds	54.3	15.2	30.5

Source: Bundesanstalt für Geowissenschaften und Rohstoffe, Hannover 1986.
* Incl. South Africa.
** Rough estimate.

particularly severe in several developing countries with only one or two minerals for export (cf. Table 8.2);
– Comparatively imperfect competition of mineral markets due to the high level of concentration on the supply side (but the degree of competition has increased in recent years);
– Limitation of access to some markets due to established trade structures;
– Uncertainty of medium and long term supply for the consumer due to a decline in exploration activities in the 1980s, production problems in exporting countries and export quotas imposed by international cartels.

This state of affairs is unsatisfactory for both the majority of producers and consumers, which is why their demands for a realignment of international mineral policy are becoming more forceful. The overall mineral policy goals of consumer countries and producer countries would appear, at first glance at least, to conflict with one another. Industrialized and industrializing nations dependent on imports of raw materials are interested in securing long-term, reliable, sufficient supplies at fair prices. In contrast, the developing countries exporting minerals are primarily con-

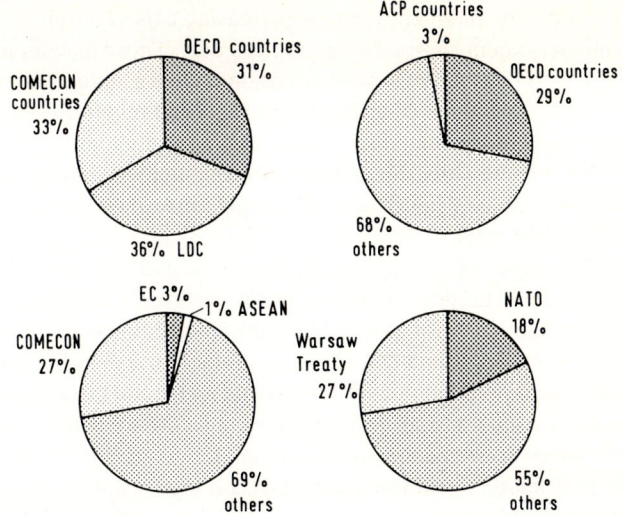

Fig. 8.1. Distribution of mineral resources in regions (BGR, Hannover 1982)

Table 8.2 Major mineral exporters among the developing countries

Country	Main mineral export	Mineral value as a per cent of country's total export value (1985)
Algeria	Petroleum	53.4
Bolivia	Natural gas	27.4*
	Tin	22.2*
Brunei	Petroleum	54.1
Chile	Copper	43.0
Ghana	Bauxite	32.4**
Guinea	Bauxite	95.0**
Guyana	Bauxite	41.5*
Indonesia	Petroleum	64.2
Iran	Petroleum	79.4*
Iraq	Petroleum	98.8*
Jamaica	Bauxite	50.2
Jordan	Phosphate rock	32.0
Kuwait	Petroleum	70.4
Liberia	Iron ore	30.1
Libya	Petroleum	98.0
Malaysia	Tin	34.1
	Petroleum	26.4
Mauritania	Iron ore	82.4
Mexico	Petroleum	40.8
Morocco	Phosphate rock	30.1**
Nigeria	Petroleum	90.7
Saudi Arabia	Petroleum	91.4
Sierra Leone	Diamonds	34.0
Togo	Phosphate rock	52.0
Zaire	Copper	33.9*
Zambia	Copper	80.5*

Sources: IBRD, World Development Report 1987, Washington, D.C., 1987.
GATT Activities 1986, An Annual Review of the Work of the GATT, Geneva 1987.
* 1983.
** 1984.

cerned with stabilizing prices, increasing export earnings from minerals, and obtaining investment capital for mining projects. Those developing countries lacking in mineral resources depend on international aid for industrialization.

An important objective of international mineral policy must be to avoid conflict between producing and consuming countries over distribution of mineral commodities by setting up a distribution system acceptable to all parties. The latter includes stable production and supply, meeting demand requirements, competitive prices and trade.

Between 1974 and 1984, however, it had become difficult to pursue this overall objective. The developing countries elevated mineral policy to the status of a major instrument to bring about the New International Economic Order (cf. Sect. 8.2.2) and it has become a central issue in the North-South Dialog. This could have highly adverse repercussions, if for example a gigantic institution were to fix prices and distribution channels as a global solution to the so-called "resource problem" and international "resource management" in "the world's interest". Such models for the strict regulation of world trade are based on the dream of optimum overall conditions and the myth of acute shortage of mineral resources. Geologists have repeatedly pointed out that none of the important minerals is in scarce supply, production being limited mainly by economic, financial, and ecological factors. The oversupply situation on many mineral markets in the mid-1980s contributed considerably to a more realistic view.

8.1.1 Objectives of Mineral Exporting Countries

The general economic policy and hence mineral policy objectives in developing countries are primarily geared to economic growth. The export of minerals, including petroleum, is viewed as the basis for financing development; foreign exchange earnings help stabilize the national budget and contribute to investment programs.

The export of minerals is accorded fundamental importance both by developing producing countries and the small group of industrialized nations rich in mineral resources (Canada, Australia, South Africa).

A priority aim is to secure stable and high prices, as exemplified by OPEC's price policy in the 1970s, with its high points in 1973/4 and 1979, although the crisis of the mid-1980s questioned the future viability of this cartel because the methods used for price policy failed in the longer run.

At the national level, the development plans for the mining and energy sectors include a whole catalog of mineral policy goals. Of particular importance are:

– Acquiring sovereignty over mineral exploitation via state control of production and marketing, altering structures of ownership (participation) or subsidizing domestic cooperatives.
– Increasing royalties and income taxes for mining.
– Increasing and rationalizing the production of minerals to maximize export earnings.
– Diversifying mine production to lessen dependence on individual mineral commodity markets.

- Increasing domestic processing of minerals to enhance net value added and contribute to industrialization.
- Protecting deposits from destructive exploitation and premature abandonment.
- Utilization of the mining sector to develop rural areas by improving the infrastructure and creating jobs outside industrial centers.
- Minimizing damage to the environment during exploitation and processing of minerals.
- Ensuring national supply of minerals (including fuels) to avoid spending foreign exchange on imports.
- Securing consistent export earnings (mainly by actively participating in international efforts).

The strategies to attain these goals are, as a rule, formulated in a general way. Mainly, exploration projects to find additional reserves are promoted, which involves geological surveys, international technical cooperation and investment of transnational mining and oil companies.

Although many countries establish a sequence of priorities, they do not always act in accordance with these priorities, because the geologic, political, economic or sociocultural conditions in individual countries mean that the aims are not always compatible with one another.

A rather astonishing political phenomenon is worth noting in this context. Although the developing countries exporting minerals are in the minority among the group of Third World countries, their demands have been supported by the others, despite the fact that those developing countries lacking in mineral resources are hard hit by an aggressive mineral price policy, as the oil price rises have clearly shown. The inequality of income distribution amongst the developing nations is further exacerbated; no adjustment mechanism is seriously being considered. The only recognizable initiative in this direction is the formation of an International Seabed Authority (cf. Sect. 8.2.2.3).

8.1.2 Objectives of Mineral Importing Countries

Just as mineral-exporting countries can be divided into a (small) group of industrialized countries and a larger group of developing countries rich in mineral resources, the importing countries can also be classified into two groups. In terms of quantity, the major importing countries are the USA, the West European industrialized nations and Japan, which depend on mineral supplies for industrial production. The other group comprises those developing countries possessing few mineral resources, whose prospects for development are influenced by international mineral policy. The COMECON countries (Eastern Bloc) are not considered.

A principal goal of the importing industrialized countries is to ensure guaranteed supplies of minerals, that of the importing developing countries, to obtain low prices.

Although for OECD countries dependent on imports mineral policy focuses primarily on supply, the following list of objectives takes other aspects into account:

- Guaranteeing the supply of minerals by international mining investments, participation in mining projects abroad, concluding long-term supply contracts or cooperation agreements and diversifying supply sources.

- Promoting domestic mineral production by supporting mineral exploration programs, subsidizing domestic mining companies and stimulating R&D in the field of substitution possibilities.
- Developing new technologies to enhance the rational and efficient utilization of raw materials.
- Increasing recycling of minerals, particularly metals.
- Minimizing environmental damage from mining and processing minerals.

The strategies and instruments to attain these objectives are varied and will be discussed in more detail below (cf. Chap. 9.1). Reference should, however, here be made to the efforts of the Western industrialized nations to arrive at common strategies, under the aegis of the EC or the OECD, for example.

The policies of mineral importing developing countries, however, concentrate on promoting domestic mineral production. They seek assistance from industrialized countries for the transfer of modern but appropriate mining technology.

8.1.3 Conflicting Aims and Possible Solutions

At first glance, the mineral policy aims of the two groups (exporting and importing countries) would seem completely irreconcilable. On closer examination, however, a number of individual goals appear by no means mutually exclusive.

To facilitate a successful dialog between the exporting developing and the importing industrialized countries, it is necessary to examine the compatibility of these goals in order to ascertain areas of common interest and potential conflict.

Those exporting countries' goals which are geared to economic growth, such as increasing, rationalizing and diversifying mineral exploitation, are quite consistent with the supply goals of the importing countries. Both groups also concur on the need to keep the damage to the environment as a result of producing and utilizing minerals to a minimum. Programs to conserve mineral resources (prevention of destructive exploitation, optimum rational use) should not give rise to any serious objections, since all policy-makers must now be aware of the limited long-term supply of these resources.

The development policy goals of the exporting developing countries who wish to increase mineral processing as part of the industrialization effort are a source of conflict. One of these is the use of processing facilities for the development of rural areas, so-called regional development. The desire to take over more of the processing stages can create difficulties for the primary industries of importing countries, depriving the traditional smelting plants, semimanufacturers and oil refineries of a part of their raw materials supply.

These conflicts cannot be resolved by the completely free market. The development goals of the exporting countries must be set against the cooperation and development policy of the importing countries, which must then take some responsibility for restructuring domestic primary industries.

An area of particular conflict remains that of commodity prices. The exporting countries adduce good reasons for high prices, while the importing countries can point to the global economic advantages of lower ones (e.g. petroleum or iron ore).

The dispute over mineral policy can thus be reduced to pricing. It is the conflict between market forces and policy concerns. As a consequence, a major issue in international mineral policy today is the search for a workable price mechanism on the commodity markets, acceptable to all parties.

8.2 Mineral Policy of International Organizations

8.2.1 The League of Nations

Mineral policy can be national or international. Although its origins can be traced back to the 16th century, when the European nations began colonizing other regions of the world, an organized systematic mineral policy did not emerge until the 20th century with the advance of industrialization. After World War I, leading mining companies concluded international agreements with governments of mineral exporting countries/territories. In 1919, the first session of the International Labor Office raised the question of trade in raw materials. As a result, the League of Nations World Economic Conference of 1927 in Geneva formulated the first principles of an international commodity policy in its Final Report. The Conference recommended that international commodity agreements should be formed by governments interested in a more methodical organization of production and a reduction of production costs.

The London Monetary and Economics Conference organized by the League of Nations between 1932 and 1933 included a subcommittee also concerned with the problems and principles of international commodity policy. The Conference adopted guidelines for the formation of international commodity agreements. The most notable were:

– In order to reestablish global economic prosperity, commodity prices were to be raised to a "fair and remunerative level".
– Better coordination between production and trade of commodities was regarded as desirable.

It should be noted that the guidelines, which were primarily concerned with the interests of the producers, consistently took those of consumers into account. A number of agreements on individual commodities attempted to comply with the guidelines, including the 2nd International Tin Control Scheme of 1934–36.

In 1937, the League of Nations appointed a commission to prepare a report on the effectiveness of mineral policy, which established that producer cartels restrict supply. The creation of bufferstocks to regulate prices in line with the market was explicitly recommended.

World economic relations were severely disrupted by World War II. The major concern was to supply the war economies and in mineral-importing countries mining was geared to nationalistic, strategic aims.

8.2.2 The United Nations

After the war, responsibility for planning and directing international mineral policy was taken over by the League of Nation's successor, the United Nations (UNO). Most conferences on commodity problems are organized by its institutions. The UN Charter of June 1945 re-affirmed the principles of international cooperation through promotion of the solution of economic, social, health, and related problems (Article 1).

A special institution devoted to the attainment of economic and social progress is the UN Economic and Social Council with its various Commissions for particular functions and regions, of which some are concerned with questions of mineral policy. On 28 March 1947, the Interim Coordination Committee for International Commodity Arrangements (ICCICA) was established to provisionally adopt responsibility for the mineral policy goals of the International Trade Organization (ITO) envisaged by the Havana Charter. As it became clear that the ITO would not become operational, the ICCICA was also dissolved at the beginning of the 1950s, not, however, without having a formative influence on mineral policy for at least the next 20 years, particularly via the conferences on commodity issues and in combination with the recommendations of the Havana Charter. The General Assembly of the United Nations also concentrated on commodity problems several times in the 1970s, particularly so during the Sixth Special Session of the UN General Assembly from 9 April to 2 May 1974 in New York, when the Declaration on the Establishment of a New International Economic Order was adopted. The Programme of Action for a New International Economic Order contains the following measures concerning commodities:

- The implementation of the Integrated Program for Commodities (cf. Sect. 8.2.2.2);
- The establishment of international commodity agreements to regulate prices and exports, guarantee sales, finance surplus production and compensate for loss of export earnings of developing countries;
- Indexation of commodity prices, i.e. linking the prices for the exports of commodities from developing countries to those for imports of industrial goods;
- Creation of producer associations on commodity markets.

General provisions of the New International Economic Order also influence commodity trade, such as:

- A code to govern the transfer of technology and to provide full national sovereignty over natural resources with the right to nationalize the mining sector without necessarily observing legal conventions on compensation;
- Unilateral access for raw materials and industrial products from developing countries to the markets of industrialized countries;
- A plan to raise the developing countries' 7% (1975) share of world industrial production to 25% by the year 2000, in compliance with UNIDO's Lima Declaration of March 1975.

In the view of the 129 Third World members (1987) of the United Nations (including China), a basic prerequisite for further development aid is the solution of structural problems. Minerals still being the most important source of foreign exchange earnings for about 30 developing countries, the latter regard mineral policy as the major focus. The rapid increase in the foreign debts of developing countries is regarded as a negative result of the present world economic system: at the end of 1986, debts amounted to US-$ 966 billion for all developing countries (except Middle East OPEC countries), debt service (1986: US-$ 130 billion) consuming a large proportion of export earnings (1986: 26% on average).

8.2.2.1 UNCTAD I–VI

In 1964 the United Nations Conference on Trade and Development (UNCTAD) was founded as a permanent organ of the UN General Assembly. It has a Secretariat in Geneva headed by the UNCTAD Secretary General and a Trade and Development Board which are in constant operation between Conferences. Amongst other things, the Secretariat is responsible for the organization of commodity conferences, where general guidelines on mineral policy are discussed or international commodity agreements are negotiated (tin, tungsten, copper).

Seven conferences have taken place to date, in Geneva (1964), New Delhi (1968), Santiago de Chile (1972), Nairobi (1976), Manila (1979), Belgrade (1983), and Geneva (1987).

Each UNCTAD Conference focused on different aspects of mineral development, depending on the world economic situation. In Geneva in 1964, basic guidelines were adopted and the following measures were regarded as urgent:

– Increasing the membership of consuming countries in international commodity agreements.
– Extending the duration of agreements.
– Providing regular adjustments of quotas and price limits.
– International financing of buffer stocks.

The Nairobi Conference in 1976 was marked by a new awareness of the commodity problem. OPEC's price policy appeared to be only in the interests of those developing countries rich in minerals: nearly all mineral prices rose substantially in 1974. International control of commodity markets on as broad a basis as possible was seen as the main aim. UNCTAD IV (1976) was prepared during various conferences, in which the developing countries played a decisive role. Worthy of note here are the UN Conference on Commodities in April 1974, where 97 Third World countries submitted the draft of the New International Economic Order, and the 8th Session of the UNCTAD Committee on Commodities on 21 February 1975 in Geneva, where the Integrated Program for Commodities was devised (Corea Plan, after the UNCTAD Secretary General Corea).

The Integrated Program for Commodities served as the basis for all the proposals relating to commodities. It was explicitly emphasized that the Program was aimed at completely restructuring international commodity trade primarily in the interests of the developing countries. The most important steps towards achieving the objectives (see Sect. 8.2.2.2) of the Integrated Program for Commodities were then stipulated:

- Establishment of a Common Fund for the financing of international commodity stocks (buffer stock).
- Setting up of international commodity arrangements.
- Harmonization of stockpiling policies.
- Indexation of prices for commodities exported by developing countries to the price of industrial goods imported from industrialized nations.
- Greater participation of developing countries in transport, marketing and distribution of commodities.

UNCTAD V in Manila designated the adoption of concrete steps to implement the Integrated Program as the central topic on the Conference agenda. The Group of 77 (77 developing countries which joined in 1964; now comprising 128 members) initiated various resolutions. The following central points need mentioning:

- The Agreement of the Common Fund for Commodities was to be adopted before the end of 1979 (actually achieved in mid-1980).
- Governments were urged to speed up negotiations on individual commodity agreements.
- The developing countries were to be supported in their efforts to secure guaranteed export earnings. In this context, proposals were made to establish a complementary facility for commodity-related shortfalls in export earnings and to promote processing of commodities in developing countries.

The latter requirement met with opposition from certain industrialized countries, including the USA and the Federal Republic of Germany. The requirement goes well beyond the maximum objectives of the Integrated Program for Commodities, because a guarantee of purchasing power is to be attained by indexation, requiring a further set of dirigistic measures, national and international production controls, regulation of investment, and mineral processing.

UNCTAD VI 1983 in Belgrade realized the interrelations in the fields of commodities, trade, money and finance, and development. The resolutions of this Conference incorporated the planned measures in the field of commodities into an overall programm. As for commodity policy, UNCTAD VI urged the signature and ratification of the Agreement Establishing the Common Fund for Commodities and terminated its promotion of developing countries in the area of processing, marketing and trading of commodities.

8.2.2.2 Integrated Program for Commodities

Since the beginning of 1974, the UNCTAD Committee on Commodities had been working on a program to regulate commodity markets, and it passed a proposal in its 8th Session on 21 February 1975 for submission to UNCTAD IV in Nairobi. The main objectives of this Integrated Program for Commodities (cf. Resolution 93/IV of the UNCTAD Conference in Nairobi) are as follows:

- Improving the terms of trade of developing countries in the field of commodities.
- Supporting commodity prices at levels which in real terms are remunerative to producers and equitable to consumers.

- Reducing excessive fluctuation in commodity prices.
- Improving and stabilizing in real terms the purchasing power of individual developing countries from export earnings.
- Expanding developing country exports of primary and processsed products.
- Diversifying production and expanding of processing.
- Assuring access to international markets.
- Increasing the participation rate of developing countries in the transport, marketing and distribution of their mineral exports.

After protracted discussions, a list of 18 commodities was agreed on (6 minerals, 5 agricultural commodities and 7 foods) of particular importance to the developing countries because they account for nearly 75% of the total value of commodity exports (excepting petroleum). The list comprises coffee, cocoa, tea, sugar, cotton, rubber, jute, hardfibres (sisal), copper, tin, tropical timber, bananas, meat, vegetable oils and oilseeds, bauxite, phosphate rock, iron ore, and manganese ore, the first ten being graded as "core" commodities. Buffer stocks are confined to these ten.

After UNCTAD V in Manila had set definite dates for negotiations on the Integrated Program for Commodities, an agreement was reached in Geneva on 27 June 1980. The Agreement Establishing the Common Fund for Commodities has been ready for ratification since 1 October 1980 and will enter into effect as soon as 90 states have given their ratification and two-thirds of the funding has been pledged. At the end of 1986, the Agreement had been signed by 114 states and ratified by 91, accounting for 58.13% of the Fund capital. The USA have not ratified the Agreement but the USSR in December 1987. UNCTAD experts estimate that the Common Fund will not begin operating before 1988.

The agreement is at present only concerned with the core of the Program, the creation of a Common Fund with two "Windows" (accounts), the first to finance international bufferstocks, the second, to finance measures to support selected commodities.

The First Account ("First Window") was provided with US-$ 400 million in the form of mandatory contributions from members. Each country is to pay a basic contribution of US-$ 1 million with US-$ 320 million owing, divided up according to a group quota system: 71% from OECD countries, approx. 13% from COMECON countries, 10% from developing countries, 5% from China, and 1% from other countries. In concrete figures, this amounts to US-$ 73.8 million for the USA, US-$ 33.7 million for Japan, US-$ 26.5 million for the Federal Republic of Germany, US-27.2 million for the Soviet Union, and US-$ 16 million for China. The voting rights in the Fund's Council, however, are allocated quite differently to the financial commitments: 47% of the votes are held by developing countries, 42% by OECD countries, 8% by COMECON countries and 3% by China.

The First Window is not intended to finance the creation of buffer stocks directly but to improve the Fund's creditworthiness or at best to help overcome temporary difficulties in liquidity. The Fund merely contributes to the financing of buffer stocks for individual commodity agreements, provided the individual agreement has signed a contract with the Fund.

The envisaged method of financing is quite complicated. The responsible organ in the individual agreement (the International Tin Council for example, see Sect. 7.2.2) stipulates the maximum financial requirements for creating a buffer stock (purchasing price for stocks and maintenance costs). A third of this sum is paid by the members of the individual agreement in cash into the First Account (Window) of the Common Fund. When funds are needed to finance stocks, this money is drawn upon and if further stockpiling is needed, a credit can be taken out, financed either from cash funds of other individual agreements or loans from the capital market on the basis of the Fund's liability capital.

As additional security, the individual agreement is also obliged to submit bufferstock warrants to the Fund. The declared aim of the Integrated Program for Commodities is to bring about international bufferstock agreements for the ten core commodities as soon as possible. In its present form, however, the Common Fund is unlikely to offer much incentive for further agreements. Probably only one of the present agreements (rubber) will make use of the new financing facility, because it offers slightly more favorable terms (lower interest rates than the International Monetary Fund).

The Second Account ("Second Window") of the Common Fund provides for "other measures" aimed at stabilizing the market in the long term. Finance is to be provided for improvements in production, marketing, and processing of commodities. These measures operate in two directions:

– Stimulating consumption by seeking new areas of application and improving the quality of industrial products.
– Stimulating production by lowering production costs, facilitating processing in the developing countries, supporting the diversification of the export structure and improving the infrastructure in exporting countries.

Initially, the Second Account is to be provided with US-$ 350 million, US-$ 280 million in voluntary contributions. In principle only those measures are to be financed that serve the interests of the commodity market as a whole, thus officially excluding projects geared exclusively to one country, though this restriction is unlikely to be practicable in every case. Those eligible to apply are international study groups engaged in promoting production or consumption of selected commodities. In the foreseeable future the Second Account is likely to be used for more commodities than the first. It is quite conceivable for example that some minerals mentioned in the Integrated Program – tin, copper, bauxite and iron ore – will profit from it. As producer associations already exist in these markets, it should be no problem to establish study groups as a precondition for financial support.

8.2.2.3 United Nations Conventions on the Law of the Sea

The Third United Nations Conference on the Law of the Sea, UNCLOS III, is viewed politically and economically as the most important conference of the 20th century and with good reason: the reform of the Law of the Sea has meant the redistribution of 70% of the earth's surface into new areas of use with offshore and ocean mining playing a growing role. At least it was the largest, longest, and most comprehensive international conference ever held.

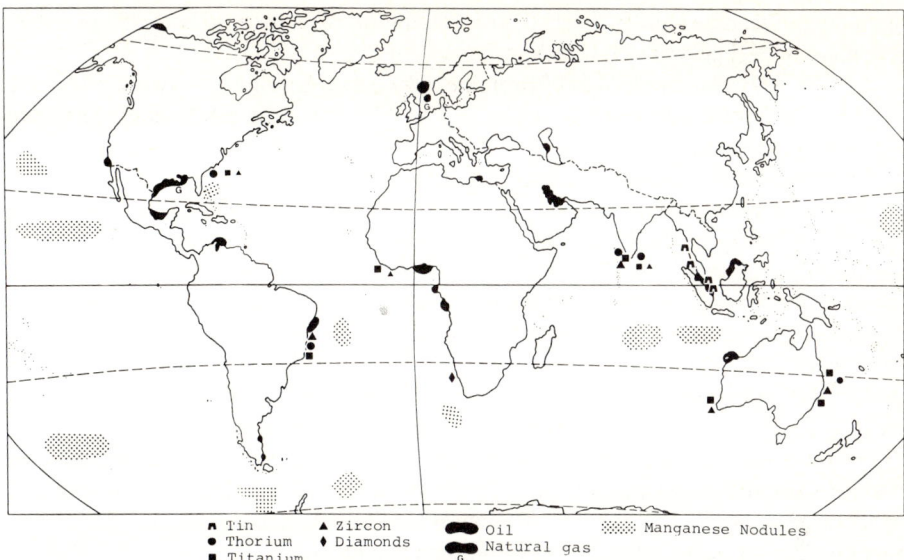

Fig. 8.2. Major marine mineral deposits (oil and natural gas)

The potential mineral resources of the oceans and its subsoils are considered very large (Fig. 8.2). At present, the main activities in ocean mining concentrate on shelf areas, within the new 200 nautical mile Exclusive Economic Zone.

The major marine mineral resources can be classified as follows:

- Dissolved elements and salts in ocean waters (magnesium, bromine, uranium as well as sodium, potassium, sulfur, boron).
- Metalliferous brines and mud with concentrations of zinc, copper, lead, and silver.
- Accumulations of industrial minerals in coastal areas (sand and gravel, limestone/shells, silica sand, semiprecious stones, phosphorite, barite, glauconite).
- Accumulations of heavy minerals in detrital placer deposits in continental shelf areas (cassiterite, rutile, zircon, monazite, magnetite, ilmenite, chromite, gold, diamonds).
- Hydrocarbons in reservoir rocks below the sea floor predominantly on the continental shelf and slope (petroleum, natural gas).
- Authigenic metalliferous nodules, encrustations and sediments with concentrations of ferromanganese, cobalt, nickel, copper, zinc (manganese nodules, cobalt crusts, ferromanganese oxides).

The exploitation of minerals from the oceans has a longer history. The production of salt started in China 4000 years ago and today, 20% of the world salt production originates from "salt farms" in India, China, Japan, USA, Israel, and Turkey. The first oil well in the sea, a few meters off the coast of Southern California, was drilled in 1896. Large-scale offshore oil production started in 1923 in the Maracaibo Bay, Venezuela. In 1907, the first offshore tin dredge operated at Tongkah Harbour, Phuket, Thailand. In 1926, the first plant to produce bromine from sea water started in California, USA. In 1930, the Palestine Potash Co. established the first potash plant at the Dead Sea and in 1941; the first magnesium plant commenced commercial operation in Freeport, Texas, USA.

In 1986, approximately 70% of bromine, 60% of magnesium, 90% of rutile, 70% of zircon, 25% of ilmenite, 24% of petroleum, 19% of natural gas, and 12% of tin in the world originated from marine deposits.

Prepared by the United Nations International Law Commission established in 1947, the First UN Conference on the Law of the Sea (UNCLOS I) met in Geneva from 24 February to 27 April 1958. UNCLOS I already challenged the traditional "freedom of the seas policy", which goes back to Hugo Grotius' doctrine of "mare liberum" (1609) and was last reaffirmed by the League of Nations' Codification Conference on questions of the territorial sea in The Hague, Netherlands, from 13 March to 12 April 1930.

For the first time, UNCLOS I provided coastal states with substantial rights of use. Of the four conventions adopted, the most important is the Convention on the Continental Shelf according coastal states the sovereign right to exploit the mineral resources on and under the sea-bed. The Convention on the High Seas guaranteed freedom of shipping, of aviation, of mineral exploration, the right to lay cables and pipelines as well as the freedom to exploit the resources of the deep sea. The Convention on the Territorial Sea and the Contiguous Zone fixed the limits of coastal waters. The limits of areas such as the Continental Shelf, the Continental Slope and the Deep Sea were, however, not precisely defined.

The Second UN Conference on the Law of the Sea in Geneva from 16 March to 26 April 1960 (UNCLOS II) was particularly concerned with reaching agreement on the limits of "territorial sea". This attempt failed because although many countries advocated a 12 nautical mile zone of coastal waters, others, above all the USA, wished to retain the 3-mile zone.

In response to the UN Maltese delegate Arvid Pardo's proposal on 1 November 1967, the UN General Assembly adopted a resolution in 1969 to call a Third UN Conference on the Law of the Sea (UNCLOS III). A declaration of principles of 17 December 1970 declared that the sea-bed and its resources beyond the limits of national jurisdiction constitute the Common Heritage of Mankind.

After preparations by a UN Committee on the Seabed, the first session of UNCLOS III began in New York. Fourteen rounds of negotiations followed and the new Convention on the Law of the Sea (UN Doc. A/CONF 62/122) was adopted on 30 April 1982 by 130 states voting in favor and 4 against, 17 abstaining.

After 9 years of difficult negotiations, the Third UN Conference on the Law of the Sea was officially concluded on 10 December 1982 in Montego Bay, Jamaica, when 117 countries (and the Council for Namibia) signed the new Convention. The USA, the United Kingdom and the Federal Republic of Germany did not sign. By the closing date of 9 December 1984, 159 states had signed the Convention, which will come into force one year after the ratification by 60 signatories; by 1987, 33 states had completed ratification.

The new Convention with 320 Articles, 9 Technical Annexes and 5 Resolutions contains important international regulations:

– Extension of the breadth of the Territorial Sea of the coastal states to a limit not exceeding 12 nautical miles, measured from a precisely determined baseline (Article 3).

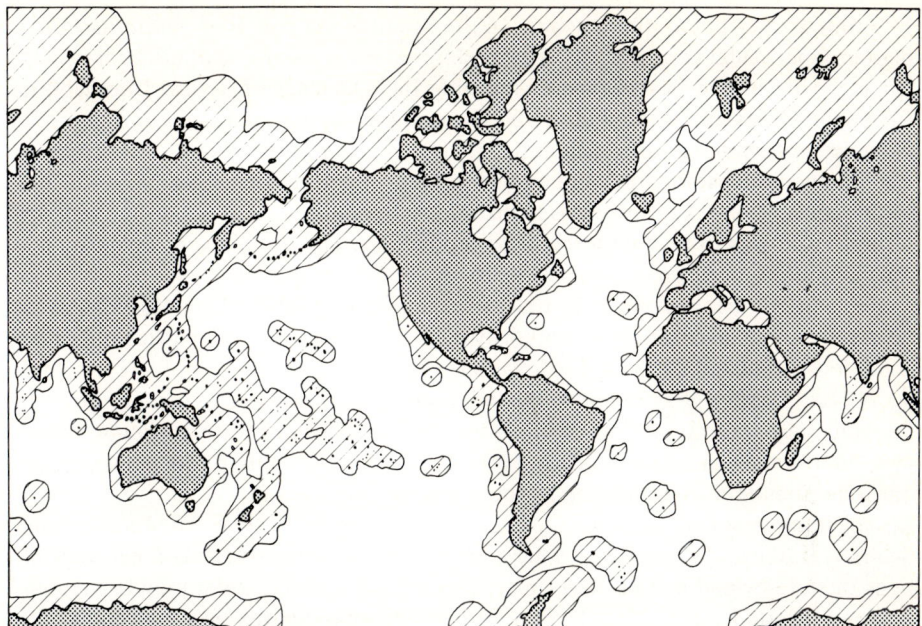

Fig. 8.3. Exclusive Economic Zone of 200 nautical miles

- Establishment of a Contiguous Zone to prevent infringements of national coastal laws, not exceeding 24 nautical miles.
- Creation of an Exclusive Economic Zone (EEZ) adjacent to the Territorial Zone or Contiguous Zone not to exceed 200 nautical miles measured from the baseline (Article 57), (cf. Fig. 8.3). Within the bounds of this EEZ, the coastal state has the sovereign right to explore, exploit, conserve and manage all natural resources in the water or on the sea-bed and its subsoil (Article 56). The prevailing national mining laws apply for the exploration and mining of minerals. However, all other states have rights referring to freedom of navigation, overflight and laying of submarine cables and pipelines (Article 58).
- Redefinition of the Continental Shelf as comprising the seabed and subsoil of the submarine areas that extend beyond the Territorial Sea throughout the natural prolongation of its land territory to the outer edge of the continental margin. On the continental shelf, the coastal state has the sovereign right to explore and exploit natural resources (Article 77).
- Revision of the Convention on the High Seas (1958) stipulating the principle of freedom but restricting the exploration and exploitation of mineral resources in the so-called Area. The Area consists of all parts of deep sea-beds beyond the EEC and the Continental Shelf. According to Article 136 the Area and its mineral resources are the common heritage of manking. The establishment of an International Sea-Bed Authority based in Jamaica is planned to govern the Area, with various organs such as an Assembly, a Council and a Secretariat to ensure a controlled utilization of minerals.
- Establishment of the Sea-Bed Disputes Chamber to function as a jurisdiction body for maritime disputes.

From the beginning, the functions of the International Sea-Bed Authority were a matter of particular controversy. The developing countries saw a comprehensive controlling body as the best means of realizing their concept of a New International Economic Order and wanted above all to ensure an equitable distribution of profits from deep-sea mining. One of their demands was thus to provide the Authority with a comprehensive exploitation and trade monopoly, the Enterprise System. According to this concept, the Authority was to dispose of an operative arm, its own mining enterprise, and was to be able to regulate production. The questions of financing and the distribution of profits (or losses) were not, however, settled.

In contrast, the group of industrialized countries suggested a franchise system for deep-sea mining, according the Authority the sole right to issue mining licenses under conditions stipulated by the United Nations.

A Joint Venture System and a Parallel System were discussed as compromises. In the latter case, enterprises would apply for exploration rights for a maximum of two mine fields which they had already explored, receiving mining-rights, however, for only one, the other to be worked by the Authority as it sees fit.

The text of the Convention is now based on the Parallel System. The International Sea-Bed Authority is entitled to issue a license to consortia of transnational corporations or state-owned mining companies, which then conduct exploration in the mineral area allotted them. After 8 years, half of the mineral area must be returned to the Authority. The Authority can, of course, dispose of the mining fields allocated to it as it wishes; its own mining firm, the Enterprise, can conduct mining operations. It is also entitled to require of the mining companies that they pay duties, submit to production restrictions and transfer modern marine technology. These latter provisions have been sharply criticized by those countries in which the technical know-how for deep-sea mining has been developed. As compensation, these countries have been accorded the status of pioneer investors who have the right to reclaim the value of the amount invested hitherto. The countries eligible for this status are Canada, the Federal Republic of Germany, France, Great Britain, India, Japan, the Netherlands, the USA and the USSR.

The redistribution of tax revenue is to be effected from a special fund to be established by the Sea-Bed Authority. The financial terms of contracts include an application fee of US-$ 500,000 and an annual fixed fee of US-$ 1 million or a production charge (royalty), whichever is greater.

Deep-sea mining laws and international agreements must now comply with the new Law of the Sea Convention, as does the Agreement on Provisional Regulations for Manganese Nodules in the Deep Sea concluded by France, the Federal Republic of Germany, Great Britain and the USA in mid-1982.

The Preparatory Committee, whose task is to establish the framework for the new law, held several sessions in 1987 in Kingston, Jamaica. The main work is on the settlement of conflicting seabed mining claims. In August 1987, the Committee decided to provide registered pioneer investor status to France, India, Japan, and the USSR.

Deep-sea mining focuses largely on the extraction of manganese nodules at depths of between 2000 m and 7000 m (average depth about 5000 m). The most promising areas, according to the latest exploration findings, are in the Pacific near the equator, for example the Clarion-Clipperton Belt. The potential resources (Ni + Cu at least 1.76%) are estimated at 100 billion t (moist material) containing 25% to 35% Mn (average 27.5%), 1% to 2% Ni (1.22%), 0.5% to 2% Cu (1.00%), 0.1% to 0.5% Co (0.25%) and small quantities of Zn, Mo, and V.

Several techniques have been designed for mining, transportation, and processing of manganese nodules. This technology has not progressed beyond the trial or pilot process. The annual capacity of the prospective first plants is estimated at 3 million t moist nodules, enabling a production of 20,000 t copper, 23,000 t nickel and 3,500 t cobalt per year per plant. The recovery of the manganese contents of the nodules remains economically controversial. The investment costs for a mining and processing unit were estimated, in 1980, at about US-$ 1 billion. The operating costs alone would amount to roughly US-$ 200 million per year. Deep-sea mining of manganese nodules is thus not feasible for the foreseeable future.

Five international consortia have shown interest in mining manganese nodules. As can be seen from a list of the members of the consortia, nine Western industrialized nations are working on the development of marine mining technology:

- A consortium made up of Sohio/Kennecott Exploration Corp. (USA 40%), Rio Tinto-Zinc Corp. (UK, 12%), Consolidated Gold Fields (UK, 12%), BP Minerals International (UK, 12%), Noranda Exploration Inc. (Canada, 12%) and the Mitsubishi Group (Japan, 12%);
- The Ocean Minerals Company (OMINCO), with participation by Billiton/Shell (Netherlands, 25%), Bos Kalis Westminster (Netherlands, 10%), Lockheed (USA, 40%) and Amoco Minerals/-Standard Oil of Indiana (USA, 25%);
- The Ocean Mining Associates (OMA), made up of Essex Minerals/US Steel (USA), Sun Ocean Ventures/Sun Oil (USA), Union Minière S.A. (Belgium) and ENI (Italy), each with 25% participation;
- The Ocean Management Inc. (OMI) with 25% participation each of Inco Ltd. (Canada), AMR-Group (Federal Republic of Germany: Metallgesellschaft, Preussag, Deutsche Schachtbau/Salzgitter), DOMCO Group (23 Japanes firms), and Sedco Inc. (USA);
- The Association Francaise pour l'Étude et la Recherche de Nodules (AFERNOD) made up of various, mainly government-owned French enterprises, such as BRGM, CNEXO, CEA, SLN, and Chantiers.

In the 1980s, the activities of the consortia have been far less extensive and enthusiastic than in the early 1970s. On the one hand, the ratification of the Convention on the Law of the Sea is still pending, and on the other, anticipated costs rose and profits have declined as a result of unexpected technological, financial, and ecological problems, and of depressed metal prices.

8.2.3 The International Energy Programme of the Organization for Economic Cooperation and Development

The Organization for Economic Cooperation and Development (OECD) comprises the 24 leading Western industrialized countries. The coordination of development aid activities is overseen by the Development Assistance Committee (DAC), established on 5 October 1961. The 18 members of the Committee (17 OECD member countries and the EC Commission) provide 99% of the funds of the Official Development Assistance (ODA) of the industrialized countries (1985: 28.9 billion US-$) and for 79% of the total World ODA. Each member is required to submit to an "examination" of its development assistance measures by the Organization.

As a reaction to OPEC's supply restrictions during the so-called oil crisis in late 1973, the USA instigated an energy conference in Washington in February 1974 in which 12 OECD countries took part. A comprehensive International Energy Programme (IEP) was worked out, which was adopted in October 1974 by the governments involved (USA, Canada, Japan, Norway and 8 EC countries, excepting France). The Program covered four areas:

– A mechanism for joint allocation of oil in crisis situations by a quota system (oil sharing).
– Ways to improve transparency of the oil market, particularly by monitoring the activities of the oil companies.
– Approaches to lessen dependence on oil, by cooperating in the rational use of energy and by developing alternative sources of energy.
– Promotion of dialogue between the oil producing and oil consuming countries.

To implement the IEP, the International Energy Agency (IEA) was founded under the aegis of the OECD. It officially commenced business in Paris in mid-November 1974, with a session of the Executive Committee in charge of administrative matters. The IEA includes a Governing Board made up of delegates (e.g. at ministerial level) from the member states and the Secretariat answerable to the Executive Director. A weighted distribution of votes and a complicated mode of voting based on the majority principle are intended to avoid possible collisions of interests. Four permanent groups are charged with detailed tasks (crisis management, oil market, long-term cooperation, relations to oil exporting countries).

Since the IEA was constituted in November 1974, the number of member countries has risen to 21.

According to the preamble in the agreement, the Agency's major goal is to secure reliable oil supplies on reasonable and equitable terms, to be attained by

– increasing self-sufficiency;
– stipulating emergency measures;
– obtaining additional information on the oil market;
– improving relations between the producing countries and consuming countries.

The Agency's major concern in the initial years was the establishment of a crisis management facility for acute bottlenecks in supply. The members were called upon to establish national buffer stocks (obligatory reserves of at least 90 days net oil imports) and to pass national legislation to restrict consumption in the event of a crisis. In December 1986 the stocks of all OECD countries comprised 440 million tons or 96 days' supply.

A joint oil distribution system for allocating rights and duties for members was also established after prolonged negotiations, but has not yet been put to the test. Over the last few years, however, the IEA has been largely occupied with supporting research and technological development programs for energy conservation and for improved patterns of energy consumption. The Intergovernmental Conference of the 19 countries then members of IEA adopted concrete energy-saving measures in October 1977 for the first time.

By 1985, members of the IEA were to reduce their oil imports to 26 million barrels per day. In addition, 12 principles of energy policy were formulated, to provide for the gradual replacement of petroleum in the generation of electricity and the sustained expansion of nuclear plant capacity. At the end of 1979, the members set new oil-saving goals. Instead of 26 million bpd in 1985, a maximum of 24.6 million bpd were to be imported by 1990 and petroleum's share in overall energy consumption was to be reduced from 52% to 40% by that time.

In May 1979, the IEA Governing Board adopted the Principles for IEA Action on Coal. These principles provided a framework for expanding world coal production, use, and trade, and established the Coal Industry Advisory Board. In December 1980, the Governing Board agreed on Lines of Action for Energy Conservation and Fuel Switching.

For the OECD countries, the International Energy Programme remains the primary international protection against tighter oil markets in the longer term and, through the IEA oil emergency allocation system, the best safety measure against disruption of oil supplies.

The main field of research of the OECD Nuclear Energy Agency, however, is concerned with special problems of nuclear power supply.

8.2.4 The European Community

The considerable dependence of EC countries on commodity imports has led them to pursue an independent EC commodity policy which is clearly geared to guarantee supplies. The situation is viewed as critical not only for petroleum and natural gas, but also for a number of metals (see Table 8.3).

In order to promote security of supply and to increase self-sufficiency in minerals, the EC supported the mining industry at home and abroad with financial and technical programs including R&D, prospecting grants and investments through the European Investment Bank (for example 52 million US-$ for the Ok Tedi goldcopper project in Papua New Guinea, or 600 million US-$ for the Carajas project in Brazil).

Table 8.3 Basic metal supply situation of the European Economic Community 1985

Metal	Import dependency*	Possibilities of:		Political risk	Supply security
		Recycling	Substitution		
Aluminum	81%	18%	Many	No	Satisfactory
Chromium	99%	15%	Some	Yes	Insecure
Copper	99%	34%	Some	No	Adequate
Tin	90%	25%	Some	No	Adequate
Iron	94%	38%	Few	No	Satisfactory
Manganese	99%	0%	Few	Yes	Insecure
Platinum	100%	50%	Few	Yes	Insecure
Tungsten	92%	11%	Some	Yes	Insecure
Vanadium	100%	2%	Some	Yes	Insecure
Zinc	76%	22%	Some	No	Adequate

* Primary products only.

Source: Eurostat, 24. Edition, Statistical Office of the EC, Luxembourg 1987.

Apart from classical supply policy aimed at cheap, secure supplies of raw materials and their rational utilization, the EC has since the early 1970s devised a new supply policy which is better adapted to the situation on the commodity markets. Of particular importance here is the cooperation with developing countries in Africa, the Caribbean, and the Pacific (ACP countries), established in multinational agreements known as Lomé Conventions (see below).

8.2.4.1 Lomé I Convention

After 19 months of negotiations in Lomé, Togo, a convention between the European Community and 46 developing countries from Africa, the Caribbean and the Pacific (ACP countries) was signed on 28 February 1975. Apart from trade agreements, this convention focuses on guaranteeing the purchasing power of export revenue in the ACP countries, many of which are among the poorest of the developing countries. The Lomé I Convention entered into force on 1 April 1976 for an initial period of 5 years, and supersedes the Agreements of Yaoundé I and II (1963 and 1969) and Arusha (1971 to 1975), concluded by the EEC and 18 and then 19 African states. The central commodity policy element in the Lomé I Convention is the so-called Stabex system which provides for financial compensation against losses on exports. The EC countries agreed to provide compensation initially for 12 commodity groups in the form of interest-free foreign exchange loans in those cases where the world market price was at least 7.5% lower than the average price for the previous 4 years. Seventy-five million ECU (European Currency Unit, equivalent to 1.12 US-$ in 1976, 1.39 US-$ in 1980, 1.10 US-$ in 1987) per annum were made available for this purpose (a total of 382 million ECU). Only one mineral (iron ore) was included in the 12 commodity groups (36 products). Stabex represented an important first step towards realizing mineral policy goals in the context of an international economic reform. In contrast to commodity agreements, Stabex can be deliberately employed to stabilize the earnings of poorer developing countries. During the Lomé I Convention, the number of members rose to 56. In subsequent conferences, negotiations were conducted to extend the list of commodities eligible for compensation and to improve the Program of Assitance. In July 1978, discussions formally commenced on Lomé II in order to ensure a smooth transition. The ACP countries demanded the inclusion of the minerals copper, bauxite/aluminum oxide, phosphate, and uranium.

In addition to the Stabex system to stabilize export revenue, the Lomé Convention also provides for important trade and development arrangements: the duty-free and quota-free access to the EC market of all industrial products from ACP countries, the duty-free importation of 94% of all ACP agricultural products into the EC, funds for trade promotion, a special quota for ACP sugar, and a special increase in the European Development Fund (EDF) for the 5-year duration of the Convention. Financial assistance for the ACP countries provided for in Lomé I amounted to a total of 3,466 million ECU.

8.2.4.2 Lomé II Convention

On 31 October 1979 in Lomé, Togo, representatives from the EC and 58 ACP countries signed a new convention (Lomé II) to remain in force from 1 January 1981 to 28 February 1985. The signatories regard it as performing a model function, as a pragmatic trade convention on the basis of equal status aimed at maintaining a balanced development policy and overall cooperation, combining the full range of aid and trade development instruments. The permanent dialog is maintained via three joint institutions: ACP-EC Council of Ministers, ACP-EC Committee of Ambassadors, ACP-EC Joint Assembly.

The central mineral-policy component in the Convention is again the stabilization of export earnings. A distinction is now drawn between the traditional Stabex, and a Mining Investment System (Sysmin). The first operates as previously, granting interest-free loans or subsidizing loss of earnings in the case of LDCs for a total of 44 agricultural commodities. The latter provides project assistance when export earnings for selected minerals decline by at least 10% as a result of technical or economic policy difficulties. In addition to iron ore, the only metal to be included in Lomé I, Sysmin covers copper, cobalt, manganese, bauxite, aluminum oxide, tin, and phosphates. Sysmin was provided with 280 million ECU (1 ECU 1981 = 1.12 US-).

Sysmin is a substitute solution to the demands of Zambia and Zaire for the inclusion of copper into Stabex. The German Federal Government played a key role in gaining acceptance for the idea of the mining system amongst the other EC partners. Disbursements from the fund are strictly tied to mining projects and are intended to eliminate the causes for losses in profitability for selected mining companies as rapidly and effectively as possible. Articles 57 to 59 stipulate the procedures for assistance to mining: the EC implements technical assistance on request in the fields of geology or mining or provides capital to support development of mining or energy projects. So far, copper mining in Zaire and Zambia has been given support. In addition, 200 million ECU were made available via the European Investment Bank for further investment projects in mining or in the energy sector, largely to counter the recent reticence to invest in Africa.

Moreover, the 35 least-developed ACP countries are to receive special support, and measures have been taken to help landlocked and island states to cope with the difficulties peculiar to their location.

The list of minerals in the Lomé II Convention is acceptable for the EC countries in terms of supply policy. The EC countries have import requirements for all the mine products mentioned and they view the future supply of some of these minerals as critical (copper, cobalt). The promotion of projects to open up new mines thus benefits both parties.

By the end of 1982, the number of participating ACP countries had risen to 63. The total value of assistance provided under the auspices of Lomé II is 5,607 million ECU.

8.2.4.3 Lomé III Convention

On 8 December 1984, again in Lomé, Togo, delegates from the 10 EC countries and 65 ACP countries signed the Lomé III Convention which came into force on 1 Mai 1986 after a long ratification process. The Convention is due to terminate on 28 February 1990.

The EC regards the Lomé Convention as the only example of the North-South Dialog in practice. The main elements in the field of development aid remain the following:

- The European Development Fund (EDF) provides grants and low-interest loans for national and regional development programs.
- The European Investment Bank (EIB) provides additional loans for development projects.
- Stabex provides cash transfers to offset losses on agricultural exports (925 million ECU).
- Sysmin provides soft loans (long-time loans with nominal interest) for mining industries (415 million ECU).
- Emergency and refugee aid provides grants for natural disasters and refugee situations (290 million ECU).

Sysmin is still the major component for the support of the mining industry in ACP countries. The loans are provided not only for the stabilization of production capacities but also for maintenance, rationalization, and diversification of mineral production. Sysmin in Lomé III covers iron ore, copper, cobalt, manganese, bauxite, tin, and phosphates. The soft loans are provided in cases where the country has earned more than 15% of total export revenue from the mineral concerned and the actual price is 10% less than the price 4 years ago.

The EC commodity policy met its major objectives: the foreign exchange earnings of developing countries in Africa, the Caribbean and the Pacific (APC) have been stabilized; the mining industry of mineral-exporting ACP countries has received financial assistance for modernization and increasing production; the mineral-import requirements of EC countries for some mining products have been better secured. In view of this experience, the Lomé Conventions are considered as a successful concept, limited only by the availability of financial resources. There is no doubt that after the termination of the present Lomé III Convention in 1990, there will be a continuation of the institution of the Lomé Conventions.

8.3 International Cooperation in Mineral Exploration and Exploitation

Coined at the beginning of the 1950s, the term "developing country" primarily refers to a nation's level of economic development. Economic indicators are the major criteria for classification. The United Nations employs the following three indicators:

- Per-capita gross domestic product (GDP)
- Share of industrial production in the GDP
- Rate of literacy.

Countries are classified by weighting these norms. The United Nations includes 128 countries (excl. China) in the developing country group and further differentiates between 3 categories:

- Least Developed Countries (LDC): underdeveloped countries, whose indicators did non exceed the following specific threshold figures when the list was drawn up in 1971: 100 US-$ GDP per capita (1986: 355 US-$), industrial production accounting for 10% of GDP and a literacy rate of 20%. In 1986, 36 countries were classed as LDCs.
- Most Seriously Affected Countries (MSAC): those countries hardest hit by the economic shocks of the 1970s, whose import (particularly petroleum) costs and indebtedness have risen drastically and whose growth prospects are limited. In 1986, 45 countries belonged to the MSAC group, of which 27 were LDCs, so that the lists overlap.
- Newly industrializing countries (NIC), i.e. those that have developed their own economic momentum, raising industrial production, per-capita income and energy consumption. In 1986, about 25 countries belonged to the NIC list, for example Brazil, Argentina, Mexico, Korea, Singapore.

The Development Assistance Committee (DAC) of the OECD has a slightly different classification and lists 158 nations and territories as developing countries.

Though each developing country has, of course, its own particular level of development, all share common features:

- Low living standard of the large majority of the population.
- Agricultural sector accounting for a large share of the GNP, though unable to produce an adequate or balanced supply of foodstuffs.
- High unemployment, much of it disguised.
- Lack of capital, low rate of savings.

Table 8.4 Offical Development Assistance (ODA) by major donors (million US-$), 1985/86

Country group	ODA	Per cent of total	Per cent of GNP
DAC total	29643	79.0	0.35
USA	9362	24.9	0.23
EC members	11775	31.4	0.51
Germany (W)	2842	7.6	0.44
France	2668	7.1	0.51
United Kingdom	1510	4.0	0.33
Japan	3853	10.3	0.29
OPEC total	3669	9.8	0.80
Saudi Arabia	2772	7.4	3.69
Kuwait	676	1.8	3.12
CMEA total	3658	9.7	0.27
USSR	3195	8.5	0.32
LDC donors*	380	1.0	
Total world	37541	100.0	

* China, India, Israel, Yugoslavia.
Source: OECD, DAC-Review, Paris 1988.

- Inadequate level of education.
- Deficient health care.
- Extremely inequitable distribution of goods and access to education.
- Growing balance-of-payments deficit and heavy foreign debt.

Despite extensive development aid granted in many forms since the 1960s (Table 8.4), by the middle 1980s the overall economic situation of the developing countries had not substantially improved.

The reasons adduced to explain this are:

- Unbridled population growth, with an annual rate of 2% and above. According to UNESCO estimates, the world's population will have risen to 6.1 billion by the year 2000, 80% thereof in developing countries.
- A continued worsening of the terms of trade with a stagnating share in world trade (1975: 20.8%; 1985: 21.2% of all imports; 1975: 23.9%; 1985: 24.3% of all exports).
- A disturbing increase in foreign debt, which has attained 1000 billion US-$ in 1987 (1970: 64 billion US-$, 1985: 912 billion US-$), the burden of debt service thus becoming prohibitive (1985: 131 billion US-$). The debts of an increasing number of developing countries have had to be rescheduled.
- Widespread political instability, leading to excessive expenditure on armaments, armed conflict causing a flux of refugees and frequent changes of government.

In the context of bilateral and multilateral development assistance, the energy sector has gained in importance since the oil price rises of 1974 and 1979 and though not accorded high priority, mining has become crucial for some developing countries.

The Resolution of the UN General Assembly of 5 December 1980 on International Development Strategy for the Third Development Decade stipulates the following goals for the energy and mining sectors:

- Pursuing the measures agreed for the Integrated Program for Commodities (cf. Sect. 8.2.2.2).
- Promoting mineral processing in developing countries.
- Supporting the establishment of institutions to explore and exploit mineral resources.
- Increasing exploitation of alternative energy sources available to lessen the dependence on hydrocarbons.

8.3.1 Cooperation *Programs of International Institutions*

Multilateral schemes in development aid are implemented by various agencies. The three major areas are:

- Technical Cooperation
- Financial Assistance
- Aid in Trade.

Over the years, a large number of institutions providing multilateral development assistance have emerged, which also directly or indirectly promote mineral projects. The major international donor organizations are listed below:

- Economic and Social Council (ECOSOC) of the United Nations with its United Nations Development Program (UNDP), which predominantly funds international Technical Cooperation projects often implemented (until 1971 exclusively) by executing agencies.
- Special Organizations of the United Nations, such as the International Atomic Energy Agency (IAEA) based in Vienna or the UN Educational, Scientific and Cultural Organization (UNESCO), which are mainly concerned with Technical Cooperation.
- Special organs of the United Nations General Assembly, such as the UN Industrial Development Organization (UNIDO) or the UN Capital Development Fund (UNCDF), which are primarily involved in Financial Assistance.
- The World Bank group which was founded in Bretton Woods, USA, in 1944 and consists of the International Bank for Reconstruction and Development (IBRD, World Bank) and its legally and financially autonomous subsidiaries, the International Finance Corporation (IFC, since 1956) and the International Development Association (IDA, since 1960), which mainly provide capital aid. Since 15 November 1947, the World Bank has been a Special Organization of the United Nations. At the end of 1985, 148 countries had subscribed 58.85 billion US-$. The largest subscribers are the USA (20.9%), Japan (6.93%), United Kingdom (6.0%) and the Federal Republic of Germany (5.83%). The total World Bank loans until 1985 amounted to 81.6 billion US-$.
- The International Monetary Fund (IMF), founded during the International Monetary Conference in July 1944 in Bretton Woods, primarily furnishes trade assistance to help improve the balance-of-payments situation of developing countries. The development activities of the IMF and the World Bank have been coordinated since 1974 by a joint ministerial committee, the Development Committee (DC) in Washington, made up of 21 members, including China.
- The General Agreement on Tariffs and Trade (GATT) established as part of the Havana Charter of 1947 with the status of a UN Special Oranization, which aims at liberalizing world trade and reducing customs barriers.
- The three large regional development banks, Banco Interamericano de Desarrollo (BID), Asian Development Bank (ADB) and Banque Africaine de Développement (BAD, BfAD), which provide capital assistance to the regional members and maintain special funds to issue credit on particularly favorable terms to the poorest member countries (Asian Special Development Fund, established in 1973 by ADB, for example).
- The Development Assistance Committee (DAC) of the Organization for Economic Cooperation and Development (OECD), which mainly coordinates and supervises Technical Cooperation projects.
- The European Community (EC) via the Lomé Convention (cf. Sect. 8.2.4), the Generalized System of Preferences of 1 July 1971, the European Development Fund (EDF) and regional cooperation agreements (of 7 March 1980 with ASEAN countries, for example).

- The OPEC countries with the OPEC Fund for International Development (cf. Sect. 7.2.1.1), which has been assisting economic development through financial cooperation since 1976.

8.3.2 The UNDP and the UN Revolving Fund

Between 1959 and 1982, the United Nations Development Program (UNDP) and its predecessors (UN Special Fund, 1958 to 1965) supported 200 projects in the mineral sector of 75 developing countries. Total expenditure amounted to approximately 200 million US-$, a mere 6% of the UNDP total in this period. The contributions of the recipient countries are estimated at an additional 140 million US-$.

A key role in United Nations natural resources development projects since 1975 has been played by the United Nations Revolving Fund for Natural Resources Exploration established in 1973 by the UN General Assembly to meet the shortage of high-risk exploration financing.

By 1985, governments of 7 industrialized and 10 developing countries had contributed 33.6 million US-$ to the Fund. The Fund received 81 proposals from developing countries for mineral exploration and geothermal energy reservoir development. Eighteen projects have been approved and ten of them have been completed. During 1985/86, the Fund for the first time attracted third-party co-financing of projects, amounting to 3.3 million US-$.

The Fund finances all exploration costs. The governments receiving assistance are obliged to make small payments to the Fund if their exploration activities lead to commercial production, e.g. 2% of the annual output value for a period of 15 years. Feasibility studies are available on a loan basis.

The UNDP Governing Council is responsible for project approvals. Project implementation is largely the job of the Natural Resources and Energy Division of the UN Department of Technical Cooperation for Development, New York.

Only government projects are sponsored. The UNDP input is provided as soon as the project is identified and planned. The largest contributions are for the provision of exploration equipment and experienced personnel. All UN experts, including short-term specialists, must have a local counterpart in order to ensure that the project is conducted jointly and that local specialized manpower receive on-the-job training.

Seven mineral exploration projects have produced good results:

- Argentina: gold and silver deposits were discovered in three areas in the Andes of Patagonia, Andean mountains (1 million tonnes of ore with 10 grammes of gold per ton).
- Benin: a kaolin deposit was found in the Ketou region, Ouémé Province (250,000 t of white kaolin).
- Congo: phosphorite was discovered in offshore areas (24 million t of phosporite concentrate).
- Ecuador: a silver deposit was explored at San Bartolomé near Cuenca.
- Cyprus: a massive sulfide deposit with base and precious metals was found in the Troodos Mountains.

- Suriname: copper and gold mineralizations for small-scale mining were explored on the Tapanahony River.
- Panama: copper and gold mineralizations of subeconomic importance were discovered in the NE part of the country.

By the year 2000 at the latest, the Revolving Fund should be completely self-financing and serve as a separate facility of the developing countries for self-help in the area of mineral and geothermal resources. It is thought that by then the repayments to the Fund will be large enough to ensure self-financing. But the projects mentioned above are not as attractive as the UNDP had hoped.

In addition to the Revolving Fund activities, the Minerals Branch of the UN Natural Resources and Energy Division is executing many Technical Assistance projects in the fields of institution building (strengthening of Geological Surveys, Bureaus of Mines etc.), mineral exploration, feasibility studies, training and editorial services. In 1987, 57 projects with a total UNDP contribution of 57.75 million US-$ were underway. Previous results included the discovery of the Mamut copper deposit in Sabah/East Malaysia, offshore tin deposits in the Heinze Basin/Burma, the iron ore deposit Mt. Nimba/Guinea, the copper deposit La Caridad/Mexico and the copper deposit Los Pelambres/Chile.

These diverse and extensive activities of regional UN organizations in the field of mineral resources can best be illustrated by the Economic and Social Commission for Asia and the Pacific (ESCAP). One of the most important committees of ESCAP is the Committee on Natural Resources which sets project priorities and draws up the 2-year working plan for the mineral sector. The executive organ is the Natural Resources Division in the ESCAP headquarters in Bangkok. This Division furnishes export advice in the fields of energy, mineral resources, and ground water development. The focus in the energy sector is on planning aid, the use of renewable energy and supplying rural areas with electricity. Assistance in the field of mineral resources is largely concerned with drawing up regional geologic and structural maps, and studies on mineral potential. Consulting work is performed via project missions of experts and by organizing seminars on specific topics.

The Natural Resources Division (NRD) also oversees four regional institutions:

- The ESCAP Regional Mineral Resources Development Center (RMRDC) originally established with funding from the UNDP and various donor countries as a subsection of the NRD in 1973 in Bangkok, granted independent status in 1978 and based in Bandung, Indonesia, from August 1979 to February 1987. As of March 1987, ESCAP suspended the activities of RMRDC. By the end of 1986, more than 260 expert advisory mission reports had been submitted and about 50 seminars or workshops conducted.
- The Southeast Asia Tin Research and Development Center (SEATRAD Center) founded with the support of the UNDP via an agreement signed by the Governments of the tin-producing countries of Malaysia, Indonesia, and Thailand on 28 April 1977. The Centre in Ipoh, Malaysia has seven departments (for geology, mining, mineral dressing and smelting, mineralogy, laboratories, documentation, and administration) which implement various research projects in the tin mining sector of the three member countries. Training courses and workshops on current problems in the tin industry are also held.

– The two coordination committees for offshore prospecting, the Committee for Coordination of Joint Prospecting for Mineral Resources in Asian Offshore Areas (CCOP/EA) and the Committee for Coordination of Joint Prospecting for Mineral Resources in South Pacific Offshore Areas (CCOP/SOPAC). Both transnational committees were established under the auspices of ESCAP in 1966 (CCOP/EA) and 1972 (CCOP/SOPAC). The former has 12 members, the latter 9. They are to provide coordination and advisory services in developing ocean mining, particularly the exploration and exploitation of heavy minerals and petroleum in the shelf regions.

8.3.3 The World Bank

In 1978, the World Bank estimated the investment needs of the developing countries in the mining sector until 1985 at 95 billion US-$, two-thirds to be met by direct investments on the part of foreign firms, the rest by means of multilateral and bilateral capital aid. The World Bank thus set itself the goal at the beginning of the 1980s of sponsoring four to six mining projects (including coal) a year, primarily by co-financing to spread risk.

The funding strategy of the World Bank is focused on minimizing lending costs and on ensuring the availability of funds. The interest rates are set from time to time at fixed spreads over the London Interbank Offered Rate (LIBOR).

The energy sector has been accorded higher priority by the World Bank since 1980. A minimum of 13 billion dollars was earmarked for a lending program from 1981 to 1985, including 1 billion US- for coal projects and nearly 4 billion US-$ for petroleum and gas projects (see Table 8.5). These loans are intended to facilitate a total investment of 57 billion US-$.

As part of this 5-year program, the following sums were pledged to the following projects (approved in 1982):

a) *Loans for petroleum projects:*

Argentina   100 million US-$ for exploration, production and pipelines construction in an Oil and Gas Project (Banco Nacional de Desarrollo).

Table 8.5 World Bank Energy Lending Programs 1981–85 (million US-$)

Energy sector	Lending program	Total project costs
Coal and lignite	840	4270
Oil and gas	3985	11760
(development)	(1020)	(2610)
(oil development)	(1755)	(5900)
(gas development)	(1210)	(3250)
Refineries	150	400
Renewable energies	625	2950
Electric power	7590	37950
Total	13190	57330

Source: World Bank, Energy in the Developing Countries, Washington, Aug. 1980.

Egypt   90 million US-$ for the Abu Qir Gas Development Project (Egyptian General Petroleum Corp.).
Ivory Coast   101.5 million US-$ for government participation in offshore exploration and development (Soc. Nat. d'Opérations Petrolières de la Côte d'Ivoire).
Romania   101.5 million US-$ for the Videle/Balaria Enhanced Oil Recovery Project.

b) *Loans for coal projects:*

Indonesia   210 million US-$ for the development of the coal deposit Bukit Asam, including the construction of a railway to the port of Panjan and the evaluation of further coal reserves in Sumatra (Coal Mining Development and Transportation Project).

The Philippines   17 million US-$ for a Coal Exploration Project (Philippine National Oil Co.).

The total of 668 million US-$ amounted to 6.5% of all World Bank credit for 1982.

In recent years, mining projects have also been funded in the following:

- Thailand   (approved 1981) 8.9 million US-$ for feasibility studies and engineering in potash deposits near Khon Kaen.
- Morocco (1982)   9.5 million US-$ for the Pilot Project for Small Scale Mining (improvement of mining and dressing technology in small lead-zinc mines).
- Liberia (1982)   11.7 million US-$ for the National Iron Ore Company Rehabilitation Project for modernizing machinery and equipment and repairing transport tracks.
- Brazil (1983)   304.5 million US-$ for the Carajas Iron Ore Project (CVRD).
- Tunisia (1984)   13.4 million US-$ for consultancy service in the phosphate industry.
- Zambia (1984)   75 million US-$ for the first phase of ZCCM copper mines rehabilitation and modernization program (co-financed by EC and BAD).
- Mauritania (1986)   20 million US-$ for the rehabilitation of iron ore mines (Soc. Nat. Industrielle et Minière).
- Zaire (1986)   110 million US-$ for a five-year rehabilitation and modernization program (1986–1990) of Gecamines copper mines (co-financed by EC, BAD, and CCCE).

It should, however, be noted that less than 1% of total World Bank credit was issued for the mining sector (period 1981–86: 545 million US-$ out of a total of 66.6 billion US-$). Nevertheless, the funding of mining projects contributed to the tendency to oversupply in some mineral markets in the 1980s.

8.3.4 The European Community

Cooperation between the EC and developing countries in the mining sector focuses on stabilizing export revenue (Mineral Fund) and facilitating trade via the measures agreed on in the Lomé Conventions (cf. Sect. 8.2.4). In addition, the European Investment Bank (EIB) also finances mineral projects. Credit is provided either from the bank's own finances or the EC budget funds it administers or a combination of the two. Here are some examples of mineral projects in ACP countries for which credit has been provided in recent years:

- Zambia received a loan at a low rate of interest (3% below market rate) of 25 million ECU in 1981 for the construction of a copper tailings processing plant in Chingola (Nchanga Cons. Copper Mines).
- Gabon was granted credit to the value of 15 million ECU in 1981 for the enlargement of the uranium ore dressing plant at Mounana Mine (COMUF).
- Ghana was granted a loan of 6 million ECU in 1983 for the modernization of the manganese mine Nsuta.
- Benin received a loan of 18 million ECU (13.5 and 4.5 million) in 1984 for the second development phase of the offshore oil field Séme.
- Sudan was granted a loan of 4 million ECU in 1985 for the development of the gold-mining district Gebeit (Sudan Minex Gold Mining Venture).
- Burkina Faso received a loan of 7 million ECU in 1985 for the development of the gold mine Pourra, West of Ouagadougou.

Notes on the Literature

Details on United Nations development policy and mineral policy can be obtained from UNCTAD (1985) in *The Least Developed Countries,* as well as from Annual Reports of the World Bank and the International Monetary Fund. A comprehensive review of the Integrated Commodity Program is provided by Behrmann (1977) in *International Commodity Agreements*. The EEC commodity policies are described in more detail by Hill (1985) and by Eisold and Hasse (1984), while more information about the OECD policies on energy can be obtained from a book published by the Internationl Energy Agency (1983): *Energy Policies and Programs of the IEA Countries*.

Chapter 9 Policies in Industrialized Countries

Most European industrialized countries and Japan depend on substantial imports of minerals. Owing to this, the supply aspect of mineral policy has in recent years been given such emphasis that the term mineral supply policy has emerged.

The EC's Office of Statistics in Luxembourg has devised a system of indicators for the supply situation covering the degree of self-sufficiency in primary products (Table 9.1), the degree of self-sufficiency incorporating domestic recycling (Table 9.2) and technological dependence on mineral imports.

The results of a study on the 21 most important minerals can be summarized as follows. The EC is not self-sufficient in any of the mineral commodities selected, importing the remainder from non member countries. For manganese, nickel, mercury, titanium, and zirconium, EC mine production plus metal recycling amounted in 1978 to between 0% and 5% of consumption, an import dependence of 95% to 100%. For cobalt, chromium, molybdenum, tantalum, and vanadium, the grade of self-sufficiency ascertained ranged from 5% to 10%, for nickel, tin and phosphate from 10% to 20%. Only for lead (69%) and fluorine (87%) was a high grade of self-sufficiency recorded.

The mineral commodity supply structure of the Federal Republic of Germany is characterized by an almost complete dependence on imports of petroleum, gas, metals and industrial minerals. Roughly half of these imports stem from industrialized countries rich in minerals, such as South Africa, USA, Canada and Australia as well as the USSR, the remainder from about 25 developing countries such as Brazil, Zaire, Chile, China, and Thailand.

Security of supply can be achieved by strategies to procure commodities and others as regards their application, particularly via the following:

Table 9.1 Metal self-sufficiency in major OECD countries 1984

Metal	EC		USA		Japan	
	Mining production	Recycling	Mining production	Recycling	Mining production	Recycling
Aluminum	15 %	18 %	3 %	12 %	–	12 %
Copper	0–1 %	33 %	46 %	20 %	3 %	29 %
Lead	7 %	43 %	42 %	43 %	13 %	32 %
Chromium	0–1 %	15 %	–	19 %	0–1 %	–
Molybdenum	8 %	5 %	37 %	–	0–1 %	–
Nickel	6 %	15 %	0–1 %	7 %	–	–
Tin	6 %	25 %	1 %	26 %	2 %	16 %
Tungsten	7 %	11 %	15 %	16 %	20 %	–

Source: Statistical Office of the European Communities, Luxembourg 1987.

Table 9.2 European Community Raw Materials Consolidated Balance Sheet, 1984

Metal (unit)	Availability					Utilization			
	Mining production	Recycling	Imports	Stocks	Total	Consumption	Exports	Stocks	Total
Al (1000 t)	956	1175	4192	49	6372	4775	1597	–	6372
Cu (1000 t)	3	1130	2235	3	3371	2659	712	–	3371
Pb (1000 t)	108	625	696	11	1440	1218	222	–	1440
Cr (1000 t)	6	102	580	–	688	644	41	3	688
Ti (1000 t)	–	2	581	6	589	383	206	–	589
Ni (1000 t)	13	36	186	–	235	178	52	5	235
Sn (t)	5216	20137	49002	4875	79230	62375	16855	–	79230
W (t)	1040	1569	11479	–	14088	9692	3134	1262	14088

Source: Statistical Office of the European Communities, Luxembourg 1987.

- Diversifying the channels of supply;
- Optimizing the use of indigenous resources;
- Developing technologies to reduce consumption;
- Intensifying recycling.

These can be complemented by foreign policy initiatives:

- To secure mineral supplies at a legal and political level.
- To preserve a mainly uncontrolled international mineral trade.
- To achieve a cooperative partnership between those industrialized nations lacking in mineral resources and the mineral exporting developing countries.
- To promote international cooperation between geologic agencies and research institutes engaged in mineral exploration and mining.

French mineral commodity policy is more global in approach. It aims at lessening the substantial fluctuations in supply to which France's industrial sector is subject by systematic stockpiling (cf. Sect. 9.1.4) and its import dependence by a selective investment policy. The Government has drawn up a list of promotion measures to stimulate mining investment abroad, particularly in developing countries rich in mineral resources, comprising direct and indirect subsidies, guarantees against risk and technology transfer. The supply program is also supplemented by extensive prospecting and exploration schemes in France's overseas territories (Guyana, New Caledonia) and in developing countries (mainly in Africa). Particularly active in this area is the Bureau de Recherches Géologiques et Minières (BRGM), which is even entitled to acquire interests in mining enterprises.

In Great Britain, a government program to safeguard mineral supplies is still in its initial phase, detailed discussion not having started before November 1979, when the

Institution of Mining and Metallurgy in London organized a symposium on the availability of strategic minerals. The first government measures were the creation of an emergency stockpile for steel refineries particularly vulnerable to shortages in supply (see Sect. 9.1.4) and a consolidation of supply links to British smelting plants and an additional scrap recycling facility for emergencies. Since 1972, however, a domestic exploration program has been receiving subsidies of up to 35% of exploration costs.

Commodity policy in the USA was for a long time implemented exclusively by means of the extensive Stockpile (cf. Sect. 9.1.4). Then in 1959, the Office of Mineral Exploration initiated a modest program to support exploration projects on the part of medium-scale mining companies. Since 1981, the Reagan administration has been giving substantial support to direct and indirect investment programs in the mining and energy sectors. Through the Economic Recovery Tax Act of 1981, mining and oil corporations enjoy the Accelerated Cost Recovery System which permits faster write-offs of depreciable assets and encourages increased investments in exploration and exploitation of minerals.

The US Tax Reform Act of 1986 retains two key provisions for US mining industry:

- The corporate tax rate is reduced from 46% to 34%;
- Two important tax incentives are maintained, e.g. percentage depletion and expensing of mineral exploration and mine development costs.

The reduction of the US corporate tax rate will have an impact on multinational companies' foreign operations. Other countries are concerned (Canada, Australia) and may change their corporate tax burden to prevent companies from relocating operations to the USA and to mining countries where the rates are lower.

In the 1960s, Japan had already begun an aggressive program on direct investment in foreign mining enterprises to secure mineral supplies by concluding long-term supply contracts. The main burden is carried by privates enterprise, supported by government measures such as financing assistance and exploration assistance, the latter via semi-government organizations, the Overseas Mineral Resources Development Co. Ltd. (OMRDC), for example.

Industrialized mineral-rich countries (Australia, Canada, South Africa) formulated a different mineral policy, taking into account the important national task of optimal utilization of the mineral resources. The latest policy approach has been initiated by South Africa where the Mineral Policy Committee presented a White Paper on the Mineral Policy in 1986. The Committee identified eleven objectives and strategies:

- Sustained mineral search (exploration strategy).
- Sound mineral resource management (resource management strategy).
- Optimal development and exploitation of mineral resources (exploitation strategy).
- Minimization of environmental damage in mineral exploitation (environmental strategy).
- Optimal utilization of manpower (manpower strategy).
- Optimum beneficiation of minerals (mineral dressing strategy).
- Maximum benefit from the export of minerals (export promotion strategy).

- Maintenance of an attractive mineral investment environment (investment strategy).
- Continuing cooperation with the states of Southern Africa in mineral matters (cooperation strategy).
- Adequate self-sufficiency in mineral requirements (self-sufficiency strategy).
- Uninterrupted operation of South Africa's mineral industry (contingency strategy).

9.1 Government Programs

9.1.1 Programs for the Preservation and Enhancement of Domestic Mineral Production

The largest industrial power, the USA, also produces substantial amounts of domestic mineral resources (including energy resources). After an oil embargo by Arab members of OPEC in 1973/1974, the US President, Richard Nixon, and the US Congress managed to enforce an American energy policy. On 7 November 1973 Nixon proposed Project Independence and urged the Congress to grant him authority to allocate crude oil and petroleum products in times of emergency supply shortage. In December 1972 the Emergency Petroleum Allocation Act was passed, which gave the President 15 days to establish a Federal Energy Agency.

Project Independence was "the most comprehensive energy analysis ever undertaken" and aimed at achieving self-sufficiency in the energy supply-demand balance by 1985 through developing domestic deposits, utilizing energy resources more effectively, and employing alternative sources of energy, such as synthetic fuels.

In April 1977 the Energy Policy and Planning Group to the Executive Office of the President issued The National Energy Plan. It became clear at this time that the USA could not attain independence from petroleum imports. In the supply-demand balances for 1985 from the National Energy Plan 6.4 million barrels of oil per day or 14% were indicated as imports (in fact, 5.3 million BOPD or 33% of the oil was imported in 1986). In response to the growing dependence on oil imports, the Carter administration developed a rather unpopular energy-saving program in 1977 and the Reagan administration has been pinning their hopes on tax acts (Economic Recovery Tax Act of 1981, Tax Equity and Fiscal Responsibility Act of 1982 and Tax Reform Act of 1986) with preferential tax concessions for small oil projects and attractive investment incentives for the mining and oil industry.

In the Federal Republic of Germany, domestic mineral resources are much more limited. The production of petroleum and gas is minimal (1986: 4.1 million t oil and 12.7 million t oil equivalent natural gas); hardly any deposits of uranium ore have been proven, though substantial reserves of lignite and coal have been discovered. By means of extensive rationalization, the lignite mining sector managed to retain its share of energy production without government support. From 1958, however, coal entered into a crisis, becoming increasingly less competitive in relation to imported petroleum and gas. Between 1958 and 1978, coal production declined from 150 million t to 83.5 million t, increasing again slightly until 1982 (88.4 million t) and went down to 80.3 million t in 1986. The number of collieries (coal mines) in operation (excluding small pits) fell from 173 to 33 (end of 1986), the number of those employed from 600,000 to 164,073 (end of 1986). In order to secure supply, government measures had to be taken to preserve a certain production capacity. In August 1965, an act was passed to promote the utilization of coal, the first "Verstromungsgesetz" (Electricity Act), followed by the second in 1966 and the third in 1974 (to terminate in 1995), all of which provided for the use of a set quantity of coal in generating electricity in order to maintain its share of domestic energy resources in the energy sector at a certain level. In April 1980 an additional agreement was reached between the mining and energy sectors to raise the proportion of coal consumed in power plants.

9.1.2 Promoting Mineral Exploration

In 1971, the Federal Ministry of Economics (BMWi) in Germany launched a program for the promotion of exploration of mineral deposits (Exploration Assistance Program). For exploration projects abroad, German firms were granted loans repayable under certain terms to cover up to 50% of the costs. The government funds provided, 7.5 million DM for 1971 and 9.0 million DM for 1972, were not, however, used up, so the program was extended to domestic projects. In 1973, 13.5 million DM were made available, in 1974, 21 million DM, and in 1975 as much as 24 million DM. After some initial difficulties, the program has proved successful: at the end of 1975, 58 projects in Germany and abroad were underway. The Exploration Program was altered as of 1 October 1975 in the following respects:

- The subsidy percentage for projects of particular economic importance was raised from 50% to 66.67% (but reduced for all foreign projects to 50% and for domestic projects to 35% as of 23.7.84).
- The list of activities eligible for support was extended.
- The repayment deadlines in the case of a find were extended.

The terms of reference were revised on 23 July 1984. The procedure for repayment was made more flexible, e. g., ten half-yearly installments 2 years after starting production. Since 1978, the program has also encompassed the exploration for nuclear fuels.

Within the general context of exploring for mineral deposits, financial support can be requested for the following activities, to be repaid only in the case of a find:

- Exploration or procurement of the results of geologic investigations conducted by others.
- The further development and testing of new methods and equipment for exploration in association with projects found eligible for support.
- Prefeasibility and feasibility studies, including the testing of pilot treatment plants.
- Acquisition of licences or concessions.
- Acquisition of interests in deposits, where development work involving a high level of risk is needed (for political reasons, normal equity participation is impermissible).
- "Extra deep" drilling for gas (only within Germany, terminated 1987).

The enterprise concerned must also guarantee that the mineral be shipped into the EC either as concentrate or metal. Its undertaking to do this must appear credible and practicable.

For 1971 to 1986 the Ministry budgeted a total of 580 million DM for the Exploration Assistance Program, of which only 475 million DM was disbursed. The subsidies were allocated to 387 projects, 318 abroad and 69 in Germany.

Domestic exploration in Germany is clearly on the decline (1985: 23 ongoing projects; 1986: 11 projects), the chances of making a find being regarded as minimal. A certain amount of success could nevertheless be achieved, particularly in areas near mines already in operation (lead-zinc mines Bad Grund and Meggen; baryte mines Dreislar, Wolkenhügel and Wolfach, graphite mine Kropfmühl).

The foreign projects were, in contrast, more effective. The major focus of exploration was the discovery of new deposits. The following projects were particularly successful:

- The tungsten mine, Mittersill, Austria, run by the Metallgesellschaft (MG).
- The lead-zinc mine, Song Toh, Thailand, with MG participation of 49%.
- The gold-copper project, Ok Tedi/Papua New Guinea, with German participation (20% DEG/MG/DEGUSSA).
- The chromite project, Llorente/the Philippines, run by Bayer AG.
- The uranium project, Yeelirrie, Australia, with a 10% participation by the Urangesellschaft.
- The magnesite project, Baymag, Canada, run by Refratechnik.

After 20 years of rather successful operation, the German Exploration Assistance Program will expire in 1990.

Three programs of financial assistance to stimulate mineral exploration for domestic reserves have been conducted by the United States Department of the Interior after World War II. Through the Defense Minerals Administration (DMA) 1950–1951, the Defense Minerals Exploration Administration (DMEA) 1951–1958, and the Office of Mineral Exploration (OME) since 1958, programs have provided for the exploration of as many as 37 minerals considered strategic or critical to national defense in the USA. The assistance of government financial participation in private industry contracts ranged from 50% to 90%, according to the mineral being explored. If a discovery was made, funds expended on the project were repayable, without interest, by a royalty on production. If no discovery was made, no repayment was required.

In 1965, the OME transferred responsibility for the exploration program to the US Geological Survey. For 27 mineral commodities the government funding amounts to 50% of the allowable exploration costs. Seventy-five per cent, however, is available to those firms searching for gold, silver, the platinum-group metals, mercury, antimony, bismuth, rutile, tantalum, and tin. The government participation in a single contract may not exceed 250.000 US-$. Between 1958 and 1974, DME had received 944 applications, approved 208 for contracts, certified 50 contracts as discoveries and spent 4.7 million US-$. The program has been successful and the estimated recoverable value of ore discovered for each dollar of government funds invested in DME contracts was about 24 US-$. In Canada, there are several exploration assistance programs run by provinces such as the Northern Mineral Exploration Assistance

Program, the Ontario Mineral Exploration Assistance Program or the Manitoba Mineral Exploration Assistance Program. The amount of money spent for exploration assistance in these Canadian programs is usually five to six times higher than that in the USA.

9.1.3 Promoting Mining Investments

One of the tools most often employed in mineral policy is the promotion of transnational mining projects. A good example is the Federal Republic of Germany, where mining and smelting firms are supported by the Government when they invest in mineral production abroad, particularly in projects in developing countries. This is done in three ways:

- Government investment guarantees.
- Government financial assistance.
- Tax incentives.

Hitherto, the only foreign investments considered eligible for support have been those in developing countries. As of January 1987 the Federal Republic of Germany has concluded investment promotion agreements with 61 of these countries, 53 of which were put into force under international law. For the Federal Government, these agreements constitute an adequate basis to guarantee investments when required to do so by the investor. In this way, the political risk is covered, not, however, the financial, though it is always difficult to demarcate the one from the other. Restrictive legislation and nationalistic economic policies in developing countries can create a gray zone hindering foreign investment and even leading to gradual expropriation. Government investment guarantees comprise two things:

- Guarantees for capital investments abroad against political risks, administered by the Treuarbeit AG for a relatively small commission.
- Guarantees for credits to finance mineral projects abroad which are not tied to deliveries from Germany, and only given for loans granted by German financial institutions.

By the end of 1986, capital investment guarantees to the value of 675 million DM had been undertaken for 14 mineral projects in ten countries, including exploration for bauxite, antimony, chromium, iron, copper, gold, niobium (columbium), and uranium ore. Credit granted for 21 mineral projects in ten countries amounted to a total of 2,373 million DM.

The financial assistance is provided by the Kreditanstalt für Wiederaufbau (KfW; Bank for Reconstruction) and the Finanzierungsgesellschaft für Beteiligungen in Entwicklungsländern (DEG; finance company for investment in developing countries).

The KfW issues long-term credit to German mining companies on market terms. The project is subjected to a financial examination to ascertain whether it is technically feasible and economically viable.

The DEG sponsors investments in developing countries involving participation or the granting of loans which are equivalent to participation. Equity was last raised in 1978 to 1 billion DM with a view to increased involvement in the primary sector. The first participation was acquired in the copper-gold project, Ok Tedi, Papua New Guinea. DEG's investments are not permanent but limited to between 8 and 15 years. A longer period is, however, envisaged for capital intensive mining projects. In addition to equity participation in new mining enterprises in developing countries, the DEG is even prepared to undertake a completion guarantee. This kind of security is of great importance for mining investment projects.

Until the end of 1981, tax incentives for mining projects abroad were provided for in a special tax law from 1963 (Entwicklungsländer-Steuergesetz) which no longer holds for investments as of 1982.

Other relevant legislation is the Foreign Investment Act of 1969 which makes particular allowance for losses incurred through foreign investment.

The Federal Republic of Germany has already signed bilateral taxation agreements with 28 developing countries in order to avoid taxation discrimination.

The governments of other industrialized countries also promote mining projects, though the emphasis varies. In France, copper is given particular support (Plan Cuivre) and the Bureau de Recherches Géologiques et Minières (BRGM) conducts mineral exploration programs, in which private enterprises can participate if prospects of a find are indicated. In Japan, the Metal Mining Agency of Japan (MMAJ) undertakes credit guarantees for Japanese firms engaged specifically in mining projects. They are also accorded tax benefits: tax-free reserves are permitted against possible losses as a result of foreign participation.

In the USA, the Overseas Private Investment Corporation (OPIC) provides insurance against political risks in developing countries, such as inconvertibility of currency, expropriation without compensation, losses due to war or revolution. Although political risk is covered to a large extent, the insurance is rather expensive, e.g. in the range of 1% and 2% of the investment. In Great Britain, the Export Credits Guarantee Department (ECGD) is responsible for these government guarantees on foreign investments.

9.1.4 Stockpile Programs

An adequate stockpile is a guarantee against short-term supply problems and unexpectedly violent price fluctuations. In addition to the voluntary or compulsory commercial stockpiles, certain industrialized countries also maintain government stocks, particularly significant being the USA Stockpile. The stockpiling of important import commodities was initially the result of legislation from 7 June 1939 and was of a military, strategic character during World War II. Within a period of 10 to 12 years after the 79th US Congress had passed the Strategic and Critical Materials Stock Piling Act on 23 July 1946, enough minerals and metals were stockpiled to cover the import requirements of the USA for at least 3 years in case of a national emergency. This National Defense Stockpile was then supplemented by a Defense Production Act Inventory in 1950 and by the Supplemental Stockpile of 1954.

The Korean War triggered the most extensive activities undertaken up to then. Despite record prices, mineral commodities valued at between 650 million US-$ and 900 million US-$ were stockpiled each year from 1951 to 1955. The cumulative costs of setting up and maintaining the National Defense Stockpile amounted to about 7

Table 9.3 US National Defense Stockpile inventory (as of March 31, 1987)

Material	Unit	Inventory	Goal
Aluminum	sh.t	2,082	700,000
Antimony	sh.t	37,107	36,000
Bauxite (Jamaica)	lg.t	12,457,740	21,000,000
Cadmium	lb	6,328,809	11,700,000
Chromite (metall.)	sh.t	2,080,710	3,200,000
Chromite (refract.)	sh.t	391,414	850,000
Copper	sh.t	29,048	1,000,000
Fluorspar (metall.)	sh.t	411,738	1,700,000
Germanium	kg	0	30,000
Lead	sh.t	601,018	1,100,000
Mica (muscovite)	lb	14,652,625	12,630,000
Nickel	sh.t	37,222	200,000
Platinum	tr.oz	452,641	1,310,000
Silver	tr.oz	121,276,707	0
Tantalum	lb	201,133	0
Tantalum minerals	lb	2,837,943	8,400,000
Tin	m.t	180,889	42,700
Tungsten (powder)	lb	1,898,831	1,600,000
Zinc	sh.t	378,316	1,425,000

Source: FEMA, Stockpile Report to the Congress, Washington, August 1987.

billion US-$, the market value as of the end of 1986 amounting to roughly 10 billion US-$ (end of 1980: approx. 15 billion).

The implementation of Stockpile Acts since 1979 has been the responsibility of the Federal Emergency Management Agency (FEMA) in Washington, which was commissioned for this purpose by the General Services Administration (GSA). For the financing of the stockpile acquisition and disposal program, a National Defense Stockpile Transaction Fund was established in 1979.

After the build-up of the Stockpile had been more or less completed in 1958, an examination was conducted to determine unjustifiably large quantities of some materials, which were then in some cases disposed of. The Stockpile Disposal Act of 16 April 1973 drastically curtailed stockpiling, completely removing some metals from the list of strategic materials, antimony, copper, molybdenum and vanadium for example. In response to a study submitted by the Federal Preparedness Agency (FPA), the National Security Council revised its stockpile policy anew. In a directive from 26 August 1976, new stockpiling quotas as of 1 October 1976 were stipulated intended to cover requirements in case of emergency for 3 years. The list now contains some 55 groups of mineral commodities including metals, ores, ore concentrates and industrial minerals (cf. Table 9.3).

Statutory provisions for the stockpiling of strategic minerals were again revised by the Carter administration on 2 May 1980, after which the Reagan administration announced a program for restocking the Stockpile in March 1981, which is planned to cost 2.5 billion US-$. The program began in 1981 by purchasing an additional 540 t of cobalt, and aims to build up stocks of many other materials like aluminum, beryllium, chromite, copper, ferrochromium, germanium, nickel, platinum, and zinc (see Table 9.3). The USA Stockpile exercises considerable influence on certain commodity markets, both purchases and disposals having affected pricing (see Sect. 7.3).

In 1987, the House Armed Services Committee voted for a transfer of the management of the National Defense Stockpile from FEMA to the Defense Secretary. Before the transfer can take place, however, it must be approved by the US Congress.

Reserves of petroleum are also being built up in the USA. At the end of 1986, the Strategic Petroleum Reserve contained 511.57 million barrels (about 75 million tons) of petroleum, representing 97 days of net petroleum imports.

Other industrialized countries have also incorporated governmental stockpiles into their national commodity policies. The creation of commercial stockpiles with government support has been under discussion in the Federal Republic of Germany since 1978. Three models are being considered:

– Mandatory reserves for the metal industry, similar to petroleum reserves.
– Stockpiling agreements between industry and a government authority.
– Tax incentives to build up material stocks in industry.

On 29 November 1978, the West German Government agreed to a proposal submitted by a group of experts to investigate possibilities for governmental strategic stockpiling, particularly for materials with a high supply risk, such as chromium, manganese, vanadium, cobalt, and asbestos. The organization of the commercial stocks, to cover a year's requirements, was to be undertaken by a stockpiling corporation, and the Deutsche Bank was to provide credit to the value of 600 million DM to finance the costs of acquisition. The Kreditanstalt für Wiederaufbau (KfW) was to be commissioned to administer the financing and discount commodity suppliers' bills to the debit of the Bundesbank. The program was suspended at the end of 1980.

The French Government decided to establish a stockpile in 1975. Initially, 250 million francs were appropriated, a further 1.600 million francs in 1979. According to unofficial sources, stocks of copper, cobalt, tungsten, chromium, lead, silver, antimony, platinum, and zirconium are held.

In early 1983, Gt. Britain began building up a strategic stockpile, initially to the value of £ 10 million for chromium, manganese, cobalt, and vanadium, sufficient to cover 3 months' requirements largely via purchases from South and Central Africa. The low prices in 1982/3 were viewed as providing a good opportunity to build up a stockpile. A more comprehensive program including phosphate, antimony, molybdenum, nickel, niobium, tantalum, and titanium would have cost over £ 200 million and thus was not undertaken.

Since 1971 in Japan, the Ministry of International Trade and Industry (MITI) has maintained varying amounts of strategic stocks and since 1976 the private Stockpile Association as well. Both stockpiles are intended to support the Japanese nonferrous metal industry, thus largely comprising the metals copper, zinc, and aluminum. A new 5-year stockpile program as of October 1982 includes metals with a high supply risk, such as nickel, chromium, cobalt, molybdenum, manganese, tungsten, and vanadium. The private sector is required to maintain enough reserves for a minimum period of 10 days, the government sector for 25, and the joint program between private and public stocks a further 25 days. The costs of the program amounted to 10 billion Yen in 1983 and to 4 billion Yen per year in the period 1984 to 1987. The stockpile planning and stockpile administration is conducted by the Stockpile Department of the Metal Mining Agency of Japan (MMAJ). In the stockpile program, the Government carries two-thirds of the burden of interest.

In Sweden, commercial stockpiles are encouraged by tax relief incentives; the efficiency of this system as a substitute for national stockpiles has, however, been called into question for some time. Small government stockpiles are now also maintained in Switzerland and India.

Spain, Korea, and Malaysia have also voiced their intention to commence national stockpiling programs for mineral commodities.

9.1.5 Promoting Mineral Trade

Trade policy instruments relate to both national and international intervention.

At a national level, imports can be restricted to protect domestic production by means of duties, import quotas and bans on imports or encouraged in order to guarantee supplies via subsidies, cheap credit, guarantees or preferential rates of exchange.

The European Community (EC) is primarily concerned to secure the free exchange of goods, free movement of labor, free payment and capital transactions, and freedom of settlement. This economic and monetary union has not as yet been fully established. Efforts in commodity policy therefore have to focus on the optimal use of the minerals available within the EC and to rely on its joint economic bargaining power in negotiations with commodity-exporting countries.

Despite heavy opposition, the General Agreement on Tariffs and Trade (GATT), concluded on 31 October 1947 in Geneva and which functions as a special organization of UNO, ought to be retained in principle, because the basic ideas of dismantling trade barriers and the principle of the most favored nation is of benefit to all countries.

The post-war monetary system is based on the Bretton Woods Agreement (1944), which originally stipulated fixed rates of exchange with the US-$ and the pound sterling as leading currencies. This system needs reappraisal now that fluctuating exchange rates have been introduced by the Washington Agreement of December 1971 and the Paris Monetary Conference of March 1973 and the reserve currencies have been virtually discarded and replaced by Special Drawing Rights (SDR) of the International Monetary Fund (IMF), now the major component of international reserves and international liquidity.

The SDR quotes have been continually increased, reaching 90 billion SDRs at the end of 1986, allocated to the 151 IMF member countries. Outstanding IMF credit amounted to 34.6 billion SDR in 1986, split up into a 6.3 billion regular facility, 6.4 billion compensatory financing facility, 6.5 billion extended fund facility, 5.3 billion supplementary financing facility and 10.0 billion enlarged access policy facility.

9.1.6 Promoting Mineral Research

In the United States of America, the Bureau of Mines in Washington, D.C., has the principal responsibility for conducting research and development programs in mineral production. The main objectives are to help secure the US mineral supply necessary to maintain economic growth and to stimulate the mineral production of mining companies. The major goals of the programs in mining research, material research, mineral information, mineral data analysis, and helium operations are the following:

- Collecting and interpreting information about minerals and mining throughout the world;
- analyzing problems relevant to USA mineral requirements;
- formulating mineral policy options and recommendations;
- providing technological advances to improve efficiency and recovery from mineral deposits to reduce environmental effects and safety hazards in mining and to improve security of using and disposing of mineral materials;
- producing and storing helium for the US Government.

In the Federal Republic of Germany, research and development programs in mineral production and utilization have been integrated in a General Program of the Federal Ministry for Research and Technology (BMFT).

The BMFT disposes of funds to sponsor scientific research and technological developments. Projects are funded in government and private research institutes, in universities as well as in industry. The BMFT R&D programs do not extend to the promotion of concrete mining projects, being confined to the development and testing of techniques and equipment.

Since 1976 a program in the field of mineral resources has been supported, which aims at improving commodity supply in the Federal Republic of Germany. It is intended to achieve this objective through the following R&D activities:

- Developing new geophysical and geochemical measuring methods;
- exploring deposits in greater depths of up to 1500m;
- investigating deposits abroad (e.g. in the P.R. of China, Canada, Indonesia, Oman, Peru, the Philippines);
- improving prospective drilling techniques;
- developing new mining methods;
- improving processing methods, especially for lateritic chromite ores.

In 1985 the Geosciences and Commodity Security Program accounted for an expenditure of 57.85 million DM. Yet with the achievement of its major objectives, the program will probably be terminated in 1990.

Other central programs of the BMFT relate to the exploration and exploitation of minerals. In the framework of the Marine Research and Technology Program a total of 78.4 million DM was allocated exclusively to ocean minerals and offshore petroleum technology in 1986. The expenditure for new coal technologies is even more substantial, especially where coal gasification and coal liquification are concerned. It was in this particular area of the central program on Energy Research and Technology that BMFT spent as much as 250.9 million DM in 1986.

A particularly costly area is marine research. The BMFT has given considerable support to the many and varied activities carried out by Germany to investigate areas in ocean mining, including the numerous voyages of the research ships Valdivia and Sonne. The BMFT now intends to promote the development and testing of suitable mining systems for manganese nodules.

9.2 Policies of Mining Companies

A free enterprise economy depends essentially on entrepreneurial initiative. In the mineral sector, mining and petroleum companies are thus expected to contribute to securing mineral supplies by international involvement in mining activities.

The activities of private enterprise are directed in two major channels. To ensure supplies for the processing sector, an active strategy of raising supply is pursued: long-term supply contracts, diversification of sources, commercial stocks and participation in mining projects, including direct investment in foreign mining enterprises. To lessen dependence on minerals with high supply risk, a more rational use of alternatives is sought.

9.2.1 Investment Strategies

The decision process starts with an attempt to achieve corporate objectives such as expansion (or maintenance) of mineral production or financial goals. In principle there are two ways to invest: via exploration (very high-risk and long-term) and via acquisition of mining rights.

Direct investments in foreign mining enterprises can be effected in various ways. They always involve the following considerations:

– Mining investments are generally a high risk.
– Mining investments constitute a long-term commitment.
– Mining investments require relatively high capital expenditure.
– The terms of investment can suddenly alter as a result of legislation; mining and investment law in developing countries is often in a constant state of flux, as the equity policy in oil exporting countries clearly illustrates.

As a rule (outside the USA), a state has the right to dispose of all the natural resources found on its territory, but it can, however, grant mining concessions to private enterprise (see Sect. 5.3).

The first licensing agreements in oil-producing countries provided foreign firms with extensive rights in prospecting and exploration, mining, and pricing. These classical agreements were of extremely long duration (50 to 99 years) and covered immense areas. An example is the concession granted to the Briton, Knox d'Arcy, by the Persian Government for a period of 60 years: in exchange for minimal royalties and limited profit sharing, he was given licences for the exploration, mining, exploitation, development, transportation and sale of petroleum, natural gas, asphalt, and native paraffin in the entire Kingdom of Persia (with the exception of the five Northern Provinces). This attractive concession was transferred to the Anglo-Persian Oil Co. in 1908, which was nationalized in 1951.

After World War II, the exporting countries became more independent and self-confident, recognizing the importance of mineral resources for their own economic development. In 1948, Venezuela developed a completely new type of agreement, establishing 50% profit-sharing participation in oil mining companies operating in the country (equity sharing). In 1968, the Algerian national enterprise, Sonatrach, was the first to attain majority participation (51%). These concessions were much shorter in duration, rarely exceeding 25 years. Minimum investments and early surrender of the land in question were stipulated for the exploration phase and a sophisticated system of levies and duties was devised.

The producing countries' efforts to gain complete control over the exploration and mining of mineral resources, however, began as early as 1963, when Indonesia signed production sharing agreements with foreign oil firms (PERTAMINA model), no longer granting them concessions but simply commissioning them on a contractual basis. The contractor undertakes the exploration work under supervision and bears the entire risk. In the case of a find, the exploration costs are reimbursed and he is given a share in production at enterprise cost. In 1966, the PERTAMINA model was adapted in Iran to a service contract. These agreements also treat the foreign company as a contractor, who is expected to bear the whole exploration risk. In the case of a find, the exploration costs are refunded but production is no longer shared; the contractor is merely permitted to purchase a part of the petroleum at preferential prices.

At present, the kind of agreement reached in the international petroleum industry depends on the estimated exploration risk. Traditional concessions are still granted for completely unexplored areas, whereas production sharing agreements or service contracts have now become usual for regions where deposits have already been discovered.

Numerous possibilities for investment are also open to mining companies: traditional mining concessions, profit-sharing agreements, equity sharing or service contracts.

Private mining investment strategies increasingly involve the formation of consortia which provide two main advantages:

– The risk is spread (see Sect. 6.2.3). This applies to all categories of risk, such as the project completion risk, political risk, reserve and production risk inherent in all mining and market risk.
– Financing is facilitated because the consortium partners raise the capital jointly and together enhance their credit worthiness and credibility for sponsors to finance projects.

An example of a consortium is the copper-gold project Ok Tedi in Papua New Guinea, where a consortium made up of the following members was formed to carry out exploration work:

– Dampier Mining Co. Ltd. (subsidiary of the Broken Hill Pty. Co. Ltd., Australia) with 37.5% participation;
– Mount Fubilan Development Co. Ltd. (formerly a subsidiary of Standard Oil Co. of Indiana, USA) with 37.5%;
– Kupferexplorationsgesellschaft (Metallgesellschaft, Degussa, Gutehoffnungshütte and Siemens, all of W. Germany) with 25%.

Another consortium was formed for the investment phase. The new Ok Tedi Mining Ltd. (OTML) based in Port Moresby, was nominated as project company. It is made up of:

– BHP Minerals, Australia with 30%
– Amoco, USA with 30%

- Papua New Guinea Government with 20%
- DEG/Metallgesellschaft, W. Germany with 20%.

The Government of PNG, however, reserved the right to take a 30% participating interest but has not yet exercised this option.

The 855 million US-$ needed to finance the initial construction phase of the project Ok Tedi (Stage I) was raised through the 4 partners contributing 256 million US-$ of equity (30%) and by means of various loans to the value of 599 million US-$. The German bank Kreditanstalt für Wiederaufbau (KfW) in Frankfurt subscribed 100 million US-$, the Export Finance and Investment Corp. of Australia (EFIC), 250 million US-$, the Export Development Corporation of Canada (EDC), 88 million US-$, the Österreichische Kontrollbank (OKB) of Austria, 50 million US-$, Citibank (USA), 150 million US-$, Lloyds Bank International/Export Credits Guarantee Department of Gt. Britain, 100 million US-$ and the Overseas Private Investment Corp. (USA), 50 million US-$.

The main objective of the concept of project financing is to limit the financial responsibility of individual mining companies. The financier is looking to the project cash flow and the assets of the project rather than to the assets of the mining companies.

9.2.2 Improvements in Mineral Utilization

Measures to improve the utilization of minerals can be taken by companies engaged in mining as well as those involved in processing, the decision-making process being at the managerial level in both cases.

Measures to enhance the efficiency of mineral utilization are:

- Technological improvement in dressing and smelting to raise the recovery rate of valuable minerals and ensure that all byproducts are also recovered.
- Rational application of minerals, for example fuels, by reducing the specific energy consumption and improving the use value. In the case of metals, changes or reductions in the contents of valuable alloy components can be substituted or reduced; molybdenum can be replaced by chromium or manganese in high-grade steel or the antimony content in battery lead can be reduced. Via modern technology in the manufacturing sector, the specific consumption of expensive metals can be reduced, tin for instance in the electrolytic production of tin-plate or steel alloys in powder metallurgy.
- Substitution of costly minerals by cheaper ones that are more readily available, replacing zinc in guttering by plastic, lead in printing letters by photo-setting, copper in oil pipelines by plastic or in electricity cables by aluminum, tin in cans by aluminum and steel in train coaches by aluminum.
- Intensification of recycling (see Sect. 7.3.1.2).

Notes on the Literature

For more details on mineral policies in the USA suggested reading is the publication of the US Congress (1976): *US Raw Materials Policy: Problems and Possible Solutions.* Specific information on US Stockpile policy can be obtained from the *Stockpile Report to the Congress,* published every 6 months by the Federal Emergency Management Agency, Washington, D.C.

Chapter 10 Policies and Special Problems in Developing Countries

In recent years, developing countries rich in mineral resources have clearly formulated their joint mineral policy goals at an international level (see Sect. 8.1.1). The major concern of each of these countries is to gain greater control over the exploitation of its natural resources and to derive maximum benefit from them. This is understandable: for many developing countries control is vital because these raw materials are indispensable as a basis for economic development. These countries depend at the same time on the assistance of the industrialized nations for the know-how and capital needed to prospect for and open up deposits. The maintenance of this assistance must thus constitute one of the goals of commodity policy in developing countries. These two goals are difficult to reconcile with one another; the conflict can only be resolved by means of a compromise. Attempts to arrive at such a compromise are evident in national mining legislation, which contains very disparate features.

Mineral policy in the developing countries is based on the principle of permanent sovereignty over natural resources. The United Nations has always been concerned to support developing countries in attaining effective control over their primary sector. Starting with the resolution of the UN General Assembly on 12 January 1952 up to the Declaration on the Establishment of a New International Economic Order on 1 May 1974, strategies to realize these objectives have been consistently developed in

Table 10.1 Nationalization of international mining companies in developing countries

Country	Mineral	Year of nationalization	Investor	Major mines or mining centers nationalized
Indonesia	Tin	1953/1958	N.V. Billiton/ Netherlands	Tin mines in Bangka and Belitung
Guinea	Bauxite	1961	Alcan/Canada	Los Islandes, Boké
Zaire	Copper	1966	Union Minière/ Belgium	Copper mines in Shaba (Katanga)
Chile	Copper	1967, 1969, 1971	Kennecott, Anaconda/USA	El Teniente, Chuquicamata, El Salvador, Exotica
	Iron ore	1971	Bethlehem Steel/USA	Atacama
Guyana	Bauxite	1970	Alcan/Canada, Reynolds/USA	Demarara
Peru	Copper, lead, zinc	1974	Cerro Corp./ USA	Cerro de Pasco
Venezuela	Iron ore	1974	Bethlehem Steel, US Steel/USA	Cerro Bolivar Orinoco district

Source: United Nations Committee on Natural Resources, New York.

Third World countries, many of which nationalized a number of mining enterprises (see Table 10.1), 32 between 1960 and 1969, and 48 between 1970 and 1976 (UN sources). During the first half of the 1980s, a private mining sector developed parallel to the state-owned mining sector in several countries such as Peru, Chile, Indonesia.

Cooperation with industrialized nations is sought on this basis, the overall legal conditions being determined by:

- General investment laws and promotion programs.
- Special provisions on mining investments as laid down in mining laws (see Sect. 5.3).

Development strategies generally place considerable emphasis on capital and technology transfer. An important criterion here is the investment climate, which is measured in terms of various "investment indices": all the factors of influence are allotted points according to which the countries in question are classified.

Of decisive importance for foreign investors are also the promotion schemes:

- Economic guarantees (e.g. against discrimination or guaranteed freedom of capital transfer).
- Tax incentives (tax concessions for a limited period, special allowances, debt balances carried forward).
- Preferential tariffs (reduction of high import duties, exemption from export duty).
- Financial assistance (preferential treatment for borrowing, currency allowance, foreign exchange allocation, provision of land).

10.1 Concession Policy

At the end of the 1960s and the beginning of the 1970s, new mining laws came into effect in many developing countries. Either prior to this or subsequently, the old license holders were wholly or in some cases partially expropriated and national mining corporations established. According to the new legislation, the ministries responsible for mining are generally empowered to grant concessions, differentiating between prospecting, exploration, and mining. Provisions to grant concessions to foreign enterprises are usually made, though their practical application differs widely. Licences granted for exploration work in particular are accompanied by conditions on minimum investment. The duration of the concessions is far shorter than previously, prospecting and exploration licenses lasting 3 to 5 years, mining concessions between 20 and 25. The operations of private license holders (foreign or national) are subject to constant control. In practice, governmental majority interest is sometimes sought after the first few years (Zaire, Zambia, Peru).

Indonesia's approach is unusual, but it could prove to be the beginning of a trend. Concessions are no longer granted; rather the Department of Mining negotiates contracts of work individually with national or foreign mining companies for all levels of project.

Where preference is given to government mining corporations or concessions granted exclusively to them, concession policy is closely related to participation policy.

10.2 Participation Policy

In mineral producing countries, two strategies are pursued to acquire control over the exploitation of natural resources:

- Progressive participation of government enterprises in foreign companies operating in the country, sometimes leading to complete state ownership, the question of compensation always being a difficult issue.
- Forming joint ventures for new mineral projects. Firms, either governmental or private, are established in developing countries on the basis of local equity participation. There are three kinds of participation: .
  - Majority joint venture (a minimum of 51%);
  - Equity joint venture (50%);
  - Minority joint venture (a maximum of 49%).

Today, the only form of participation in many developing countries is that of the minority joint venture. A form of investment which has now become a kind of venture in its own right is the "fade out" joint venture, in which a majority interest is agreed upon to be converted into a minority interest after an amortization period of between 7 and 20 years. In some cases the foreign investor withdraws entirely (phase out provisions).

In the 1960s, the OPEC countries pursued a policy of active participation (see Sect. 7.2.1.1). The concession agreements were renegotiated, and the oil companies had to establish new local subsidiaries within the country, in which government oil companies then participated, in some cases taking them over. Even Saudi Arabia acquired all of Aramco's share in 1979, after Iraq (1971), Venezuela (1975), Kuwait (1977) and Qatar (1979) had completely nationalized.

These also served as precedents for non-OPEC countries. Particularly Latin American countries with petroleum deposits, such as Mexico, Brazil, Argentina, Bolivia, or Peru, pursued a similar path of nationalizing the oil industry.

The mining industry displays two contradictory trends. On the one hand, various developing countries are increasing government participation and on the other, some are again inviting increased foreign participation.

The following are examples of increased government participation:

- In Malaysia, the financial capital of the Malaysia Mining Corporation (MMC) is wholly owned by the government group Pernas. The MMC took over large sections of tin mining in 1980 and also acquired interests in domestic tin smelting (see Sect. 7.1.1).
- In Nigeria, the state-owned Nigeria Mining Corporation purchased a substantial number of shares in tin mining and acquired interests in tin smelting in Jos.
- In Togo, West Africa, the Government acquired a majority interest in the phosphate company, Compagnie Togolaise des Mines de Benin from W.R. Grace (USA) and French shareholders.
- In Angola, the majority interest in Diamag was acquired from De Beers in 1977.

Majority interest by foreign enterprises is again admissible in the following countries:

- Chile, where the Government sold shares in the Disputata copper mine to Exxon and now accepts majority participation in new projects, such as the Quebrada copper project (Falconbridge, Canada), the El Indio project (St. Joe, USA) or the Andacollo copper project (Noranda, Canada).
- Peru, where the Government has abandoned its intention of acquiring majority participation in the Southern Peru Copper Corp. (52.3% ASARCO, USA and other US firms), even though Southern Peru's Toquepala and Cuajone mines have now become the largest copper and molybdenum producers in the country.
- Papua New Guinea, where the PNG Government and Investment Corp. acquired a mere 20.2% interest in Bougainville Copper Ltd. and 20% of the Ok Tedi Mining Co., to which the Government intends to confine itself.

The new mining projects in developing countries are characterized by numerous joint ventures, with all three types of participation in evidence. Recently, interesting variants of facilities to finance governmental participation have been devised:

- Carried interest: the government of the host country does not pay for its shares in a given enterprise until production has commenced, either partly or at full scale. This sort of agreement was reached by the Nigerian Government for an uranium project, for example.
- Free interest: the government of the host country is permitted to participate without having to invest or to do so in return for tax concessions. The Government of Zaire, for instance, demanded a 20% interest in the Tenke-Fungurume copper project; Liberia acquired 50% participation in an iron ore project in return for exempting the project from corporate taxes.
- Production sharing: the foreign contractor undertakes initial full responsibility for equipment and management, for which he receives contractually stipulated quantities from current production (ore concentrates, petroleum, metals, petrochemicals).
An example is the Dumai oil refinery project on Sumatra, a joint venture concluded between PERTAMINA (Indonesia) and Sumitomo Shoji (Japan) in 1967, where the latter operates as the prime contractor and organizer of a Japanese syndicate of companies and receives fixed quantities of petrochemicals.

Examples of fade out joint ventures in mining are contracts in Indonesia (Rio Tinto Zinc, Freeport) and in Panama (Texasgulf), according to which the foreign interests are surrendered after an agreed period of 10 to 20 years.

The participation policies in developing countries have five main objectives:

- Maximizing government revenues and foreign exchange earnings. Revenue from mining operations can be claimed by various fiscal measures like royalty, company tax-free interest equity, carried interest equity, resource rent tax or additional profit tax.
- Achieving direct control over mineral production and marketing of minerals. Transnational companies (TNC) wish to maximize their international profits and use global market strategies as well as transfer pricing to do this. By acquiring equity the government is in a position to monitor the performance of the foreign investor closely.

- Exercising management control in mining companies in order to supervise decision-making on employment, investment, procurement, local equipment purchases, local processing of minerals, environmental protection, or production levels.
- Enhancing the transfer of technology and know-how to indigenous engineers and managers. Direct collaboration with foreign experts can have educational effects benefiting not only the operation of the nationalized mining company, but also the national economy and administration.
- Realizing the political ideals of the ruling government with respect to implementation of full national sovereignty over mineral resources or asserting national identity in former colonies. New concepts of cooperation between foreign investors and national mining companies have been developed taking into account the political pressure to acquire sovereignty while transferring financial risk and management responsibility to the TNC (production sharing contracts; risk service contracts).

10.3 Fiscal Policy

The main aim of mining taxation policy in developing countries is to earn as much government revenue as possible from the exploitation of mineral resources, preferably in the form of foreign exchange (see also Sect. 6.3.4.2). The mining firm concerned must, of course, be able to rely on a minimum return: 15% internal rate of return on capital is currently often considered the threshold figure.

Development policy goals can also be pursued by taxation: promotion of industrial development, protection of national investors from foreign competition, regional development of rural areas, diversification of mining production. These goals are not accorded equal priority in every country.

All companies are, of course, subject to corporation tax, though to varying degrees and in some cases graduated according to the type of mineral and duration of project. Two categories of taxes are common: net profit taxes and royalties based on volume of output. Many authors regard income or profit taxes as superior to royalties.

Payment of royalties has, however, a long tradition in the mining industry. This duty on each ton of ore mined or on the value of mine output is usual in the petroleum industry and ore mining. Only few countries have dispensed with this form of government revenue (Zambia). Although they provide the government with reasonably constant revenue, they can deprive it of proportionate participation in rising profits, which is why a profit tax is generally imposed as well, usually in graduated form. An interesting approach is the linkage of the rate of taxation on profits to the rate of return on investment, as introduced in Peru, where it has proved an effective investment incentive.

Another example is the production levy for bauxite mining in Jamaica imposed in 1974. This levy is based on the average realized price of aluminum ingot of the three major US producers (Alcoa, Kaiser, Reynolds). The introduction of the Bauxite Production Levy – promoted also by the International Bauxite Association (see Sect. 7.2.1.3) as a main tool for its indirect pricing policy – led to considerably higher government revenues in Jamaica and other bauxite-exporting countries.

Some countries also introduced a resource rent tax (RRT) on mining resources, taking into account the nature of mineral deposits as a national resource. The basis of assessment for corporation tax can be difficult where the transnational corporations display close vertical integration, as in the petroleum or aluminum industry and profits in oil and bauxite mining can be kept to a minimum in order to benefit the processing sector abroad. For this reason, some countries have introduced additional profit taxes (APT). In Indonesia, for example, an investor must pay a windfall profits tax in addition to the normal rate (35% in the first 10 years and 45% after that), when earnings exceed 15% of total investment.

In many developing countries other forms of tax are levied, such as high land rent, annual concession fees, bonus payments when contracts are signed or when production starts. In some mineral exporting countries, special duties, usually graduated, are imposed on exports of commodities. Higher rates are charged for petroleum and ore concentrate than for petrochemicals or metals in order to stimulate manufacturing in the country of origin.

Indirect components of taxation policy are import duties on mining equipment, split exchange rates (higher rates for commercial import goods or less favorable rates for company earnings), or statutory provisions on the reinvestment of company profits.

10.4 Marketing Policy

In order to lessen their dependence on traditional trade relations or transnational corporations, various developing countries have evolved schemes to establish independent marketing systems for minerals. They have adopted various approaches:

- A government enterprise, such as the national petroleum corporation, has the exclusive right to export minerals and seeks to negotiate direct supply contracts with foreign firms, as is the case in Iran, where since 1978 the National Iranian Oil Co. has been selling roughly 25% of its petroleum exports directly to Japan, India and other European countries in accordance with intergovernmental agreements, or in Jamaica, whose government bauxite corporation has concluded long-term agreement supply contracts with new aluminum smelters in Algeria and Venezuela.
- Government marketing enterprises are set up to market the product of state-owned mining companies as well as that of private firms; in Peru for example, Minero Peru Commercial (MINPECO) markets the entire production of the large government mining corporations CENTROMIN PERU and MINEROPERU.
- Government mining banks undertake to sell the product of small mines or precious minerals such as gold, silver, or diamonds. In Bolivia, the Banco Minero de Bolivia was entrusted with these functions. Since then, small mines have been compelled to sell their product at fixed prices.
  A directive was issued on 17 April 1978 by the Philippine Government instructing all gold mining enterprises in the country to sell their entire product to the Central Bank, which pays them in local currency.

10.5 Importance of Small-Scale Mining in Developing Countries

The United Nations Institute for Training and Research (UNITAR) is in favor of expanding small-scale mining in developing countries. Related development policy recommendations were directed at the multinational and bilateral development aid organizations at the first international conference on the special problems of this sector of the mining industry, which was held in December of 1978 in Jurica/Mexico.

The Mexican conference revealed clearly, however, that views differ widely on the significance, usefulness and risks of small-scale mining. One of the chief reasons for this is that country-specific and resource-specific criteria play an important role in determining the limits of this sector of the mining branch.

Although small-scale mining is not restricted to developing countries, the problems and the need for aid are significant in those areas. The organizational form most commonly preferred in developing countries is the co-operative. Government aid programs to create such mining cooperatives have been implemented with varying degrees of success in a large number of countries. As a rule they are only attractive when coupled with financial assistance via a mining bank.

Although the United Nations attempted a general assessment of the value of small-scale mining and concluded that this sector accounts for an estimated 10% of world mining production (F. Skelding 1972), some very clear distinctions must be made when assessing its importance. Only a few mineral resources are suitable for this type of operation and only a few countries have been able to maintain the tradition of small-scale mining. As a result, the importance of this type of mining is somewhat limited, but in terms of development policy its role can be so significant that it is appropriate to analyze all the implications and to determine whether or not it is worthwhile supporting this type of mining, bearing in mind the overall impact on the mineral supply situation.

10.5.1 Definition and Characteristics of Small-Scale Mining

There is no international classification that defines exactly what constitutes small-scale mining, although a clear definition would be important from the point of view of national mining legislation and support measures.

There are various criteria that might be used to define this sector, and ideally they should be used not singly but in combination:

- Quantities of run-of-mine ore (ROM) produced or processing capacity.
- Number of employees.
- Size of the mining concession.
- Amount of fixed assets.
- Amount of revenue from sales (gross income).
- Annual profit.

The current practice in Peru will serve as an example here. According to the revised Mining Act of 1978 (Ley de la Pequena Mineria) all mines are classified as small-scale operations if their annual gross income is between 500,000 and 10 million Soles (= 2,000–40,000 US-$, October 1979), their dressing plant has a maximum capacity of 200 tons ROM ore/day and they have acquired a concession of not more than 1,000 ha. The body representing this sector of the economy, namely the Sociedad Progreso de la Pequena Mineria, has, on the other hand, called for the lower limits of the "medium-scale mining" sector to be put at 200 tons of ROM ore per day (not dressing capacity), 100 employees, 5 million US-$ worth of fixed assets (not expressed in Soles because of the high inflation rate in the country) and sale revenues of 3 million US-$.

Another definition is used in the Philippines, where small mines have less than 25 employees. The United Nations consider a mine as small when less than 50,000 tons of ore per day are produced.

For the purpose of comparison it is interesting to consider the views held in the industrialized countries. The journal Mining Magazine publishes each year a list of the medium-sized and large mining operations prepared by Mining Journal Research Services and states that according to the "international standard" all mines producing less than 150,000 tons of ROM ore per year should be categorized as small.

It is also customary to draw a minimum size limit – small-scale mining does not include individual miners (tributers, dump pickers, pirquinieros, garimpieros), who work on a contract basis and are obliged to deliver their ore to mining companies or to state mining banks.

The general characteristics of small-scale mining in developing countries include in particular:

– Low capital investment because the ore is extracted chiefly by human labor.
– Extremely arduous working conditions, particularly in underground mines and in remote mountain regions.
– Unsystematic exploitation of visible ore minerals because inadequate geologic knowledge is available on the distribution of the minerals in the deposit.
– Inadequate social security, if any at all, for the miners, because the mining regulations regarding minimum wages or accident prevention are ignored. The companies do not set up their own schools or hospitals and no disability pension or old age pension benefits exist.

10.5.2 Production of Small-Scale Mining

Although no overall statistics exist on the share contributed by small-scale mining to the overall output of minerals, an attempt should be made to quantify this contribution. From the point of view of economic geology, two pre-conditions must first of all be met:

– Deposits must occur at or close to the surface.
– The deposit should contain minerals which can be identified with the naked eye or by means of simple concentration processes.

The size of the deposit plays an indirect role. If the above-mentioned conditions are met, then small-scale mining concentrates chiefly on minerals which occur in countless small deposits because such naturally determined distribution patterns impose tight technical and economic limits on mechanization.

The above-mentioned deposit-related conditions result in small-scale mining being linked, to a great extent, with specific mineral resources. Among the ore minerals, three non-ferrous metals, two steel alloying elements and one precious metal stand out particularly clearly (cf. Table 10.2).

Table 10.2 Minerals exploited by small-scale mining

	Mineral	Per cent of total world
Metals	Antimony	approx. 25 %
	Tin	approx. 15 %
	Tungsten	approx. 15 %
	Mercury	approx. 15 %
	Chromium	approx. 10 %
	Gold	5 – 10 %
	Titanium (rutile)	
	Ilmenite	
Minerals industrial	Semiprecious stones	75 – 80 %
	Diamonds	10 – 15 %
	Mica	approx. 20 %
	Quartz	15 – 20 %
	Sulfur	approx. 10 %
	Graphite	approx. 25 %
	Fluorspar	20 – 25 %
	Gypsum, kaolin, rock salt	20 – 30 %

Source: Gocht (1980).

Table 10.3 Small-scale mining in selected developing countries

Country	Major minerals for small-scale mining
Latin America	
Bolivia	Tin, antimony, lead, tungsten, gold
Peru	Lead, antimony, tungsten, gold, zinc, silver
Brazil	Semi-precious stones, beryllium, gold, ilmenite, chromium, diamonds, mica (vermiculite)
Mexico	Pyrochlore, mercury, fluorspar, tin, silver, opal
Chile	Copper, sulfur, gold
Africa	
Nigeria	Tin, columbite, gold
Zaire	Diamonds, tin
Zimbabwe	Chromium, gold, lithium, precious stones
Mozambique	Semiprecious stones, tantalite, mica
Sierra Leone	Diamonds
Asia	
China	Tin, antimony, tungsten (iron ore), coal)
India	Manganese, mica, gypsum (coal, iron ore)
Turkey	Chromium, antimony, mercury, lead, zinc
Thailand	Tin, antimony, tungsten
Burma	Tin, tungsten, antimony
Malaysia	Tin, ilmenite, monazite
Indonesia	Tin, gold

Source: Gocht (1980).

Table 10.3 shows the regional distribution of the mineral producing countries in which small-scale mining plays a significant role. In Latin America these countries are above all Bolivia, which has 2,000–3,500 small mines in production (antimony, lead, tin, tungsten, gold), Peru with 2,000–3,000 mines (antimony, lead, tungsten, gold, zinc), Brazil with more than 4,000 small mines (semiprecious stones, beryllium, gold, tin) and Mexico with about 2,500 mines (mercury, fluorspar, tin, silver, opal).

In Africa, Nigeria (tin, columbite), Zaire (diamonds, tin, gold), Zimbabwe (chromium, gold), Mozambique (semi-precious stones), and Sierra Leone (diamonds) all have a considerable number of small mining operations.

Finally, in Asia, the chief countries that deserve mention are China, India, Turkey, and Thailand. In China, small-scale mining has been encouraged deliberately, and it is therefore widespread not only for tin, antimony, and tungsten, but also for iron ore (approx. 7% of the country's output) and coal (approx. 10%). In India, too, there are several hundred small coal mines, about 200 small iron ore mines, approximately 400 small manganese mines and an estimated 700 small mica mining operations. Turkey can look back on a long tradition of small-scale mining (chromium, antimony, lead-zinc), even though a process of consolidation has taken place in the last 15 years. In Southeast Asia this sector is economically important. In Thailand approx. 1,000 small-scale mines are in operation (1985), in Malaysia approx. 200, in Indonesia approx. 500, and in Burma approx. 200.

The statistical data always relate to active mines whose number may fluctuate considerably at short notice, depending on the market situation.

10.5.3 Advantages and Disadvantages of Small-Scale Mining

The negative and positive effects of small-scale mining will be described in socially related categories, paying particular attention to the problems of the developing countries. When the economic aspects are considered, both the advantages and disadvantages are apparent. From the point of view of a developing country, it is certainly positive that small-scale mining requires low capital investment and consumes relatively small amounts of energy because this sector of the mining industry is generally labor-intensive and only modest mechanization has taken place.

This results in another advantage, namely that capital goods (e.g. sluice-boxes, gravel pumps, screens, trams and carts, simple tools, etc.) can be produced by local workshops. The decisive economic disadvantage, however, is the low output which means that small-scale mines are generally low-efficiency, high cost operations and thus require government assistance in the form of low-interest loans and tax relief. These remarks will be illustrated by a review of tin mining in Southeast Asia (see Table 10.4).

Relative to the output, the amount of labor required in small tin mines is almost twice as high as in medium-sized mines and 25 times as high as in large-scale operations. At the same time, the degree of mechanization – and thus also the consumption of energy – is much lower.

It is interesting to note from Table 10.4 that there is a point at which mechanization of medium-sized mines is no longer cost effective. In a gravel-pump mine, much more horsepower is installed than in dredges, but the annual production is only about

Table 10.4 Factors of production in typical tin mines in Southeast Asia (1985)

Mining method	Dredge (large mine) Malaysia	Gravel pump mine (medium-size) Thailand	Sluicebox mining (small mine) Thailand
Number of employees	80	75	18
Installed HP capacity	3,200	450	8
Annual production (t, conc.)	750	45	7
Hours worked per 100 t concentrate (1,000 h)	30	460	750
HP consumed per 100 t concentrate (10,000 HP)	320	720	80

7%. Special attention must be paid to this disadvantageous fact when introducing modern technology into small mines.

Thus, small-scale mining can be regarded as a source of foreign currency earnings to a limited extent only and even less so as a source of tax revenues. Nevertheless, it does help to reduce unemployment, especially in mostly rural areas where there are few alternatives to agriculture. Although this advantage is generally cited as the most important reason for supporting small-scale mining, it does involve negative social aspects such as the often very poor working conditions and the associated high accident rate, below-average wages and totally inadequate training opportunities for the miners.

It is particularly in the least developed regions such as remote mountain valleys or almost inaccessible jungle areas that miners are members of the most socially underprivileged class. In such areas, where the government and its agencies have insufficient influence, the workers are not normally given contracts, frequently not even the minimum wages are paid, even the most basic accident-prevention rules are ignored and child labor is widespread. As a result, it is a common practice among miners to work only part of the time in the small mines and to hire on, for example, as farm laborers during the harvest season. Of the 70,000 persons employed in almost 3,000 small mines in Peru (compared with 55,000 in all the large and medium-sized mines), only about a third work regularly as miners all the year round.

From the point of view of socio-economic development, the subsector "small-scale mining" does not play a significant role because practically no transfer of technology occurs and very little is contributed towards improving the regional infrastructure, although these are the two areas in which mining projects are normally of benefit in the context of development policy.

Only the larger of the small mines have constructed access roads to and from their production sites. Often enough, the products are still transported by donkey, mule, lama, elephant, camel or on the backs of human porters; valuable products such as diamonds, gold, or tin are also sometimes carried along smugglers' trails (Peru, Brazil, Zaire, Burma, Thailand) into neighboring countries. Usually no attempt is made to undertake other infrastructural measures, such as the construction of hospitals and schools.

The disadvantages also outweigh the advantages in the area of mineral policy. The risk of destructive exploitation is particularly great in the sector of small-scale mining because, as a rule:

- High-grade blocks of the deposit are mined selectively.
- The recovery of minerals is low.
- No attempt is made to extract valuable by-products.

Antimony mining in Peru can serve as an example here. In Puno province a large number of small mines produce a total of 8,000 to 10,000 tons of antimony ores per annum. Mining is carried out in narrow drifts using picks and shovels, and only ores containing more than 10% Sb are taken, whereas the cutoff-grade in mechanized mines is around 4% Sb. The recovery achieved in the primitive dressing plants, where hand picking is used, is only about 40%, whereas a modern plant can achieve 80%. Peruvian antimony production thus amounts to only about 450 tons of metal instead of 700–800 tons. Finally, the molybdenite contained in the ore is not extracted because this requires additional processing.

The small tin mines of Southeast Asia, the small tungsten, antimony or lead mines of Bolivia, and the small fluorspar or mercury mines of Mexico run similar risks of being exploited destructively.

Finally, the ecological aspects should be mentioned. Small-scale mines cause much less environmental damage than large mines, although the opposite may sometimes be the case if the low recovery rate leads to toxic concentrations of metals in the tailings which are discharged without supervision into streams and rivers.

10.5.4 Forms of Technical Cooperation in Small-Scale Mining

The disadvantages affecting small-scale mining, to which reference has been made above, can be minimized to a considerable extent by providing various forms of multilateral and bilateral Technical Cooperation. There are six main areas in which such assistance can be considered:

- Planning and implementing exploration programs to detect mineral deposits which are suitable for small-scale mining.
- Technical advice on the improvement of traditional mining and processing methods in order to increase the recovery of valuable minerals, thereby increasing the profitability and reducing destructive exploitation.
- Evaluating mineral deposits.
- Improvement of infrastructures in rural areas where small-scale mining is common, in order to guarantee energy supplies, water supplies, transportation routes, medical services and educational facilities.
- Establishment of training facilities for the miners and managerial staff of small mines.
- Organizational assistance in setting up co-operatives, central workshops, sales offices and regional mining banks.

Without doubt, one of the main weaknesses of small-scale mining is that deposits are not systematically exploited. Since no exploration activity of any kind is carried out, the discovery of new ore reserves is left to chance. Only the government geologic surveys are in a position to provide assistance, but in most developing countries these are either still in the course of being established or they must concentrate on supporting the large state mining companies. Therefore the only assistance available to small mines at present comes in the form of special development aid programs. In such cases, the most effective form of assistance can be provided in the areas of mining and ore dressing technology. By consulting with and advising the operators of small mines, there ist a very good chance that sociologically adapted technologies can be developed and disseminated.

Finally, the importance of training in the areas of small-scale mining should be emphasized. This sector of the economy has the potential to provide impetus for industrialization and regional development in rural areas, but the realization of this potential depends on whether small-scale mining can break out of the bonds of traditional technical rigidity which currently hamper it. Training of the miners at all levels from basic to advanced is essential in this regard. Such training can be given in apprentice workshops, vocational schools, or on the job.

10.6 Effects of Mining in Developing Countries

The mining sector is of considerable significance for a well-defined group of 25 to 30 developing countries, though also in these cases the differences in the economic structures are evident. Structural characteristics manifest themselves in the predominant position of the mining sector in economic planning as well as in the role attributed to this sector with regard to economic growth and the industrialization process. Furthermore, these structural differences within the respective economies appear in the fiscal, foreign trade, and investment policies because their national systems of taxation and regulations for export and for the activities of multinational companies are influenced by the mining sector. Investigations of the effects of mining activities in developing countries (Gocht 1984) have pointed to the predominant significance of tax revenues from mining activities which account for a considerable portion of the government budget.

A less favorable situation has been observed in typical mining countries in the Third World. The dualistic structure of both the employment *income* and the dissemination of technologies is more distinct; inflation and unemployment rates tend to be higher and agriculture suffers from a stagnation which leads to increased food imports.

The sectoral development plans in some important mining countries reflect their growth-orientated objectives and strategies. After agriculture the mining sector has a high national priority. Among the primary objectives of the mining sector, as defined in most of the countries, are:

- Increasing production and efficiency in the mines.
- Diversifying mine production with special reference to energy resources such as coal and lignite, as well as industrial minerals.
- Intensifying the processing of ore concentrates to increase the domestic value added.

In addition to these frequently pursued objectives further goals appear in some of the mining countries, for example:

- Reducing environmental hazards.
- Assessing national mineral potential.

Some of the strategies to attain these objectives are:

- The expansion of the Geological Survey, especially of departments and laboratories for the exploration and evaluation of mineral deposits.
- Intensification of exploration activities of government institutions in order to minimize the risk and create incentives for investors.
- Expansion of the transfer of technology in the fields of mining geology and mining in order to improve the exploration techniques, the mining methods, and mineral dressing processes.

10.6.1 Primary Effects

The primary effects of a mining project are direct costs and benefits for the mining company and are influenced by company strategies. It may here be assumed that not only private but also government enterprises generally have profit-orientated objectives. The primary effects may be classified in two economic groups:

- The investment interests with priority given to achieving the best return on investment possible;
- The mineral production interests with the aim of securing the supply of company-owned smelters or the extension of technical know-how.

For a quantification of profits or of the return in investment several reliable methods are available (see Sect. 6).

10.6.2 Secondary Effects

The spectrum of effects and impacts of a mining project is not confined to the individual enterprise. Especially mines, which are often situated in remote areas, frequently exert diverse and long-term influence on the project region and even on the national economy of a developing country. This means that effects arise which are either of economic benefit or give rise to social costs. These impacts will here be designated as "secondary effects", which term has in the meantime been widely adopted and is also the most appropriate as a contrast to "primary effects".

In general, secondary effects are not intended by the mining company. It is still a matter of controversy whether the company is or is not capable of exerting an influence on them. Some of the secondary effects can be influenced by the mining company, such as certain ecological and infrastructural impacts. Government measures, however, do exert an influence on secondary effects and it is therefore an essential part of a responsible development policy to enhance the advantages and to minimize the disadvantages of mining projects.

Prior to an analysis of the main secondary effects of mining projects it should be pointed out that even exploration projects show considerable impacts, which can be classified into the following four groups:

- Mineral economic effects through an improvement of the knowledge of regional mineral potentials as well as through incentives for mining investments.

- Effects with regard to professional qualification of geologists, miners, and mine managers.
- Institution building effects through the acquisition of know-how by the competent authorities, e.g., Geological Survey, Department of Mineral Resources, and others.
- Infrastructural effects through the construction of roads and air strips.

a) *Infrastructural Effects*

As specific mining operations have no free choice of location, the necessary material infrastructure and personnel infrastructure must be created and brought into line with the project, from which additional expenses of up to 50% of the total investment costs may arise (see Table 10.5). To begin with, the mine must be connected to the local traffic network, which implies the construction of new roads or railway connections. The next step is to secure the supply of water and energy through appropriate installations. Finally, the personnel structure of the plant necessitates the construction of housing and the recruitment and training of personnel. Mining projects are thought to play a significant part in the development of remote regions because they force the companies to build up a relevant infrastructure. Yet as long as these new infrastructures are formed and used merely in accordance with individual mining projects, they serve only partially as an impetus to regional development. Additional government measures, however, can strengthen these effects. Developing countries should therefore integrate their mining projects, from the planning period onwards, into an overall concept of regional economic structures in order to make the maximum economic profit from mining infrastructures.

b) *Effects on Employment and Income*

The effects on income are considered to be especially relevant for measuring the success of investments with regard to the national economy. Mining projects create new income in the following three fields:

Table 10.5 Investment on infrastructure (as per cent of total mining investment)

Project	Investment			
	Transport	Energy	Social	Infrastructures total
Copper Mine Cuajone, Peru	16.9	6.0	18.1	43.7
Gold-copper Mine Ok Tedi, PNG	27.4	8.6	14.2	50.2
Iron ore Mine Bong Range, Liberia	12.5	13.3	10.3	36.1
Copper Mine Bleida, Morocco	8.0	12.0	30.0	50.0
Lead-zinc Mine Strathcona, Canada	14.5	6.1	33.1	53.7

Source: Schippers (1982).

– Employment income in the mine;
– Capital income for domestic shareholders;
– Expanded labor and capital income in linked industries.

Mining on an industrial scale is basically a capital-intensive activity and thus only major projects or others with special preconditions show considerable employment effects. An OECD study (1975) gave an impressive illustration of this fact, comparing the contribution of the mining sector to the GDP in developing countries with its share in the total labor force. In Zambia the mining sector (copper) accounted for 25.4% of the GDP, but only 4.2% of the national labor force was employed in this sector; the respective percentage in Indonesia was 9.3% and 0.2% and in Suriname (bauxite) 26.2% and 7.0%. According to more recent statistics published by the United Nations (1983) and the International Labor Office (1981) the data for Botswana (diamonds) are 25.9% share of mining in GDP and 6.8% employment in the mining sector or for Bolivia (tin) 9.9% and 5.6% respectively. Direct employment effects and thus a contribution to the reduction of unemployment rates in developing countries are therefore significant only in a few cases.

The lignite mine in Mae Moh, North Thailand, (opencast mine and power plant) is one example. In its first project stage as much as 1,463 jobs were generated in the open-pit section and 1,039 in the power plants. Only 17% of these jobs are held by local inhabitants (Schützdeller 1984). The expansion of capacity which is already under way will lead to the creation of 5,000 additional jobs in 1988.

The income on investment of domestic shareholders in developing countries has up to now been modest, as on the one hand, domestic private investment occurs to any significant extent only in a few LDCs (Thailand, Philippines, Brazil), and on the other hand, the profits made by these mining companies are often relatively small.

For suppliers and processing plants the induced effects on employment and income in the mining sector tend to be lower than the world-wide industrial average. In a series of investigations the income multiplier of mining projects was reckoned at only 1.5 to 3.3. Furthermore, these multiplier effects depend largely on the stage of development in the country. Brazil for instance offers much better preconditions for domestic industry to hold a large share in the manufacturing of spare parts and equipment than Zaire or Burma.

c) *Effects on Industrialization*

In large mines, there is a considerable demand for services and supply of material. Enterprises in structurally less developed regions try to obtain a high degree of self-sufficiency by establishing company-owned repair shops. But in any case, a process is initiated whereby small industries and transport contractors settle in the area and offer their services to the mine. These backward linkages are often small but labor-intensive plants. They carry out not only maintenance, repair work, and transport of all kinds, but also the production of material for mines such as timber, bricks, and charcoal is set in motion.

On the other hand, mining production creates forward linkages, utilizing the extracted minerals. Coal mining, for example, enables enterprises with a demand for low-priced thermal energy to dry agricultural goods (grains, tobacco, coffee, etc.) or to produce industrial goods (cement, fertilizers, etc.) to settle in this region. For processing ore concentrates, however, the only suitable way is to smelt them on an industrial scale. This first step towards a vertical diversification was successfully taken in numerous mining countries in the 1960s and 1970s. Zinc and bauxite/aluminum are impressive examples of this diversification. While in 1960 only 29% of the zinc supply from developing countries entered the world market in the form of metals, this percentage rose to 42% in 1970 and to 61% in 1980. The share of aluminum on the world market increased from 3% in 1960 to 10% in 1970 and 22% in 1980.

d) *Transfer of Technology and Know-How*

The expansion of mining plants in developing countries always implies contact with modern technologies and, in the case of cooperation with industrialized countries, with modern management. It is in the undisputed interest of the developing countries to integrate a systematic transfer of technology and know-how into the concepts of their industrialization processes. For this purpose, the LDCs should create the essential conditions, because the transfer of technology depends largely on initial personnel, socio-economic and administrative conditions, and cannot merely be enforced by law.

The transfer of technology and know-how is first of all a problem of personnel. The scientists and engineers working with the technology user should not only be in a position to apply the technology but also to adapt it to local conditions and finally to develop it. The more training the personnel receive, the more effective the transfer in mining projects becomes. In the developing countries there is a demand for training at all levels, whereby it is often observed that the individual mine has to train a larger number of personnel than it actually requires because of the considerable migration rate of qualified personnel. Here, there results a special multiplier effect of mining projects located in remote areas.

e) *Fiscal Effects*

Traditionally, all kinds of mining operations are taxed. The royalties have proved to be of particular fiscal attractiveness. These are taxes which are collected regardless of the profit and are imposed on the crude ore tonnage. The royalties, together with corporate income taxes, export taxes, licence fees, and special mineral taxes, contribute essentially to the fiscal effects of a mining project and constitute one of the most outstanding project effects from the point of view of the national economy.

Tax regulations are therefore extremely important for the evaluation of mining projects. The government will always try to take a substantial share of the profits made by the mining companies. It did, however, emerge that fiscal instruments should be dealt with carefully so as not to be harmful to the entire mining sector. Examples of a flexible tax policy are found in the tin mining industry in Thailand and Malaysia, where the governments even reduced taxes when the profitability of small and medium tin mines was seriously threatened. Because of the high risks involved in mining production, private investors regarded as fair a rate of return on investment of 15%.

f) *Effects on the Balance of Payments*

Mining projects in developing countries are generally related to foreign trade, which manifests itself in:

- The export of minerals and metals.
- The reduction of mineral imports by domestic mining production.
- The importation of mining equipment and material.

The role of mines as foreign-exchange earners for the developing country is often overestimated, taking into account only the export of mining products and not the imports necessary for mining projects. The import of capital goods such as machines, vehicles, electrical systems, or spare parts and also the import of materials such as diesel oil, explosives, or chemicals can reduce the foreign exchange earnings considerably.

The extent to which the positive effects on the national balance of payments make themselves felt depends largely on the degree of industrialization of the individual developing country. To a considerable extent, Brazil, Korea (South) and India are already in a position to produce mining equipment and spare parts locally. Burma, Jamaica, and Sierra Leone, on the other hand, are compelled to import nearly all of their mining equipment and additionally a major part of the fuels. Considering the market interest rate, however, the foreign exchange account of a mining project may even be negative, because the investments must take place several years before the production starts.

Yet other mining projects, which reduce the import of minerals and thus contribute to the saving of foreign currency, may show positive effects on the balance of payments. Coal or lignite mines in developing countries may serve as an example. The savings on foreign currency through the substitution of imported crude oil can clearly increase the internal rate of return of such investments from an overall economic point of view. (The lignite mine Mae Moh in North Thailand achieved an increase of from 16.8% to 28%).

g) *Ecological and Social Effects*

Only some evident impacts out of the range of ecological and social effects of mining projects shall be mentioned here:

- The effects on the environment through the pollution of water and air by mining activities or the damage to the environment by larger opencast operations (see also Sect. 6.3.4.3).
- The impacts on the socio-cultural environment when economic oases are created or migrant workers are employed.
- The exploitation of limited mineral resources.

The production of mineral resources on an industrial scale always entails environmental problems. Modern mining technology is trying to solve these problems, though their solutions are sometimes costly. In the meantime the governments of most LDCs have developed a stronger sense of responsibility for their environment and have firmly embodied the protection of the environment in law. The widely held view that developing countries give incentives for the settlement of industries by emphasizing less strict environmental regulations is not true – at least for the mining sector. Yet developing countries always meet with difficulties in checking the observance of the mine regulations because of a lack of efficient laboratories and experienced inspectors. In this field there is a demand for cooperation with industrialized countries.

Numerous investigations have revealed that, especially when there are foreign shareholdings, larger mining companies pay relatively high wages and salaries and offer special social benefits. Mine workers in rural areas thus come to occupy a special social position, which can lead to conflicts with the rest of the rural population. This results in a migration of labor into the project region, which causes a series of problems, among them the isolation of ethnic minority groups and the separation of the immigrants from their families.

It is, however, possible to minimize the negative secondary effects of mining projects by taking appropriate measures. Before the necessary political steps can successfully be undertaken a detailed analysis of all project effects must be carried out. For this work, contributions made by experienced mineral economists are essential.

Notes on the Literature

Readers interested in more details on mineral strategies in developing countries are referred to Bosson and Varon (1977): *The Mining Industry and the Developing Countries,* and to Radetzki and Zorn (1979): *Financing Mining Projects in the Developing Countries.* Additional information can be obtained from Wälde (1984): *Third World Mineral Development: Current Issues,* as well as from Sims (1985): *Government Ownership Versus Regulation of Mining Enterprises in Less-developed Countries.* Those interested in mining impacts should read the article by Gocht (1984): *Effects of Mining Projects in the Developing Countries.*

References

For suggested reading, please refer to "Notes on the Literature" at the end of each chapter.

Adelman J, Adelman F (1959) The dynamic properties of the Klein-Goldberger Model. Econometrica 27
Adelman MA, Houghton JC, Kaufman G, Zimmerman MB (1983) Energy resources in an uncertain future: Coal, Gas, Oil, and Uranium Forecasting. Ballinger, Cambridge, Massachusetts
Allum JAE (1969) Photogeology and regional mapping. Oxford
American Geological Institute (1972) Glossary of geology
Association of Iron Ore Exporting Countries (APEF) Iron ore news, various issues, Genf
Australian Mineral Economics Pty. Ltd. (1980) The world tin industry. Sydney, p 278
Baldwin WL (1983) The world tin market. Durham, NC, p 273
Behrmann JR (1977) International commodity agreements. An evaluation of the UNCTAD Intergrated Commodity Program, Washington
Bender F (ed) (1983) New path to mineral exploration. Stuttgart
Bender F (ed) (1985) Geo-resources and environment. Proceedings of the fourth international symposium Oct 1985, Hannover, p 144
Bennet MK, and Associates (1968) International commodity stockpiling as an economic stabilizer. New York
Birnie RW, Francica JR, Jr (1981) Remote detection of geobotanical anomalies related to porphyry copper mineralization. Econ Geol 76, pp 637–647
Bischoff G, Gocht W (eds) (1981) Das Energiehandbuch, 4th edn Braunschweig, Wiesbaden
Blondel F, Lasky SG (1970) Concepts of mineral reserves and resources. UN Surv, New York
Bosson R, Varon B (1977) The mining industry and the developing countries. Oxford, p 292
Brealey R, Myers S (1984) Principles of corporate finance. McGraw-Hill, New York
Brobst DA, Pratt WP (eds) (1973) United States mineral resources, US Geol Surv Prof Paper 820. Washington, p 722
Brown R (1986) New mining codes: salient features. Paper presented at a conference on mining ventures in developing countries. Institute for Foreign and International Trade Law, Frankfurt/Main, 20 June 1986
Buchanan LJ (1981) Precious metal deposits associated with volcancic environments in the southwest. Arizona Geol Soc Digest vol IX, pp 237–262
Burn RG (1984) Exploration risk. CIM Bull 77, 870:55–61
Calkins JA, Keefer EK, Ofsharick RA, Mason GT, Tracy P, Atkins M (1978) Description of CRIB, the GIPSY retrieval mechanism, and the interface to the General Electric MARK III Service. US Geol Surv Circular 755–A, p 47
Cansier D (1987) Besteuerung von Rohstoffrenten. Berlin, p 199
Carnegie RH (1986) Outlook for mineral commodities. Group of Thirty, New York, p 35
Chaffee MA (1976) Geochemical exploration techniques based on distribution of selected elements in rocks, soils and plants, mineral butte copper district. Pinal County, Arizona. Geol Surv Bull 1278
Chamber of Mines of South Africa (1986) Statistical tables, Johannesburg, p 42
Chao Hung-Po (1979) Economics with exhaustible resources. New York
Chapman PF, Roberts F (eds) (1983a) Metal resources and energy. London
Chapman PF, Roberts F (1983b) Metal resources and energy. London, p 238
Charles River Associates (1978) The economics and geology of mineral supply: An integrated framework for long-run policy analysis. Report N.327. Charles River, Boston, Massachusetts
Chaussier JB, Morer J (1987) Mineral Prospecting Manual. Elsevier, Amsterdam, New York, p 300

CIPEC Quarterly Reviews, various issues, Paris
Cobalt Development Institute, Cobalt news, various issues, Brussels
Compton DW, Gjostein NA (1986) Materials for ground transportation. Sci Am 255 (4):92–100
Cox DP (1983) US Geol Surv – Ingeominas mineral resource assessment of Colombia: Ore deposit models. US Geol Surv Open File Report, pp 83–423
Cox DP, Singer DA (1986) Mineral deposit models. US Geol Surv Bull 1693:379
Crane-Engel M, Schanze E (1988) Multilateral exploration assistance: the United Nations programs. In: Tilton JE, Eggert RG, Landsberg HH (eds) World mineral exploration: Trends and economic issues. Resources for the Future, Washington, DC
Cranstone DA (1988) The Canadian mineral discovery experience since world war II. In: Tilton JE, Eggert RE, Landsberg HH (eds) World mineral exploration: trends and economic issues. Resources for the Future, Washington, DC
Crawford JT III, Hustrulid WA (eds) (1979) Open pit mine planning and disign. Am Inst Min, Metallurgical, and Petroleum Engineers, New York, p 367
Cronin DS (1980) Underwater minerals. Academic Press, London, p 362
Crowson P (1986) Minerals handbook 1986–87. Stockton, New York, p 331
Crowther J (1985) Feasibility studies. Bulletin of the Proceedings of the Australasian Institute of Mining and Metallurgy 290 (5):14–18
CRU Consultants Inc. (ed) (1985) Copper to 1995. Challenges for an industry in transition, vol 3, New York
Cumming JD, Wicklund AP (1985) The diamond grade handbook, 3rd edn. J.K. Smith & Sons Diamond Products, Toronto, p 541
Cummins AB, Given IT (1973) SME mining engineering handbook. Am Inst Min, Metallurgical, and Petroleum Engineers, New York
Daniel P (1985) Minerals in independent Papua New Guinea. Policy and performance in the large-scale mining sector. Working paper No 85/10, Australian National University, p 67
Dasgupta PS, Heal GM (1979) Economic theory and exhaustible resources. Cambridge
David N (1977) Geostatistical ore reserve estimation. Elsevier, Amsterdam, p 364
Davis JC (1973) Statistics and data analysis in geology. Wiley, New York
Davis JC (1986) Statistics and data analysis in geology. 2nd edn. Wiley, New York, p 646
DeGeoffroy JG, Wignall TK (1985) Designing optimal strategies for mineral exploration. Plenum, New York, p 364
Demler FR (1983) Beverage containers. In: Tilton JE (ed) Material substitution: Lessons from the tin-using industries. Resources for the Future, Washington, DC, pp 15–35
Department of Energy, Mines and Resources (1975) Department terminology and definitions of reserves and resources, Ottawa/Canada
Department of Mineral and Energy Affairs (1985) South Africa's mineral industry 1985, Johannesburg, p 224
Derry DR (1980) World atlas of geology and mineral deposits. John Wiley, New York, p 110
Desai M (1967) An econometric model of the world tin economy, 1948–1961. Econometrica 34:105–134
De Verle PH (1984) Mineral resources appraisal. Oxford geological sciences series. Oxford, p 445
Dixon CJ (1984) Atlas of economic mineral deposits. Chapman & Hall, London, p 143
Earney CF (1980) Petroleum and hard minerals from the sea. Halsted, New York, p 291
Eckstrand OR (1984) Canadian mineral deposit types: A Geological Synopsis. Geol Surv Can, Econ Geol Rep 36:86
Economic Geology (1983) An issue devoted to techniques and results of remote sensing: 78: pp 573–770
Edwards R, Atkinson K (1986) Ore deposit geology. Chapman & Hall, London, p 466
Eggert RG (1986) Changing patterns of materials use in the US automobile industry. Mater Soc 10 (3):405–431
Eggert RG (1987a) Metallic mineral exploration: An economic analysis. Resources for the Future, Washington, DC
Eggert RG (1987b) Exploration and access to public lands. In: Cordes JA (ed) Public policy and the competitiveness of US and Canadian metals production. In preparation
Eisold H, Hasse R (1984) Time for reorientation in Lomé III. Inter Econ 19 (2): pp 78–83

Ely N (1964) Mineral titles and concessions. In: Robie EH (ed) Economics of the mineral industries. 2nd edn, Am Inst Min, Metallurgical, and Petroleum Engineers, New York, pp 81–130
Emerson C (1984) Mining taxation in ASEAN, Australia and Papua New Guinea, Kuala Lumpur/Canberra, p 107
Engineering and Mining Journal, Library of operating handbooks. vol 1: Operating handbook of mineral processing (1977); vol 2: Mineral surface mining and exploration (1978); vol 3: Mineral underground mining (1978)
Eveleth RW (1979) New methods of working an old mine – Case history of the Eberle group, Mogollon, New Mexico. Summarized in Mining Engineering, pp 138–139
Federal Emergency Management Agency (1986) Stockpile report to the congress. Washington, DC, p 61
Fettweis GB (1981) Die internationale Einordnung von Mineralvorräten (The international classification of mineral resources). Erzmetall 34, pp 400–406; pp 465–469
Financial Time (ed) (1986) Mining international year book 1986. London
Fischmann LL (ed) (1980) World mineral trends and US supply problems. Washington
Fisher FM, Cootner PH, Baily MN (1972) An econometric model of the world copper industry. Bell J Econ Manage Sci 3:568–609
Fitch AA (ed) (1979) Developments in geophysical exploration methods. London
Fletcher WK (1981) Analytical methods in geochemical prospecting. In: Govett GJS (ed) Handbook of exploration geochemistry, vol 1. Elsevier, Amsterdam, p 255
Forrester JE (1949) Principles of field and mining geology. New York
Fortescue and Hornbrook (1967)
Fox W (1974) Tin – the working of a commodity agreement. London, p 418
Frame A (1983) The copper industry in crisis; prices and the supply/demand balance. In: Copper '83. Preprint book 1, Conference 1–2 Nov 1983, Royal Lancaster Hotel. London
Francis AA (1981) Taxing the transnationals in the struggle over bauxite. Kingston
Gilbert JM, Falk CE Jr (1986) The geology of ore deposits. Freeman, New York, p 985
Gocht W (1969) Der metallische Rohstoff Zinn. Berlin
Gocht W (1980) The importance of small scale mining in developing countries. In: Natural resources and development, 12:7–18
Gocht W (ed) (1981) Proceedings of the seminar on exploration and evaluation of tin deposits in Southeast Asia, Intertechnik, Bd. 21. Aachen
Gocht W (1982) Schutz natürlicher Ressourcen durch Vermeidung von Raubbau. In: Internationale Kooperation, Bd. 23. Baden-Baden, pp 60–82
Gocht W (1983) Wirtschaftsgeologie und Rohstoffpolitik, 2nd edn. Springer, Berlin Heidelberg New York, p 295
Gocht W (1984) Effects of mining projects in developing countries. Geol Jahrb A 75:167–175
Gocht W (1985) Handbuch der Metallmärkte, 2nd edn. Springer, Berlin Heidelberg New York, p 444
Gocht W, Jütte-Rauhut J (eds) (1986) Proceedings of the seminar on importance of primary tin mining in Southeast Asia, Intertechnik, Bd. 28. Aachen
Gordon-Ashworth F (1984) International commodity control. London, pp 103–273
Govett GJS (1983) Rock geochemistry in mineral exploration. Handbook of exploration geochemistry, 3. Elsevier, Amsterdam, p 440
Griffiths OH, King RF (1980) Applied geophysics for geologists and engineers. Pergamon Press, New York, p 230
Guigues J, Devismes P (1969) La prospection minière à la Bateé dans le Massif Americain. Mémoires du BRGM, 71, p 171
Guilbert JM, Park CF Jr (1986) The geology of ore deposits. WH Freeman Company, New York, p 985
Günther R (1972) Remote Sensing in der Geologie. BMBW-Forschungsbericht W72-28, Clausthal-Zellerfeld
Gupta S (1982) The world zinc industry. Lexington/Toronto
Gy P (1982) Sampling of particulate minerals. Elsevier, Amsterdam, p 530
Hansen J (1978) Guide to practical project appraisal: Social benefit cost analysis in developing countries. UNIDO project formulation and evaluation series, 3. New York
Harris DP (1984) Mineral resources appraisal. Oxford University Press, Oxford, p 445

Harris DP, Agterberg FP (1981) The appraisal of mineral resources. Econ geol 75:879–938
Harris DP, Skinner BJ (1982) The assessment of long-term supplies of minerals. In: Smith VK, Krutillia JV (eds) Explorations in natural resource economics. John Hopkins University Press for Resources for the Future, Baltimore, pp 247–326
Hazen SW Jr (1967) Assigning an area of influence for an essay obtained in mine sampling. US Bureau of Mines Report Investigation 6955:75
Hill T (1985) Wither Lomé? A review of the Lomé III negotiations. Third World Q, pp 661–681
Hill VG, Ostojic S (1986) Primary aluminium: Its raw materials situation and implications for bauxite exporting countries. In: Proceedings of the 5th Jugoslav international symposium on aluminium, 23.4.–25.4.1986, Mostar
Hohmann, GW, and Ward SH (1981) Electrical methods in mining geophysics. In: Skinner BJ (ed) Economic geology 75th Anniversary Volume. pp 806–828
Hood PJ (ed) (1977) Geophysics and geochemistry in the search for metallic ores. Geol Surv Can, Econ Geol Rep 31:811
Hoppe R (ed) (1978) E/MJ Operating handbook of mineral surface mining and exploration. McGraw-Hill, New York, p 450
Horikoshi E, Sato T (1970) Volcanic activity and ore deposition in the Kosaka mine. In: Tatsumi T (ed) Volanism and ore genesis. University of Tokyo Press, pp 181–195
Hoskins J, Green W (1977) Mineral industry costs, 2nd edn. Spokane
Howarth RJ (ed) (1983) Statistics and data analysis in geochemical prospecting. Handbook of exploration geochemistry, 2. Elsevier, Amsterdam, New York, p 436
Howe CW (1979) Natural resource economics. New York
Hustrulid WA (ed) (1982) Underground mining methods handbook. Am Inst Min, Metallurgical, and Petroleum Engineers, New York, p 1754
Institution of Mining and Metallurgy (1971) A pricing and marketing of metals. Trans IMM, London
International Bauxite Association (1975) Agreement establishing IBA. Kingston
International Energy Agency (1983) Energy policies and programmes of IEA countries. OECD, Paris, p 416
International Monetary Fund (ed) (1986), Primary commodities – market developments and outlook. Washington, DC, p 74
International Primary Aluminium Institute (1986) Statistical summary
International Tin Council (ed) (1979) Tin production and investment. London, p 185
International Tin Council (ed) (1987) Monthly Statistical Bulletin and Quarterly Reports, 1975–1986. London
Joralemon IR (1973) Cooper, the encompassing story of mankind's first Metal. Howell-North, Berkeley, California
Journel AG, Huijbregts CJ (1981) Mining Geostatistics. Academic Press, Londono, p 600
Jütte-Rauhut J (1986) Project financing in the mining industry. In: Gocht W, Jütte-Rauhut J (eds) Proceedings of the seminar on importance of primary tin mining in Southeast Asia, Intertechnik, Bd. 28. Aachen, pp 266–274
Kaufman A (1976) Mineral disposal systems. In: Vogely WA (ed) Economics of the mineral industries, 3rd edn. Am Inst Min, Metallurgical, and Petroleum Engineers, New York, pp 644–647
Kaufman PJ (1984) Handbook of futures markets: Commodity, financial, stock index, and options. John Wiley, New York
Keefer EK, Caulkins JA (1977) Description of individual data items and codes in CRIB. US Geol Surv Circular 755-B: p 32
Kennedy P (1985) A guide to econometrics, 2nd edn. MIT Press, Cambridge, Massachusetts
Krauskopf KB (1979) Introduction to geochemistry, 2nd edn. McGraw-Hill, New York, p 617
Krige DG (1962) Statistical applications in mine evaluation. J Inst Mine Surv South Africa 12 (2, 3)
Kumar R, Radetzki M (1987) Alternative fiscal regimes for mining in developing countries. World Dev 15 (5):741–758
Labys WC, Field FR, Clark J (1985) Mineral models. In: Vogely WA (ed) Economics of the mineral industries, 4th edn. Am Inst Min, Metallurgical, and Petroleum Engineers, New York, pp 337–379
Lasaga M (1981) The copper industry in the Chilean economy: An econometric analysis. Lexington Books, Lexington, Massachusetts
Law AD (1975) International commodity agreements. Lexington
Lawson CW (1984) The future of east-south trade after UNCTAD VI. Third World Q, pp 145–154

Lee TD (1984) Planning a mine feasibility study – an owner's perspective. In: Mine Feasibility – Concept to Completion (short course notes available from Northwest Mining Association, 414 Peyton Building, Spokane, Washington USA 99201)
LeFon SJ (ed) (1983) Industrial minerals and rocks, 5th edn. Am Inst Min, Metallurgical, and Petroleum Engineers, New York, p 1674
Leshy JD (1987) The mining law: a study in perpetual motion, Resources for the Future. Washington, DC
Levinson AA (1974) Introduction to exploration geochemistry. Applied Publishing, Wilmette
Levinson AA (1980) Introduction to exploration geochemistry, 2nd edn. Applied Publishing, Wilmette, p 924
Lowell JD (1968) Geology of the Kalamazzo orebody, San Manuel district, Arizona. Econ Geol 63: pp 645–654
Lowell JD, Guilbert JM (1970) Lateral and vertical alterationmineralization zoning in porphyry ore deposits. Econ Geol 65: pp 373–408
Lydon JW (1983) Chemical parameters controlling the origin and deposition of sediment-hosted stratiform lead-zinc deposits. In: Sangster DR (ed) Sediment-hosted stratiform lead-zinc deposits. Min Assoc Can Short Course Handbook: 8: pp 175–250
MacDonnell LJ (1987) Government mandated costs: the regulatory burden of environmental, health, and safety standards on US metals production. In: Cordes JA (ed) Public policy and the competitiveness of US and Canadian metals production. In preparation
Mansbach TB (1985) OPIC insurance programmes for the mining sector. In: Tinsley CR (ed) Finance for the minerals industry. New York, pp 457–460
Matheron G (1971) The theory of regionalized variables and its applications. Fontainebleau, Cahiers Centre
McCammon RB (ed) (1975) Concepts in geostatistics. Springer, Berlin Heidelberg New York
McCammon RB (1974) The statistical treatment of geochemical data. In: Levinson AA (ed) Introduction to exploration geochemistry, 1st ed. Applied Publishing Ltd, Wilmette, pp 469–508
McCammon RB (1980) The statistical treatment of geochemical data. In: Levinson AA (ed) Introduction to exploration geochemistry, 2nd edn. Applied Publishing, Wilmette, pp 469–843
McGill S (1985) Financing the Ok Tedi mine – case study of the process from a government perspective. In: Tinsley CR, Emerson ME, Eppler WD (eds) Finance for the minerals industry. Am Inst Min, Metallurgical, and Petroleum Engineering, New York
McGill S, Crough G (1987) Indigenous resource rights and mining companies in North-American and Australia. Natural Resources Forum 11, 1: pp5–26
McKelvey VE (1973) Mineral resource estimates and public policy. US Geol Surv Prof Paper 820:9–19
McKelvey VE (1986) Subsea mineral resources. US Geol Surv Prof Paper 1689-a:106
McKelvey VE, Soregaroli AE, Balla JC (1987) 1985 Mineral and exploration statistics, United States and Canadian companies. Econ Geol 82, pp 257–263
McKinstry HE (1957) Mining geology. New York
McLaren DJ, Skinner BJ (1987) Resources and world development. John Wiley, Chichester, p 940
Megill R (1979) An introduction to exploration economics, 2nd edn. Tulsa
Metallgesellschaft AG (ed) Metallstatistik, various issues. Frankfurt/Main
Mikesell RF (1979) The world copper industry, structure and economic analysis. John Hopkins University Press for Resources for the Future, Baltimore, p 298
Mikesell RF (1984) Foreign investment in mining projects. Oelschläger, Gunn & Hain, Cambridge, Massachusetts
Miller S, Emerick JC (1985) The secondary effect of mineral development. In: Vogely WA (ed) Economics of the mineral industries, 4th edn. Am Inst Min, Metallurgical, and Petroleum Engineers, New York
Mine Development Bimonthly (1985) vol III no 3 (December 31)
Minister of Supply and Services, Canada (1985) Canadian minerals yearbook 1983–1984. Min Rep 33, Ottawa
Müller-Ohlsen L (1981) Die Weltmetallwirtschaft im industriellen Entwicklungsprozeß. Tübingen, p 265
Nash CR (1980) Photogeology and satellite imagery interpretation in mineral exploration. Min Sci Eng 12, pp 216–243

NCSS (1976) Government and the natural resources. Washington
Nevitt PK (1984) Project Financing. Euromoney Publications, London
Newendorp PD (1975) Decision analysis for petroleum exploration. PPC Books, Tulsa, Oklahoma
Northern Miner (1987) Quebec study shows flow-through value. March 2:1, 19
Northwest Mining Association (1984) Mine feasibility-concept to completion. (Short course notes available from Northwest Mining Association, 414 Peyton Building, Spokane, Washington USA 99201)
OAPEC (ed) (1983) OAPEC in brief. Kuwait, p 41
Office of Technology Assessment, Congress of the United States (1979) Management of fuel and nonfuel minerals in Federal land: current issues and status. Government Printing Office, Washington, DC
Ohle EL, Bates RL (1981) Geology, geologists, and mineral exploration. In: Skinner BJ (ed) Economic Geology 75th Anniversary Volumem, pp 766–774
OPEC (ed) (1980) The statute of the organization of the petroleum exporting countries. Wien
Overstreet WC, Marsh SP (1981) Some concepts and techniques in geochemical exploration. In: Skinner BJ (ed) Economic geology 75th Anniversary Volume, pp 775–805
Paley HN (1984) The joint NASA/Geosat test case project. AAPG Bookstore, Tulsa
Paley HN (1985) The joint NASA/Geosat fest case project. AAPG Bookstore, Am Asso of Petroleum Geologists
Parasnis DS (1973) Mining geophysics. Elsevier, Amsterdam New York
Parasnis DS (1975) Principles of applied geophysics. London
Parr CJ (1982) Environmental considerations. In: Hustrulid WA (ed) Underground mining methods handbook. Am Inst Min, Metallurgical and Petroleum Engineers, pp 155–181
Patterson JA (1959) Estimating ore reserves follows logical steps. Eng Min Jour 160, p 115
Paterson NR (1983) Exploration geophysics: airborne. In: Woakes M, Carman JS (eds) AGID Guide to mineral resources development. Assoc Geosc Intern Dev, Bangkok, pp 121–151
Pawlek F, Fischer R (1982) Rückgewinnung von NE-Metallen aus Schrotten und Rückständen – wirtschaftliche und technische Entwicklungsrichtungen. In: Metall, 36, pp 428–431
Pearl RM (1973) Handbook for prospectors. New York
Peters WC (1987) Exploration and mining geology, 2nd edn. John Wiley, New York, p 685
Petersen U, Maxwell SR (1979) Historical mineral production and price trends. Min Eng:25–34
Petrick A Jr (1985) Mineral investment and finance. In: Vogely WA (ed) Economics of the mineral industries. Am Inst Min, Metallurgical, and Petroleum Engineers, New York, pp 227–301
Petroleum Economist (1985) July. London
Pickett J, Robson R (1986) Manual on the choice of industrial technique in developing countries. OECD, Paris, p 128
Pindyck RS, Rubinfeld DL (1981) Econometric models and economic forecasts, 2nd edn. McGraw-Hill, New York
Pollard DE (1984a) Law and policy of producer's association. Oxford, pp 109–115
Pollard DE (1984b) The IBA – the first ten years. IBA-Q Rev 10, 1
Post AM (1983) The implications of change in mining finance and participation. Erzmetall 36, 7/8:338–391
Pralle GE (1985) The effect of host government attitude upon foreign investment in mining. In: Tinsley CR (ed) Fianance for the minerals industry. New York, pp 353–358
Primary Tungsten Association: Monthly Bulletin, various issues. London
Prost G (1983) Mineral exploration with Skylab photography in central Colorado. Econ Geol 78:633–640
Radetzki M (1980) Changing structures in the financing of the minerals industry in LDCs. Dev Change 11:1–15
Radetzki M (1982) Regional development benefits of mineral projects. Resour Policy 8, 3
Radetzki M (1985) State mineral enterprises: An investigation into their impact on international mineral markets. Resources for the Future, Washington, DC
Radetzki M, Zorn S (1979) Financing mining projects in developing countries, Min Jour Books, London, p 189
Ramsley JB (1981) The economics of exploration for energy resources. London
Rassan GN, Gravesteijn J, Potenza R (eds) (1987) Multilingual thesaurus of geosciences. Pergamon, New York

References

Reedman JH (1979) Techniques in mineral exploration. App Sci, London, p 533
Reeves RD, Brooks RR (1978) Trace element analysis of geological materials. Wiley, New York, p 421
Rensburg WCJ von, Bambrick S (1978) The economics of the world's mineral industries. Johannesburg
Robertson W (1982) Tin: Its production and marketing. London, p 210
Robinson EL (1985) Changing in the financial structure of the mining industry over time. In: Tinsley CR, Emerson ME, Eppler WD (eds) Finance for the minerals industry. New York, pp 10–16
Rose AW, Hawkes HE, Webb JS (1979) Geochemistry in mineral exploration, 2nd edn. Academic Press, New York, p 657
Rosenkranz RD, Boyle EH Jr, Porter KE (1983) Copper availability – market economy countries: a mineral availability program appraisal. US Bureau of Mines Information Circular 8930. Government Printing Office, Washington, DC
Rudawsky O (1986) Mineral economics: development and management of natural resources. Elsevier, Amsterdam, p 192
Sabin AE (1985) Tin availability – market economy countries: a minerals availability appraisal. Proceedings of the 1st International Tin Symposium Dec. 1985. Tin International, pp 79–103
Sato T (1974) Distribution and geologic setting of the Kuroko deposits. Soc Min Geol Japan Spec Issue 6, pp 1–9
Sawkins FJ (1984) Metal deposits in relation to plate tectonics. Springer, Berlin Heidelberg New York, p 325
Schenck GHK (1985) Methods of investment analysis for the minerals industries. In: Tinsley CR, Emerson ME, Eppler WD (eds) Finance for the minerals industry. Am Inst Min, Metallurgical, and Petroleum Engineering, New York, pp 77–93
Shell International Petroleum (1986) London
Schlitt WJ (ed) (1982) Interfacing technologies in solution mining. Am Inst Min, Metallurgical, and Petroleum Engineers, New York, p 370
Schneider GJ (1981) In situ neutron activation analysis. In: Premining investigations for hardrock mines. US Bureau of Mines Information Circular 8891, pp 46–54
Schützdeller R (1984) Sekundäreffekte von Bergbauprojekten unter besonderer Berücksichtigung vom empirischen Untersuchungen im Braunkohlenbergwerk Mae Moh in Thailand, Internationale Kooperation, Bd. 25. Baden-Baden
Shanks WC III (ed) (1983) Cameron volume on unconventional mineral deposits. Am Inst Min, Metallurgical, and Petroleum Engineers, New York, p 246
Short NM (1982) The Landsat tutorial workbook: basics of satellite remote sensing. NASA Reference Publication 1078, Washington, p 553
Siegel FR (1974) Applied geochemistry. New York
Siehl A, Thien J (1978) Geochemische Trends in der Minette (Jura, Luxemburg/Lothringen). Geol Rundschau 67, pp 1052–1077
Sillitoe RH (1973) The tops and bottoms of porphyry copper deposits. Econ Geol 68, pp 799–815
Sims R (1985) Government ownership versus regulation of mining enterprises in less-developed countries. In: Natural Resources Forum, New York:265–280
Sinclair AJ (1976) Applications of probability graphs in mineral exploration. Association of Exploration Geochemists, Toronto. Special Volume 4:95
Sisselman R (ed) (1978) E/MJ Operating handbook of mineral underground mining. McGraw-Hill, New York, p 440
Skinner BJ (1979) The frequency of mineral deposits. Geol Soc South Africa Trans, annecture to 82 (16):12
Skinner BJ (ed) (1981) Economic geology 75th Anniversary Volume. Economic Geology Publishing Co., Texas, p 964
Skinner BJ (1986) Earth resources, 3rd edn. Prentice Hall, Englewood Cliffs, NJ, p 184
Smith D, Wells L (1975) Mineral agreements in developing countries: structure and substance. Am J Int Law 69:560–590
Snow GG, Mackenzie BW (1981) The environment of exploration: economic, organizational, and social constraints. In: Skinner BJ (ed) Economic geology 75th Anniversary Volume, pp 861–896
Stermole FJ (1987) Economic evaluation and investment decision methods. Investment Evaluations, Golden, Colorado

Stobart CH (1985a) Effect of government involvement on the economics of the base-metals industries. In: The institution of mining and metallurgy (ed) Role of government in mineral resources development. London, pp 96–102
Stobart CH (1985b) How can copper be promoted effectively? 6th International Conference of the Commodities Research Unit Oct. 1985. London
Stokex E, Zeckhauser R (1978) A primer for policy analysis. W.W. Norton, New York
Stone TA, Birnie RW, Zantop H (1982) Landsat mapping of rocks associated with copper mineralization, northern Bahia State, Brazil. Proc Int Symp Remote Sensing of Environment, Second Thematic Conference: Remote Sensing for Exploration Geology, Dec 6–10, 1982, Fort Worth, TX, pp 935–946
Sutulov A (1983) Veränderungen der Weltbergbaustruktur. Metall 37, pp 281–282
Takeuchi K, Strongman JE, Maeda S, Tan CS (1987) The world copper industry: Its changing structure and future prospects. World Bank Staff Commodity Working Paper no 15. World Bank, Washington, DC
Tantalum-Niobium International Study Center: TIC quarterly bulletin, various issues. Brüssel
Telford WM, Geldart, Sheriff RE, Keys DA (1976) Applied geophysics. Cambridge University Press, London, p 860
Thoburn J (1981) Multinationals, mining and development. A study of the in industry, Westmead, Farnborough. Hampshire UK, p 183
Thomas LJ (1979) An introduction to mining, 2nd edn. Halsted, New York, p 421
Thomas R (ed) (1977) E/MJ Operating handbook of mineral processing. E/MJ Mining Informational Services, MaGraw-Hill, p 426
Thrush PW (ed) (1968) A dictionary of mining, mineral, and related terms. US Bureau of Mines, Washington, DC, p 1269
Tilton JE (ed) (1983) Material substitution: lessons from the tin-using industries. Resources for the Future, Washington, DC
Tilton JE (1985) The Metals. In: Vogely WA (ed) Economics of the mineral industries. Am Inst Min, Metallurgical, and Petroleum Engineers, New York, pp 383–416
Tilton JE (1987) Changing trends in metal demand and the decline of mining and mineral processing in North America. In: Cordes JA (ed) Public policy and the competitiveness of US and Canadian metals production. In preparation
Tilton JE, Millett J, Ward R (1986) Mineral and mining policy in Papua New Guinea. Institute of National Affairs, Port Moresby, Papua New Guinea
Tilton JE, Eggert RG, Landberg HH (eds) (1988) World mineral exploration: Trends and economic issues. Resources for the Future, Washington, DC
Tin International (1987). London
Tinslex CR, Emerson ME, Eppler WD (eds) (1985) Finance for the minerals industry. Am Inst Min, Metallurgical, and Petroleum Engineering, New York
Tong Kooi Ong, Kam Cheng Eng (Jan. 1986) The tin market crisis and the future. Malays Tin 13, 1:3–7
UNCTAD (1976) Manila declaration and programme of action. New York
UNCTAD (1982) Lending policies and criteria for the Second Account of the Common Fund for commodities. Genf
UNCTAD (1985) The least developed countries. Introduction to the LDC's and the substantial new programme of action for them. New York, p 157
United Nations Institute for Training and Research (1980) The future of small scale mining. Washington
Uranium Institute (1986) The Uranium Institute: The first ten years 1985–1985. London
US Bureau of Mines (ed) Mineral commodity summaries, various issues. Washington, DC
US Bureau of Mines: Mineral Yearbooks
US Congress, Congressional Budget Office (1976) US raw materials policy: problems and possible solutions. Washington, DC
US Department of State (1980) Global 2000, Report to the President, vol 1 and 2. Washington, DC
US Department of Energy (1987) Annual energy review 1986. Government Printing Office, Washington, DC
Van Landinghan SL (ed) (1983) Economic evaluation of mineral property. Benchmark papers in geology/67. Hutchinson Ross, Stroudsburg, PA, p 385

Verly G, David M, Journel AG, Marechal A (1983) Geostatistics for natural resources characterization, vol 1 and 2. D. Reidel, Dordrecht, p 1092
Vietor RHK (1984) Energy policy in America since 1945. New York
Wagenhals G (1984) The world copper market: structure and econometric model, ol. 233. In: Beckmann M, Krelle W (eds) Lecture notes in economics and mathematical systems. Springer, Berlin Heidelberg New York
Wälde TW (1984) Third world minerals development: current issues. Columbia J World Business XIX, 1:27-35
Wallace SR (1975) The Henderson ore body – Elements of discovery, Reflections. Min Eng, June 1975, pp 34-36
Wallace SR, MacKenzie WB, Blair RG, Muncaster NK (1978) Geology of the Urad and Henderson molybdenite deposits, Clear Creek Country, Colorado, with a section on a comparison of these deposits with those at Climax, Colorado. Econ Geol 73, pp 325-368
Walrond GW, Kumar R (1986) Options for developing countries in mining development. Houndmills/Basingstoke/Hampshire/London, p 190
Ward SH (1981) Gamma-ray spectrometry in geologic mapping and uranium exploration. In: Skinner BJ (ed) Economic geology 75th Anniversary Volume, pp 840-849
Warnecke SJ (1984) Stockpiling of critical raw materials, Chatham House Papers No 5. Roy Inst Intern Affairs, London
Wellmer FW (1986) Rechnen für Lagerstättenkundler und Rohstoffwirtschaftler. Clausthal-Zellerfeld
White paper on the mineral policy of the Republic of South Africa. Pretoria, p 19
White L (1980) Mining in Mexiko. E/MJ 32, pp 62-194
Whiteley RJ (ed) (1981) Geophysical case study of the Woodlawn ore body, New South Wales, Australia. Pergamon, Oxford, p 588
Wiebmer JD (1979) The small scale miner-industry's silent partner. Min Eng 31, pp 135-142
Wolfe JA (1984) Mineral resources: a world review. Chapman & Hall, New York, p 293
World Bank (ed) (1986) The world bank annual report 1986. Washington, p 236
Wright NP (1981a) Gravity and magnetic methods in mineral exploration. In: Skinner BJ (ed) Economic geology 75th Anniversary Volume, pp 829-839
Wright PN (1981b) Seismic methods in mineral exploration. In: Skinner BJ (ed) Economic geology 75th Anniversary Volume
Wyllie RJM, Argall GO (eds) (1975) World mining glossary of mining, processing, and geological terms. Miller-Freeman, San Francisco, p 432
Zantop H, Nespereira J (1978) Heavy-mineral panning techniques in the exploration for tin and tungsten in northwestern Spain. Geochemical exploration 1978. In: Watterson JR, Theobald PK (eds) The Association of Exploration Geochemists, pp 329-336
Zwartendyk EJ (1981) Economic issues in mineral resource adequacy and in the long-term supply of minerals. Econ Geol 76: pp 999-1005

Subject Index

AAS 52, 64
Abu Dhabi Fund 149
ACP countries 208–210, 217
acquisition strategy 96
ADB 213
additional profit tax 239
aerial photographs 27, 32, 33
aeromagnetic survey 38
AFERNOD 205
AFMAG 40
African Petroleum Producers Association 150
Agreement on Provisional Regulations for Manganese Nodules 204
aid in trade 212
airborne electromagnetic survey 40
airborne geophysical survey 36
airborne geophysics 24, 29, 35
airborne magnetic survey 37, 44
airborne survey 44
Alcan Aluminium 138
Alcoa 137
alumina production 138
Aluminium Company of America 137
aluminium market 137
aluminium oxide plant 137
aluminium smelter 137, 138
Alusuisse 138
Amalgamated Metal Corporation 136
American Zinc Institute 158
Anaconda 141
analytical methods 51
Anglo American Corp. 139, 140
anomalies 36, 38, 42, 45
APEC 145
APEF 155
APPA 150
Asian Development Bank 213
assaying ore 61
Assimer 159
Association des Pays Exportateurs de Minerai de Fer 115
Association Francaise pour l'Etude et la Recherche de Nodules 205
Association Internationale des Producteurs des Mercure 159
Association of Tin Producing Countries 144, 164

Atlantic Richfield 141
atmospheric survey 50
atomic absorption spectrometry 52, 64
ATPC 164
auger drill 59

backward linkages 249
backwardation 175
BAD 213
balance of payments 251
Banco Interamericano de Desarollo 213
Bandoeng Pool 143
Bangka drill 59
bank loans 127
Banque Africaine de Développement 213
bauxite mining 137, 138
Bauxite Production Levy 238
bedrock 51
benefit-cost analysis 102
benefits 96
BfAD 213
BID 213
biogeochemical survey 49
biogeochemistry 45
Bismuth Institute 159
Bouger anomaly 43
brands 175
BRGM 220, 226
British Mining Act of 1880 91
British Mining Law 90
British Petroleum Co. 141
Buchanan model 20
Buffer Stock Manager 163
bufferstock 143, 162, 163, 195, 199
building materials 5
bulk mining methods 77
bulk sampling 24
Bundesanstalt für Geowissenschaften und Rohstoffe 23
Bureau de Rècherches Géologiques et Minières 23, 220, 226
Bushfeld Complex 11
business risk 103

Cadmium Association 158
Canadian Export Development Corporation 126

capital investment guaranty   225
carbon-in-pulp-recovery   86
cartel   133
cash flow   97, 105, 106
   analysis   99, 115, 118, 120, 121
   table   117
cash price   175
causal forecasting method   185
CBT   177
CDA   158
Centre d'Information de Métaux Non Ferreux   159
Centre d'Information du Cobalt   159
ceramics   86
channel samples   63
chemical assay   63
Chicago Board of Trade   177
churn drills   59
CIMNF   159
CIPEC   150–153
civil law   90
Clarion-Clipperton Belt   204
classical statistical calculations   66
classification of reserves   68
classification of resources   68
clastic sedimentary rock   14
cluster analysis   54
colorimetry   52
COMECON   170, 193, 199
COMEX   174, 176
COMIBOL   136, 163
commercial stocks   168, 169
Committee for Coordination of Prospecting for Minerals   216
commodity agreement   144, 160, 167
commodity association   196
commodity cartel   144
commodity policy   221
Common Fund   198–200
common law   90, 91
competition   133
competitive market   133
competitive price   166, 174
completion risk   104
composite material   86
concentration of elements   8
concessions   235
   policy   235
Conference on the Law of the Sea   202
conglomerates   14
Conseil Intergouvernmental des Pays Exportateurs de Cuivre   151
Consolidated Tin Smelters   136
contact metasomatic hydrothermal deposits   16
contango   175
continental crust   8–10
continental shelf   203

Convention on the High Seas   203
Copper Development Association   158
copper export   152
copper market   132
core drilling   59, 60
Corea Plan   197
corporation tax   238
costs   22, 77, 96
   of exploration   22
   structure   114
crustal evolution   10
crystallization   11
cutoff grade   113

DAC   205, 211, 213
debt   99, 126, 127
decision-tree   107
   analysis   106
deep sea mining   204, 205
DEG   225, 226, 233
demand risk   103
dendrogram   55
density separation   80
depletion allowances   99
deposit development   85
depreciation   99
detailed exploration   111
detailed mapping   24
detailed target evaluation   119, 123
developing countries   89, 127–130, 134, 190, 191, 196, 198, 210–213, 220, 234
Development Assistance Committee   205, 211, 213
Development Committee   213
development costs   25
development funds   128
development goals   194
development of a mineral project   24
development phases   23
diamond drill   59
direct investment   231
discount rate   97, 99, 101, 102, 104, 105
discovery methods   88
discovery of a mineral project   24
discriminant analysis   56
discount factors   117
dispersion halos   45–47
dispersion variance   67
distribution of mineral deposits   189
diversification strategy   96
drill hole geophysics   60
drill hole logging   35
drill samples   62
drilling   24, 29, 58
dump leaching   78

earth's crust   8–11
east-west trade   170

Subject Index

EC 194, 207, 213, 217, 219, 229
ecological effects of mining 251
econometric models 180, 183
economic analysis 117, 118, 121
Economic and Social Commission for Asia and the Pacific 215
Economic and Social Council 213
economic evaluation 94
economic geology 2, 94
economic resources 73
economic risk 103
EDF 208, 210, 213
EEZ 203
effects on income 248
effects on industrialization 249
effects on the balance of payments 251
EIB 210, 217
electric survey 24, 30, 38
electromagnetic anomalies 40
electromagnetic log 60
electromagnetic survey 24, 30, 39, 40
emission spectroscopy 52
empirical ore deposit models 18
ENAF 136
energy consumption 172
energy resources 190
energy savings 172
engineering model 182
environmental protection 118
environmental regulations 123, 124
EOSAT 31
equity 99
   financing 126, 129
   joint ventures 236
   sharing 232
ERS-1 31
ESCAP 215
eurodollar market 130
European Community 207, 213, 217, 229
European Development Fund 208, 210, 213
European Economic Community 207
European Investment Bank 210, 217
European Zinc Institute 158
evaluating mineral deposits 245
evaluation 22, 109
   of mineral projects 24
exclusive economic zone 201, 203
Ex-Im Bank 126, 130
expected value 106
exploitation 231
exploration 25, 26, 85, 98, 109, 231
   cost 24
   expenditure 29
   financing 126
   geology 2
   method 22, 23, 29, 30, 87
   model 18, 20

   program 245
   risk 232
   sequence 24
   strategy 26
   tool 22
Exploration Assistance Program 223
export control 162
export credit 127, 130
Export Credits Guarantee Department 226
export license 144
export quota 162, 190
extension variance 67
external fund 125
Exxon 141

feasibility analysis 94
feasibility study 24, 26, 119, 123, 127
financial assistance 212, 235
financing 22, 124
first account 199
first window 199
fiscal effects 250
fiscal policy 238
fiscal regime 122
floor price 162
flotation 81
flow-through share 126
follow-up examination 24
follow-up exploration 27
forecasting 184
foreign-exchange risk 103
formation of mineral deposits 10
forward linkages 249
forward pricing 175
French Mining Law 90
froth flotation 81, 86
fungible goods 166
future market 175

gamma radiation 41
gamma-ray spectrometry 60
gamma rays 31
GATT 213, 229
General Agreement on Tariffs and Trade 213, 229
General Mining Union Corp. 139
General Prussian Mining Law 90
General Services Administration 170, 227
genetic ore deposit model 18
geochemical anomalies 45, 88
geochemical exploration 45, 47
geochemical log 61
geochemical method 30
geochemical survey 30, 53
geochemistry 29, 87
geologic cycles 10
geologic exploration 34, 46

geologic factors   74, 181
geologic log   61
geologic mapping   34–36, 38
geologic model   17, 110
geological survey   89
geometric method   64
geometric ore reserve estimation   64, 65
geophone arrays   43
geophysical anomalies   36, 61, 88
geophysical exploration   34
   method   36, 37
geophysical logging   60, 61
geophysical method   29, 30
geophysical survey   27
GEOSAT   32
Geosciences and Commodity Security
   Program   230
geostatistical evaluation   66
geostatistical method   67
geostatistical ore reserve calculation   67
geothermal grandient   8
German Association of Metallurgical and
   Mining Engineers   112
German Exploration Assistance Program   224
Gold Exchange of Singapore   176
Gold Fields of South Africa   139
Gold Institute   160
gold loans   128
gold market   138
gold mining companies   140
government assistance   127
government equity participation   123
government policy   181
government stockpile   170
grade of reserves   113
gravimetric survey   24, 30, 41, 42, 44
gravity anomaly   42
gravity profile   39
gravity survey   39
ground geophysical survey   44
ground geophysics   24, 29
group of 77   198
GSA   170, 227

Havana Charter   196
heap leaching   78, 86
heavy mineral survey   30
hedging   176
helium exploration   50
Hg detector   30
high-tonnage metal deposit   76
Hongkong Commodity Exchange   176
hydrogeochemical survey   48
hydrogeochemistry   45
hydrothermal solution   11, 15

IAEA   157, 213
IBA   153, 154

IBRD   213
ICCICA   196
ICP   52, 64
IDA   213
IEA   206, 207
IEP   206
IFC   213
IMA   158
IMF   139, 213, 229
import dependency   207
income tax   122, 238
indexation of prices   198
indicated reserves   70
indirect benefit   102
indirect costs   102
induced polarization   39
   logs   60
   survey   39, 40, 44
inductively coupled plasma spectrometry
   52, 64
industrial minerals   5, 190
industrialized countries   130, 196, 206, 219
inferred reserves   70
inflation   100
infrastructural effect   248
initial target evaluation   111
institution building effect   248
insurance premium   175
Integrated Program for Commodities   155,
   196–198
interest rate   117, 130
Interim Coordination Committee for International Commodity Arrangements   196
internal fund   97, 99, 125, 126, 127
internal rate of return   100, 120
International Atomic Energy Agency   157, 213
International Bank for Reconstruction and
   Development   213
International Bauxite Association   153, 238
international cartel   167, 190
International Commodity Agreement   143,
   160, 196, 197
International Copper Cartel   151
International Development Association   213
International Energy Agency   206
International Energy Programme   205, 206
International Finance Corporation   213
International Labor Office   195
International Lead Zinc Research Organi-
   zation   158
International Magnesium Association   158
international mining companies   134, 141, 142
International Monetary Fund   139, 213, 229
international oil companies   141
International Primary Aluminium Institute
   155, 158
International Sea-Bed Authority   203, 204

Subject Index   267

international study group 143
International Tin Agreement 143, 161
International Tin Control Scheme 143, 195
International Tin Council 161, 199
International Tin Study Group 143, 164
International Trade Organization 196
International Tungsten Agreement 165
investment 100, 107, 122, 129
   analysis 96
   decision 99, 101, 108
   fund 125
   guaranty 225
   strategy 231
IP 39, 40, 44, 60
iron and steel alloys 190
iron ore association 155
iron ore prices 156
IR-photography 31
IRR 100, 118, 120
ITA 161, 162
ITC 161, 162
ITO 196

joint ventures 236
judgmental forecasting method 185

Kaiser Aluminum Chemical Corp 138
Kennecott 141
KfW 225, 228
KLTM 177
Kreditanstalt für Wiederaufbau 225, 228
Kuala Lumpur Tin Market 176, 177
Kuroko-type volcanogenic massive sulfides 15, 16
Kuwaiti Fund for Development 149

labour requirement 118
Landsat I 31
LDC 211
leaching 78
Lead and Zinc Study Group 158
Lead Development Association 158
League of Nations 195
least developed countries 211
legal considerations 118
levy 238
LIBOR 216
Lima Declaration 196
LME 163, 164, 174–176
   tin market 163
   tin price 164
logging 60
Lomé Convention 208–210, 213
London Gold Futures Market 176
London Interbank Offered Rate 216
London Metal Exchange 163, 164, 169, 174–176, 182

long-term debt 125
longwall mining method 77

magma 10, 11
magmatic process 11
magnetic anomaly 38
magnetic log 60
magnetic separation 80
magnetic survey 24, 30, 37, 41
magnetic susceptibility 35
majority joint venture 236
Malaysia Mining Corporation 136
manganese nodules 78, 205
marine mineral deposit 201
marine mining 78, 205
Marine Research and Technology Program 230
market 132
   analysis 133, 168
   force 195
   model 179
   organization 142
   risk 103, 232
   structure 133, 134
   studies 160, 174
marketing 22
   policy 239
McKelvey box 71
measured reserves 70
medium-sized deposits 25
merger strategy 96
metal concentration 9
metal exchanges 174
Metal Market and Exchange Company 175
Metal Mining Agency of Japan 226, 228
metallic mineral resources 5
metallogenic concept 5
metamorphic process 16
mine development 127
mine financing 128
mineral
   demand 87, 173
   deposits 5, 8, 11
   development 83, 87, 95
   dressing 2
   economics 2, 94
   endowment 72, 73
   exploration 29, 95, 194, 223
   exploration and development 85
   exploration projects 214
   exporters 191
   exporting countries 144, 192
   importing countries 193
   market 1, 83, 132, 190
   market model 179
   model 183
   policy 1, 189, 190

potential 57
price 1, 181
pricing 166
processing 80, 85
research 229
reserves 168
resources 1, 5, 7, 8, 69, 189–191, 201, 242
resources computer file 19
supply 85, 87, 168
trade 229
utilization 233
minimum acceptable reserves 111, 112
minimum price 144, 161
mining 8, 74, 79, 85, 117, 231
   companies 231, 234
   concession 232, 240
   engineering 2
   geology 2
   in developing countries 246
   investment 225, 231, 248
   law 90, 122
   lease 26
   legislation 144
   method 244
   production 219, 220
   project 23, 237
Ministry of International Trade and Industry 228
Minor Metals Traders Association 160
minority joint venture 236
Mississippi Valley lead-zinc deposits 16
MITI 228
MMAJ 226, 228
MMTA 160
Molycorp 141
monopolistic producer 167
monopoly 133
   price 166
Monte Carlo simulation 108
MOS-1 31
most seriously affected countries 211
MSAC 211
MSS 31
multinational mining 89
multispectral scanning 30, 31
multivariate statistics 53

Nairobi Conference 197
National Defense Stockpile 226
nationalizations 134
natural resources 196, 231, 234
net cash flow 98, 105
net present value 98–100, 117
net profit tax 238
neutron activation 61
New International Economic Order 196, 234
New York Commodity Exchange 174, 176
newly industrializing countries 211

non-oil-exporting countries 148
nonrenewable natural resources 189
Northern Mineral Exploration Assistance 224

OAPEC 145, 150
Ocean Management Inc. 205
Ocean Minerals Company 205
Ocean Mining Associates 205
oceanic crust 10
ODA 206, 211
OECD 193, 194, 199, 205–207, 211, 213
Office of Mineral Exploration 221, 224
Official Development Aid 205, 206, 211
oil companies 89, 145
oil embargo 222
oil market 206
oil price 1, 149, 193
   crisis 146
   policy 146
oligopolistic market 133
oligopolistic producer 167
oligopoly 133
   price 166
O-mode analysis 55
O-mode factor analysis 56
OPEC 141, 145, 192, 197, 214, 222, 236
   Fund for International Development 214
   government revenues 148
   national quotas 147
   Special Fund 148
operating costs 118
OPIC 130, 226
opportunity cost 97, 99
   of capital 98, 99
optimal scale of operation 113
ore 5, 8
   deposits 8, 10, 112
   enrichment factors 9
   forming process 10
   grade 8, 9
   reserve calculation 69
   reserve estimation 64–66
   reserves 65
   to-waste ratio 75
Organisation for Economic Cooperation and Development 205, 213
Organization of Arab Petroleum Exporting Countries 150
Organization of Petroleum Exporting Countries 145
Overseas Mineral Resources Development Co. 221
Overseas Private Investment Corporation 226

parallel price 167
participation 192, 231
   policy 236
pathfinder elements 46

Subject Index

payback period 100, 101
Pechiney Ugine Kuhlmann 138
percussion drill 59
PERTAMINA model 146, 232
petroleum export policy 147
petroleum market 132
petroleum price 149
photogeologic interpretation 34
photogeology 24
photography 31
pitting 58
placer deposit 75
plastics 86, 87
point samples 63
political risk 103, 104, 207, 225, 232
porphyry copper deposits 11, 13, 14, 17, 18, 88
porphyry copper models 20
possible reserves 70
posted price 144, 167, 177, 178
potential mineral resources 144
precambrian metamorphic belts 16
precious metal vein 17
precious metals 190
prefeasibility study 26, 111, 117–120
preferential tariffs 235
present value ratio 100
Preussag 136
price 174, 180
  analysis 169
  cartel 157
  fixing 133, 177
  fluctuation 160, 189
  policy 193, 197
  trend 167
pricing 166, 167, 195
primary effects 247
Primary Tungsten Association 156, 166
probable reserves 70
producer association 143, 144, 158
producer cartel 143, 144
producer price 177, 178
production costs 114, 115
production levy 238
production rate 114
production risk 232
production sharing contract 238
production trend 168
profit sharing 231
  agreement 232
  participation 231
program design 26, 109
project description 117
project financing 130
project risk 232
prospecting 25, 29, 88
prospector 87
  system 20

Prospector and Developers
  Association of Canada 127
proven reserves 70
PTA 156
P.T. Timah 136

quantity adjustment 133
quantity fixing 133

Radar Sat 31
radiometric survey 24, 30, 41, 44
rate of return 98, 101
reconnaissance exploration 24, 26, 27, 94, 110
reconnaissance mapping 24
recovery rate 233
recycling 144, 171–173, 207, 219, 220, 233
regalian law 90, 91
Regional Mineral Resources Development
  Center 215
relative price 166, 167
remote sensing 24, 27, 30, 32
renewable energies 216
reserves 8, 69, 70, 88, 190
  estimation 64
  risk 104, 232
resistivity 35
  log 60
  survey 40
resources 8, 69, 70, 88
  base 73
  rent tax 239
retrenchment strategy 96
Reynolds Metals 138
rift zones 10
ring 175
risk analysis 102, 110, 111, 207
risk service contract 238
R-mode analysis 54, 56
R-mode factor analysis 55
rock-forming 10
rock geochemical survey 51
rock geochemistry 45
room-and-pillar method 77
rotary drill 59
royalty 122, 146, 154, 192, 231, 238
RRT 239
run-of-mine 240

sample preparation 63
samples 47, 48
sampling 61, 62
  density 47
Saudi Fund of Development 149
SDR 229
Sea-Bed Disputes Chamber 203
seafloor exploration 8
SEATRAD Center 215

second account 200
second window 155, 200
secondary effects 247
sedimentary basin 15
sedimentary-exhalative ore
  deposit 13, 17
sedimentary process 14
seismic method 36
seismic survey 24, 30, 43, 44
selective mining 77
Selenium-Tellurium Development Association 159
sensitivity analysis 105, 106, 118, 120
service contract 232
short-term debt 125
Silver Institute 159
simulation modelling 108
skarn deposit 11
small-scale mining 240–243
social effects 251
socialist economies 90
socio-economic evaluation 94
Sohio 141
soil geochemical anomalies 49, 51
soil geochemical surveys 48
soil geochemistry 45
solution mining 78
Southeast Asia Tin Research and Development Center 215
spanish mining ordinance 90
Special Drawing Rights 229
spectral photometry 30
spectrometry 31
SPOT 31
spot market 175, 177
Stabex system 208, 209
state-owned mining companies 89
statistical forecasting techniques 184
statistical method 66
statistical ore reserve calculation 69
STATPAK 53
steel crisis 156
steel market 156
stereoscopes 33
Stillwater Complex 11
stockpile 221, 226–228
  price 170
  program 226
stockpiling 168, 169
  agreement 228
  policy 198
stoping 77
strategic minerals 221, 227
Strategic Petroleum Reserve 228
strategic stockpiles 170
stratiform deposit 15
stratiform sulfide deposit 15

stream sediment geochemical survey 47
stream sediment geochemistry 24
strip mining 76
structural analysis 133
submarine volcanism 15
substitution 174, 180, 181, 207, 233
sulfide deposits 14
supply factor 168
surface mineralization 87
surface mining 75
Sydney Future Exchange 176
Sysmin 209, 210
systematic sampling 57

Tantalum-Niobium International Study Center 159
target drilling 94
target selection 94
tax incentives 225, 235
taxes 119
technical cooperation 212, 245
technical log 61
technical risk 103, 104
technological innovations 174
tectonic model 10
tectonic process 10
terminal market 175
THAISARCO 136
TIC 159
time value of money 97
Tin Control Scheme 143
tin market 135, 163
tin mining enterprises 136
Tin Producers' Association 143
Tin Study Group 161
TNC 237
trademark 175
transfer of technology 196, 250
transnational corporations 237, 239
trenching 58
trust 133
Tungsten Association 156
Tungsten Committee 165

United Nations 196, 201
  Capital Development Fund 213
  Committee on the Seabed 202
  Committee on Tungsten 165
  Conference on Trade and Development 1, 151, 155, 165, 197
  Convention on the Law of the Sea 200
  Development Program 90, 127, 213, 214
  Institute for Training and Research 240
  Revolving Fund for Natural Resources Exploration 90, 127, 214
UNCLOS 200, 202

Subject Index

UNCTAD 151, 152, 155, 165, 197, 198
   Committee on Commodities 198
   Committee on Tungsten 165
   Preparatory Working Group 165
underground exploration 61
underground mining 77
underground stoping 76
UNDP 213, 214
UNESCO 213
UNIDO 196, 213
Union Oil 141
UNITAR 240
univariate statistics 53
Uranium Association 157
Uranium Institute 157
uranium market 157
US Bureau of Mines 229
US Export-Import Bank 126
US General Mining Law of 1872 91, 92
US Geological Survey 23, 224
US Leasing Act of 1920 91
US National Defense Stockpile 227

US Overseas Private Investment Corporation 130
US Treasury Department 140

vapor geochemical survey 50
vapor geochemistry 45
variograms 67
vein deposit 11, 13
volcanogenic copper-lead-zinc deposit 13
volcanogenic massive sulfides 15, 17, 88

Winnipeg Commodity Exchanges 176
World Bank 213, 216
   Energy Lending Program 216
world mineral production 6

X-ray fluorescence 52
XRF 52

Zinc Development Association 158
Zinc Institute 158

International Mineral Economics

provides an integrated overview of the important concepts. The treatment is interdisciplinary, drawing on the fields of economics, geology, business, and mining engineering.

Part I examines the *technical* concepts important for understanding the geology of ore deposits, the methods of exploration and deposit evaluation, and the activities of mining and mineral processing.

Part II focuses on the *economic* and related concepts important for understanding mineral development, the evaluation of exploration and mining projects, and mineral markets and market models.

Finally, Part III reviews and traces the historical development of the policies of international organizations, the industrialized countries, and the developing countries.